Computational Nanotoxicology

Computational Nanotoxicology

Challenges and Perspectives

edited by

Agnieszka Gajewicz
Tomasz Puzyn

JENNY STANFORD
PUBLISHING

Published by

Jenny Stanford Publishing Pte. Ltd.
Level 34, Centennial Tower
3 Temasek Avenue
Singapore 039190

Email: editorial@jennystanford.com
Web: www.jennystanford.com

British Library Cataloguing-in-Publication Data
A catalogue record for this book is available from the British Library.

Computational Nanotoxicology: Challenges and Perspectives

ISBN 978-981-4800-64-8 (Hardcover)
ISBN 978-0-429-34137-3 (eBook)

Contents

Preface

Innovations in nanotechnology and nanomaterials design have changed the landscape of industrial and consumer products. They have opened up promising application prospects in numerous areas of human daily life (e.g., drug delivery, cosmetics, electronics, the energy sector, etc.). However, several nanoparticles (NPs) have been identified as being toxic to humans and the environment. Thus, prior to marketing new NPs, their impact on living organisms along their entire life cycle should be thoroughly assessed. Unfortunately, our knowledge of mechanisms driving (eco)toxicological activity of NPs is still limited. From this perspective, we are especially interested in to what extent the physicochemical properties and specific structural features of NPs that can be controlled at the stage of synthesis are correlated with the observed toxicity endpoints. Therefore, the development of computational (in silico) methods to be used for supporting human health and environmental risk assessment of engineered nanomaterials is nowadays of high interest. This is mainly because the application of such methods enables us to fill the existing knowledge gaps in the relationships between the structure, properties, and toxic activity. Nevertheless, when considering the existing high degree of complexity and multifunctionality of engineered nanoparticles, in silico methods originally developed for regular chemicals cannot always be applied explicately for nanoparticles.

This book focuses on critically discussing the state of the art and future challenges to be faced by computational nanotoxicology, with the aim of making the material accessible to readers primarily interested in the development and application of such tools in risk assessment. Divided into four sections, the book addresses a wide array of methodological aspects related to achieving scientific, regulatory, and industry acceptance of computational methods for nanotoxicology. Opening with an introduction to the field of computational nanotoxicology, in view of regulatory requirements and applications, the book continues with an account of the theory and applications of computational chemistry (i.e., quantum mechanics, density functional theory, molecular mechanics, molecular dynamics) and chemoinformatic methods (i.e., quantitative structure–activity relationships modeling, grouping, and read-across). The third part of the book addresses the challenges related to recent trends in the validation of computational models and discusses how to cope with the uncertainty of predictions. The final section focuses on the key challenges and roadblocks that have to be overcome in the next decade in order to accelerate progress in nanotechnology research, development, and manufacturing, as well as to support environmental, health, and safety activities more efficiently and cost-effectively. This—in a longer perspective—should lead to designing nanomaterials that are safer for humans and the environment.

We believe that with the excellent contributions from all authors, this book will promote the possibilities and benefits of applying in silico methods in risk assessment. Moreover, it will enhance the dialogue between experts in nanotechnology, computational chemistry, and/or regulation.

Finally, we wish to express our gratitude to the editors at Jenny Stanford Publishing for their excellent cooperation throughout the editing process and, last but not least, to our families for their friendly support and friends for their valuable comments.

Agnieszka Gajewicz

Tomasz Puzyn

2020

Chapter 1

Modeling of Nanomaterials for Safety Assessment: From Regulatory Requirements to Supporting Scientific Theories

Lara Lamon, David Asturiol, Karin Aschberger, Jos Bessems, Kirsten Gerloff, Andrea-Nicole Richarz, and Andrew Worth
European Commission, Joint Research Centre, Ispra, Italy
andrew.worth@ec.europa.eu

In this chapter, we describe different theories relevant to understanding the behavior, fate, and effects of manufactured nanomaterials (NMs). In the first part, background information on regulatory requirements related to NMs is reported, along with an overview of risk assessment as an approach to address risks posed by exposure to NMs. The second part is dedicated to the identification of key physicochemical properties that are relevant for characterizing and understanding the behavior (fate and biological effects) of NMs. An understanding of the scientific basis of NM behavior is important in the development and application of standard and alternative approaches to animal testing.

Computational Nanotoxicology: Challenges and Perspectives
Edited by Agnieszka Gajewicz and Tomasz Puzyn

ISBN 978-981-4800-64-8 (Hardcover), 978-0-429-34137-3 (eBook)
www.jennystanford.com

1.1 Introduction

Manufactured nanomaterials (NMs) are being increasingly included in a variety of goods and products because of their novel physical and chemical characteristics [1]. There are concerns, however, that the very same characteristics may also lead to environmental and human health risks. In the European Union (EU), NMs have been defined in legal terms. Legally binding definitions are included in the Cosmetic Products Regulation (EC) 1223/2009, the Biocidal Products Regulation (EU) 528/2012/EC, and Regulation (EU) 1169/2011 on Provisions of Food Information to the Consumers (FIC Regulation). In addition, in 2011, the EC adopted a recommendation (2011/696/EU) on the definition of the term "nanomaterial" with the goal to promote consistency in the interpretation of said term for legislative and policy purposes in the EU [2]. This definition is broadly applicable across different regulatory sectors but not legally binding. The EC definition applies to all particulate NMs regardless of their origin: natural, incidental, or manufactured. It refers to a size range of 1–100 nm and also establishes a threshold of 50% or more particles <100 nm in the number size distribution, which in specific cases can be lowered to 1%. This size range has been proposed in several definitions, including ISO/TS 12805:2011 by the International Organization for Standardization (ISO).

The NM definition in the Biocidal Products Regulation is based on the EC definition, while the definitions for cosmetic products and food were implemented previously and contain some relevant differences. Although the size range is comparable (<100 nm) the main difference is the restriction to intentionally manufactured (or produced) materials, which is further restricted for cosmetic products to insoluble or biopersistent materials. This size range has been proposed in several definitions, including ISO/TS 12805:2011 by the ISO. The revision of the EC definition foresees the consequent harmonization in cosmetic products and food regulations [3].

To evaluate and manage the environmental and health impacts of NMs, chemical risk assessment (CRA) is applied to assess the potential risks caused by NMs. CRA is a process by which scientific and regulatory principles are applied in a systematic approach to address qualitatively and/or quantitatively the likelihood that humans or environmental species may be harmed due to potential exposure to chemicals. Regulation 1907/2006 concerning the Registration, Authorisation and Restriction of Chemicals (REACH) [4, 5] requests registrants to demonstrate the safe use of chemical substances, including those in nanoform. REACH aims to ensure chemical safety while promoting innovation and competitiveness as well as reducing the use of in vivo testing through the use of nonanimal alternatives, such as in silico, in vitro, and in chemico methods; read-across; and weight of evidence.

This chapter describes regulatory information requirements for NMs, with a special focus on REACH, as well as the scientific and technical basis for understanding their behavior (fate and effects). The REACH regulation (Section 1.2) and the risk assessment approach (Section 1.3) are briefly introduced, and an overview is given (Section 1.4) of key physicochemical (PC) properties that are relevant for characterizing and understanding the behavior (fate and biological effects) of NMs. This chapter also touches on the scientific basis of NM behavior (Section 1.5), since this understanding is important in the development and application of standard (Section 1.6) and alternative approaches to animal testing (Section 1.7).

1.2 Information Requirements for Risk Assessment: Legal Provisions and Guidance

In the EU, substance- and sector-specific pieces of legislation provide a binding framework to ensure the safety of substances and products on the market (manufactured or imported). NMs are implicitly covered or explicitly addressed (e.g., cosmetics and biocidal products), depending on the applications and applicable legislation. The "substance" definition of REACH applies to chemicals irrespective of size, shape, and physical state, and thus its provisions apply also to NMs, even if there are currently no provisions that explicitly refer to NMs [5, 6].

1.2.1 Chemical Substances under REACH

According to the Second Regulatory Review on Nanomaterials [7], REACH sets the best possible framework for the risk management of NMs when they occur as substances or mixtures. More specific requirements may be introduced through the possible revision of some of the REACH annexes.

The provisions of REACH contain extensive obligations for manufacturers to generate and assess data on chemicals (including PC properties, manufacture, uses, and hazardous properties) and to demonstrate that risks can be adequately controlled during their use. All chemicals manufactured or imported in quantities higher than 1 ton/year (t/y) have to be registered with basic information requirements. In addition, for all chemicals manufactured or imported in quantities higher than 10 t/y, a chemical safety assessment has to be performed and documented in the chemical safety report.

Under REACH, different forms of one substance (e.g., solids, suspensions, powders, and NMs) are considered within a single registration of a substance [7]. However, the registrant must ensure the safety of all included forms and provide adequate information to address the different forms in the registration. This means that more than one endpoint study for different forms may be required or different forms within one registration may have different hazard classifications. The REACH approach to hazard assessment and risk characterization, with its built-in flexibility, makes it suitable for NMs [7]. On the basis of REACH Implementation Project on Nanomaterials (RIPoN) reports [8], the European Chemicals Agency (ECHA) has published specific guidance for NMs [9–11]. In addition, ECHA has set up an NMs working group to give advice on scientific and technical issues in relation to NMs under REACH.

The information collected or generated under REACH is used, for example, for priority setting, classification and labeling, chemical safety assessment, and PBT/vPvB[1] assessment. It needs to be adequate for both classification and labeling and for chemical safety assessment if the latter is required (triggered by classification or PBT assessment). REACH Article 10 defines the information requirements

[1]PBT, persistent, bioaccumulative, and toxic; vPvB, very persistent and very bioaccumulative.

to be submitted for registration, and Annex I outlines the general provisions for assessing substances and preparing chemical safety reports. Annexes VII to X more specifically detail the standard information requirements. Information on intrinsic properties is mainly dependent on the tonnage at which the substance is brought on the market. Column 2 of each of these annexes (VII to X) defines the specific rules for adaptation of the standard information requirements as defined in Column 1, that is, when a specific step (test) does not need to be conducted or further studies may be considered if the results are positive. Guidance notes on fulfilling the requirements of Annexes VII to X are presented in REACH Annex VI. This annex also contains the information requirements that are independent from the tonnage, that is, general registration information, identification of the substance, manufacture and use, classification and labeling, and guidance on safe use. REACH Annex XI defines the general rules for adaptation of the standard testing regime set out in Annexes VII to X.

The bases for the use of alternative approaches under these rules are laid down by Articles 13 and 25. Article 25(1) states that "testing on vertebrate animals shall be undertaken only as a last resort." According to Article 13(1):

> Information on intrinsic properties of substances may be generated by means other than tests, provided that the conditions set out in Annex XI are met. In particular for human toxicity, information shall be generated whenever possible by means other than vertebrate animal tests, through the use of alternative methods, for example, in vitro methods or qualitative or quantitative structure–activity relationship models or from information from structurally related substances (grouping or read-across).

As described in Annex XI, testing can be waived or replaced by other information and methods when (i) testing is not scientifically necessary, (ii) testing is technically not possible, or (iii) substance-tailored exposure-driven testing (i.e., exposure-based waiving) is appropriate. The types of tests that can be waived or replaced by alternative tests are dependent on the tonnage and the requirements of the respective annex. Guidance is provided in the ECHA (2011) Guidance on Information Requirements and Chemical Safety

Assessment, Chapter R.5: Adaptation of Information Requirements [23].

REACH Annex XI also explains when alternative approaches, such as qualitative or quantitative structure–activity relationship (QSAR), grouping and read-across, in vitro methods, and weight of evidence can be used instead of testing to indicate the presence or absence of a certain dangerous property (see Section 1.7). Results obtained from valid QSARs may be used to indicate the presence or absence of a certain dangerous property if (i) the scientific validity of the QSAR model has been established, (ii) the substance falls within the applicability domain of the QSAR model, (iii) the results are adequate for the purpose of classification and labeling and/or risk assessment, and (iv) adequate and reliable documentation of the applied method is provided (REACH Annex XI, 1.3). Substances whose PC, toxicological, and ecotoxicological properties are likely to be similar or follow a regular pattern as a result of structural similarity may be considered as a group or "category" of substances; and PC, human health effects, and environmental effects or environmental fate may be predicted from data for reference substances within the group by interpolation to other substances in the group (read-across approach) (REACH Annex XI, 1.5) (see also Section 1.7.4). Guidance on how to assess QSARs if conditions are met and on the technically and scientifically justified methodology for the grouping of substances is provided in the ECHA (2008) Guidance on Information Requirements and Chemical Safety Assessment, Chapter R.6: QSARs and Grouping of Chemicals [202].

The current minimum standard information requirements as described in the REACH annexes may not address all NM-specific PC properties and characteristics; especially the PC properties may not be sufficient to adequately characterize the NM and discriminate it from other forms. However, it is already possible to document relevant information in the International Uniform Chemical Database (IUCLID), which is the electronic tool used for submissions of REACH dossiers to ECHA. IUCLID included the 13 OECD Harmonised Templates, which cover NM-specific endpoints (updated in April 2016) (http://www.oecd.org/ehs/templates/harmonised-templates-physico-chemical-properties.htm).

The impact assessment on the possible amendment of the REACH annexes to more specifically address NMs has been finalized, and the potential elements were presented as "Informal Considerations" (nonpaper) at the May 2014, March 2016, and March 2017 meetings of the Subgroup of the REACH and CLP Competent Authorities on Nanomaterials (CASG Nano). The proposed amendments include, for example, more detailed information on substance identity and highlight the existence of nanoforms of the same substance. The proposal also includes a definition of nanoform that is in accordance with the European Commission recommendation of October 18, 2011, [2] on the definition of an NM.

1.2.2 Cosmetic Products

Certain ingredients of cosmetics require authorization on the basis of a scientific risk assessment and inclusion in Annexes IV (colorants), V (preservatives), and VI (UV filters) to Regulation (EC) No 1223/2009 [12]. Some of these substances may be nanoparticles (NPs). The EC provides a definition of NMs (insoluble/partially soluble, or biopersistent and intentionally manufactured) as well as mechanisms for notification, labeling, and safety evaluation of cosmetic products containing NMs. It also imposed a ban on testing finished products and ingredients on animals as well as on marketing such products and ingredients (from March 2013).

The EU Scientific Committee on Consumer Safety (SCCS) released a guidance "Safety Assessment of Nanomaterials in Cosmetics" [13], where it is recommended that nano-specific PC properties be considered in a safety assessment. It highlights the need for special considerations in relation to the safety of NMs, in view of the possible distinct properties, interactions, and/or effects that may differ from those of the conventional forms of the same materials. It covers and gives recommendations addressing the main elements of risk assessment, that is, material (PC) characterization, exposure assessment (e.g., likelihood and extent of translocation of NMs across biological barriers), and toxicological evaluation (e.g., consideration of nano-related aspects, such as particulate form, or possible interaction with biological entities). The guidance presents an overview of available methods of toxicological evaluation of NMs for each endpoint and concludes that at present, in view of current limitations, an approach using in vitro assays only is too premature to be applied for risk assessment. Concerning read-across, the SCCS considers that in the absence of a sufficient knowledge base on NM properties, behavior, and effects, a category approach to risk assessment is currently not feasible for NMs and risk assessment of each NM needs to be carried out on a case-by-case basis. The SCCS is, however, open to considering the application of read-across and mathematical models that will enable a category approach to conduct risk assessment of NMs in the future, when new knowledge will increase the understanding of the key parameters that drive the properties, biological interactions, and toxicological effects of NMs [13].

Further guidance on NM data submission is provided in the SCCS's "Notes of Guidance for the Testing of Cosmetic Ingredients and Their Safety Evaluation," where it is stressed that specific properties be taken into consideration in safety assessment [14]. Furthermore, in its memorandum "Relevance, Adequacy and Quality of Data in Safety Dossiers on Nanomaterials" [15], the SCCS highlights the importance of submitting data that are relevant, adequate, and of good quality in support of risk assessment. This also highlights that when NMs are submitted under the same dossier, the possibility to read across should be justified.

1.2.3 Biocidal Products

The risk assessment and authorization of NMs in biocidal products is addressed by the Biocidal Products Regulation (EU) No 528/2012 [16]. It states that for the approval of NMs as active substances and for subsequent product authorization, the test methods applied to the NMs shall be accompanied by an explanation addressing their scientific appropriateness, taking into consideration the specific characteristics of each NM. The information requirements for active substances are outlined in Annex II of the regulation, whereas those for the biocidal product are given in Annex III. The general rules for the adaptation of the data requirements for biocides are presented in Annex IV of Biocidal Products Regulation (EU) No 528/2012. These are in accordance with Annex XI of the REACH regulation [3].

1.2.4 Plant Protection Products

The Plant Protection Products Regulation (EC) 1107/2009 [17] does not make any explicit reference to NMs. The data requirements for active substances and plant protection products are set out in Annexes II and III to Directive 91/414/EEC [18]. Guidance on how to assess the risk of NMs in plant protection products is provided by European Food Safety Authority's (EFSA's) Guidance on the Risk Assessment of the Application of Nanoscience and Nanotechnologies in the Food and Feed Chain [19].

1.2.5 Food Production

The use of nanotechnology and NMs in food production is currently covered by EC No 258/97 concerning "novel foods" and "novel food ingredients" [20], which addresses food not consumed to any significant degree in the EU prior to May 1997. The proposal for the revised regulation [21] addresses NMs and nanotechnology more explicitly by covering:

> Foods modified by new production processes such as nanotechnology and nanoscience and food or vitamins, minerals and other substances containing or consisting of 'engineered nanomaterials'.

A premarket approval (safety assessment and authorization) is required for novel food as well as for food additives.

In 2011 the EFSA issued "Guidance on the Risk Assessment of the Application of Nanoscience and Nanotechnologies in the Food and Feed Chain" [19], providing a practical approach for assessing potential risks arising from applications of nanoscience and nanotechnologies in food additives, enzymes, flavorings, food contact materials, novel foods, feed additives, and pesticides. This guidance is now being updated. This document provides guidance to applicants for data generation on the PC characterization and on testing approaches to identify and characterize hazards arising from the properties of NMs. It describes six exposure scenarios, depending on the possible transformations of the NM before and after ingestion of the food/feed, and for each of these scenarios, the type of data needed to conduct a risk assessment is specified. When no exposure to NMs is verified, for example, by data indicating no migration[2] from food contact materials or by demonstration of complete degradation/dissolution with no possible absorption of NMs, the information requirements can be reduced. Sometimes, there is direct use of NMs in food or feed, where NMs are transformed to non-nanoforms in the food/feed matrix and before ingestion can be treated as and follow the guidance for non-nanoforms. For such NMs only local effects in the gastrointestinal (GI) tract and possible absorption before transformation should be considered. A comparison should indicate whether the nanoform has increased, less, or similar hazard as compared to the non-nanoform.

1.3 Risk Assessment

The three major steps in a risk assessment (or to prepare a chemical safety assessment under REACH) are (a) hazard assessment, (b) exposure assessment, and (c) risk characterization (REACH Annex I; ECHA, 2009; European Parliament and Council, 2006a).

The objective of a hazard assessment is to identify the hazards of the substance; assess their potential effects on human health and the environment; and determine, where possible, the no-effect levels (threshold levels). If a substance meets the criteria for classification as dangerous or is identified as PBT or vPvB, it will undergo exposure assessment and risk characterization.

An exposure assessment serves to measure or estimate the dose or concentration of the substance to which humans and the environment are, or may be, exposed in specific exposure scenarios that cover all identified uses and life stages of the substance. An exposure assessment is required only for hazardous substances in REACH.

For risk characterization, derived no-effect levels extrapolated from the in vivo no-effect levels are compared with the levels of exposure to assess whether the risk is controlled. When a no-effect level cannot be quantitatively determined, a qualitative or semiquantitative approach is used.

[2]The guidance document does not include details on analytical or mathematical methods for detecting migration of NMs to food and feed; the analytical approach should be well justified by the applicant.

Although the necessity of an exposure assessment and risk characterization depends on the results of the hazard assessment, exposure information may play an important role in influencing it. In situations where human or environmental exposure is absent or so low that additional effects information is not requested to improve the risk management, exposure-based waiving of hazard information can be considered (Reach Annex XI, Section 3, Substance-Tailored Exposure-Driven Testing [5]). Reducing/minimizing exposure is, therefore, an important risk management option in the development and production of NMs and NM-containing products, especially when there are still scientific uncertainties in the hazard assessment.

In contrast to exposure-based waiving, additional (targeted) testing can be triggered if the risk characterization indicates a need to investigate further the effects on humans or the environment of exposure situations. Exposure assessment also informs hazard assessment in terms of which routes of exposure (e.g., inhalation) and which concentrations are the most relevant to test. Exposure-based adaptations (EBAs) are an integral part of the ECHA Guidance on Information Requirements and Chemical Safety Assessment, Chapter R.5: Adaptation of Information Requirements [22]. According to this guidance, EBAs can be considered when they can be justified on the basis of no-release, strictly controlled conditions, absence of exposure, or no significant exposure. In practice, EBAs of the standard hazard testing regime under REACH require reliable information on use and exposure, which is often not available for NMs.

The applicability of current risk assessment methodologies for NMs has been assessed by the EC scientific committees, including the Scientific Committee on Emerging and Newly Identified Health Risks (SCENIHR), the SCCS, the EFSA, and the European Medicines Agency. The SCENIHR [23] concluded the following:

> Risk assessment methodologies for the evaluation of potential risks of substances and conventional materials to man and the environment are widely used and are generally applicable to nanomaterials. However, specific aspects related to nanomaterials still require further development. As there is not yet a generally applicable paradigm for nanomaterial hazard identification, a case-by-case approach for the risk assessment of nanomaterials is still warranted.

Similar conclusions were made by the scientific committee of the EFSA for NM applications in food and feed [24].

The Scientific Committees as well as the Organisation for Economic Co-operation and Development (OECD) have identified that for NMs, further development of some standardized and validated methods is required, for example, for sample preparation and characterization, exposure data and models, and some endpoints in toxicological test guidelines (TGs) (Section 1.6). Harmonization and standardization of measurement and test methods in support or risk assessment of NM is being promoted through the OECD and by a commission mandate to the European Standards Organizations [7].

Several (regulatory) risk assessments on NM-containing products have been completed, and various products in different sectors have been authorized (second regulatory review 2012: 20 medicines, 3 food contact materials, and 4 cosmetic products) [7]. The SCCS (and its predecessor committees) has performed a number of risk assessments on NMs used as cosmetic ingredients (Table 1.1).

Table 1.1 SCCS risk assessment on nanomaterials used as cosmetics

Nanomaterial	Document
Titanium dioxide as UV filter, as pigment	SCCNFP/0005/98
Zinc oxide in sunscreen	SCCP/1147/07
TiO_2	SCCS/1516/13 + SCCS/1539/14
ZnO	SCCS/1489/12 + SCCS/1518/13
ETH 50 (nano and non-nano)	SCCS/1429/11
MBBT	SCCS/1460/11
Silica	SCCS/1545/15
Carbon black	SCCS/1515/13
Hydroxyapatite	SCCS/1566/15
Additional coatings for TiO_2 (nanoform) as UV filter in dermally applied cosmetic products	SCCS/1580/16
Titanium dioxide (nanoform) as UV filter in sprays	SCCS/1583/17 (in finalization)

Note: Documents available at https://ec.europa.eu/health/scientific_committees/consumer_safety/opinions_es

1.4 Properties That Drive NM Behavior (Fate and Toxicity)

Table 1.2 provides a list of properties that have been identified as relevant for understanding the fate and effects of NMs. These properties have been identified from the discussions of the OECD Working Party on Manufactured Nanomaterials (WPMN) [32], Member States for REACH implementation, the SCENIHR (2009), and the Scientific Committee on Consumer Safety (2012) as well as the scientific literature. Information requirements for NMs under REACH are at the moment not yet finalized; inclusion of relevant properties in REACH submissions is under discussion [13, 23]; such properties include crystalline phase, dustiness, crystallite size, zeta potential, pore density, radical formation potential, surface chemistry and functionalization, specific surface area (SSA), dispersibility, (photo)catalytic activity, particle size distribution, and agglomeration/aggregation states [25, 32].

Properties in Table 1.2 have been organized following the scheme proposed by Sellers et al. (2015) into chemical identity, particle characteristics, fundamental transport and behavior, and reactivity/activity, which is consistent with the properties suggested in ECHA's voluntary practical guide on grouping NMs for read-across [37, 50]. In the Sellers proposal, the chemical identity group comprises an NM that has impurities and surface coating and information on crystal structure. The particle characteristics group may include shape, size-related properties, and porosity. The transport and behavior group covers properties that influence the movement of an NM in the environment or within the organism and may depend on the NM or on the environment, like water solubility, dispersibility, and zeta potential. The reactivity/activity group comprises properties that are exclusively related to the reactivity of the NM as a whole, like radical formation potential, catalytic activity, protein binding properties, hydrolysis, and dissociation constant.

Table 1.2 Physicochemical properties and their relevance in (eco)toxicological assessment

Category	Term	Definition (unit of measure [um])	Relevance in (eco)toxicological assessment	REACH requirement or OECD recommendation
Chemical identity	Chemical composition	It is the chemical identity and atomic arrangement of the nanomaterial; it shall consider also impurities and additives (purity is expressed as the percentage of the intended NM present).	The type of elements involved in the core as well as the form or valence of those elements are relevant in toxicology.	The definition is applicable to NMs [25, 28].
	Crystal type[a]	It refers to the specific arrangement of a chemical group (crystallite) in the 3D space. In some cases more crystal types of the same NM may observed: in this case the relevant information is the fraction of the different crystalline forms that are present [25].	This is a recognized indicator of the biological impact of NMs [29, 30].	It is considered a regulatory-relevant endpoint [31].
	Crystallite size	It is the size of the crystal or the grain.	This is important for ensuring comparability between tests and for interpreting data from the tests.	For crystalline materials it is considered a regulatory-relevant endpoint [31].
	Surface chemistry	It refers to the chemical nature, including composition, of the outermost layers of the nano-object.	It allows the recognition of the various modifications of the surfaces of NMs that will lead to numerous potential interactions and will play a key role in determining fate in natural aqueous systems, colloidal stability, and exposure [25, 32].	It is a relevant regulatory parameter [26, 33].

[a]Crystalline phase in the OECD [25]

Category	Term	Definition (unit of measure [um])	Relevance in (eco)toxicological assessment	REACH requirement or OECD recommendation
Particle characteristics	Surface functionalization, coating	Functionalization is mainly related to the introduction at a surface of chemical groups intended to subsequently react with other species [34]. Surface modification, functionalization, and coating are considered in RIPoN 1 [35] as synonymous with surface treatment.	Surface treatment is considered to affect environmental fate and health safety of NMs [32].	It is considered as surface chemistry [26].
	Basic morphology	NMs with the same composition may have different shapes (i.e., spheres, rods, tubes, fibers, and plates), which may be related to different physical, chemical, and biological properties. Shape is defined by the OECD [25] as the (semi)qualitative geometrical description or dimensionless term of the extremities of the particle, collections of particles, and the agglomerates or aggregates of the particles of the material under investigation. Recognized shape descriptors are sphericity, circularity, aspect ratio, convexity, and fractal dimension. The ISO [36] defines three levels of detail in describing shape, described by the geometric form of the external boundary of the particle and referring to convexity, perimeter, and circularity at different scale levels (macro-, meso-, and microshape descriptors; um: m/m)[b].	The elongation (or aspect) ratio (together with rigidity) is an important determinant of the potential health effects of fibrous materials, including NMs.	It is considered as shape; it is considered in the EC definition and required by regulation [33].

[b]Macrodescriptors are defined from size measurements made on the particle silhouette; mesodescriptors are morphological mathematical descriptors (computing robustness and the largest concavity index), a concavity tree (providing general insight into the organization of concavities and their complexity), and angularity descriptors (fractal dimension, Fourier descriptors, and bending energy); microdescriptors determine the roughness of shape boundaries using fractal dimension and higher-order Fourier descriptors/coefficients for surface and textural analysis [36].

(*Continued*)

Table 1.2 *(Continued)*

Category	Term	Definition (unit of measure [um])	Relevance in (eco)toxicological assessment	REACH requirement or OECD recommendation
	Particle size	It is the physical dimension of the smallest discrete form of a substance under specified measurement conditions. If particles in a group are of differing sizes, they may be described by particle size distribution [25].	Once the particles are internalized, particle size may also affect the distribution within the body and the toxicity both at the point of entry and distally. Size distribution is not a static parameter; it may also change during the course of (environmental) toxicity testing (as well as during the life cycle of the material) due to, for example, partial dissolution, interaction with test media, or preferential absorption of smaller particles [37].	
	Pore density	Pore density is defined as the measure of the void (i.e., "empty" spaces) in a material and is a fraction of the volume of voids over the total volume (g/cm^3).	Pore density has relevance in the fate of NMs as it influences their kinetics (i.e., sedimentation rate and size distribution as suspended material) in different media [38, 39].	It is considered relevant in the regulatory context [26].
	Specific surface area (SSA)	SSA is the surface area of the particle surface area (SA) [m^2] per unit mass m (g) SSA = SA/m (m^2/g) [40]. It is an intensive property and hence allows expressing the information related to the SA in a comparable way, independent of the quantity of the available material. The area of the exposed surface of a single particle plays an important role in influencing the physical and chemical interactions of the particle in the media. SCENIHR [23] and ECHA [9] consider also the concept of volume-specific surface area (VSSA, m^2/cm^3) $\text{VSSA} = \frac{\text{SA}}{\text{V}} = \text{SSA} \cdot \rho$,	Chemical reactions take place at surfaces; hence a high SSA represents high reactivity. SSA is relevant for a number of parameters for (eco)toxicological and risk assessment. It will dictate the surface charge density in cases where NMs are surface-functionalized. This, in turn, has direct consequences for (i) NM interactions (i.e., agglomeration) with other naturally occurring particulate matter (i.e., contaminant vectors), (ii) route of exposure as a function of surface ligand–biological interface (i.e., bioaccumulation pathway and bioavailability), and (iii) mechanisms of toxicity (e.g., dose response curves normalized for SA may indicate different results compared to results presented on a per mass basis).	It is considered relevant for the definition of NMs [2].

Category	Term	Definition (unit of measure [um])	Relevance in (eco)toxicological assessment	REACH requirement or OECD recommendation
		where ρ is material density and V [cm^3] is the volume of the material. VSSA is proposed as a measurement that can be used to distinguish dry solid nanostructured material from non-nanostructured material at VSSA ≥ 60 m^2/cm^3 [40].		
Fundamental transport/behavior	Dispersibility	Dispersibility is the degree to which a particulate material can be uniformly distributed in another material (the dispersing medium or continuous phase). A dispersion is a suspension of discrete insoluble particles in a fluid that may falsely have the visible appearance of a solution (i.e., the product of the conversion of a solid substance into a liquid form by mixture with a solvent). A dispersion of an insoluble material may elicit a different response from that anticipated from the classical molecular or the elemental toxicity expected from the chemical composition. Dispersion stability is an important parameter to assess in the context of sample preparation [8].	It may influence the ability of the NMs to reach and enter the cell [25].	The definition is under discussion [25, 26].

(*Continued*)

Table 1.2 *(Continued)*

Category	Term	Definition (unit of measure [um])	Relevance in (eco)toxicological assessment	REACH requirement or OECD recommendation
	Water solubility	Water solubility/dispersibility refers to the mass proportion of a given sample of NM that is held in a water solution or as a colloidal suspension in water as a function of time or where the sample of NM loses its particulate character as it changes from a particle form to a molecular form [25]. (um: μg/L, mg/L, g/L, g/cm^3, kg/m^3, ppb, and vol%)	If an NM is soluble, it is likely to be presented to the in vitro/in vivo test system in a molecular or ionic form and can be expected to elicit the same response as more usual chemical forms of the material (with different dissolution rates compared to the bulk substance). If the NM under investigation is insoluble in biological or environmental media, then it will be presented to the test system in a particulate form and might elicit a different response from that expected on the basis of the chemical composition.	The definition is already required in REACH [5, 9] but needs revision for NMs [26].
	Dissolution rate	According to experimental evidence, the rate of dissolution of soluble materials increases with decreasing particle size and differs in different media; the dissolution rate is relevant as a process when dealing with NM behavior [41].	The dissolution rate is considered in several studies as a measure for NM biodurability; it is then related to NM (bio)persistence [41, 42].	The definition is not required in REACH.
	Hydrophobicity/ Hydrophilicity	Hydrophobicity refers to the "water-avoiding" behavior of a nonpolar molecule or group that has low affinity for water molecules. Hydrophilicity refers to the "water-liking" behavior of a polar molecule or group that has high affinity for water molecules.	Influences protein binding to NM [43, 44], and characterizes inhibitory activity towards various enzymes [45]. Hydrophilicity is one of the most relevant descriptors in uptake models [46].	The definition is not required in REACH.

Category	Term	Definition (unit of measure [um])	Relevance in (eco)toxicological assessment	REACH requirement or OECD recommendation
	Dustiness	It is the propensity of a material to generate airborne dust during its handling. The measure of interest is the degree to which a given NM can remain in the air column before settling. Interactions of NMs with other common airborne particulate matter should be studied.	It provides a basis for estimating the potential for inhalation exposure [47].	The definition is relevant as a regulation requirement [31].
	Zeta potential (surface charge)	It is defined as the charge at the particle interfaces. It is a repulsive interparticle force. Because a colloidal, charged system is a stable system [48], therefore, ionization enhances the particle stability. In fact, a charged surface attracts counterions in its vicinity. Therefore, when two particles approach each other, their diffuse layers will overlap and the resultant repulsive force may outweigh the attractive van der Waals attraction, rendering the suspension stable [mV]. Van der Waals forces are expressed by the Hamaker constant, which is related only to the properties of the interacting bodies and the medium and consists of the summation of intermolecular interactions [49]. This constant could be useful in determining the degree of agglomeration and sorption [50].	Zeta potential can be related to the stability of colloidal dispersions. Zeta potential indicates the degree of repulsion between adjacent, similarly charged particles in dispersion. For molecules and particles that are small enough, a high zeta potential will confer stability, that is, the solution or dispersion will resist aggregation. When the potential is low, attraction exceeds repulsion and the dispersion will break and flocculate. In nanotoxicology, zeta potential (surface charge) plays a key role in determining the (i) degree of colloidal interaction, which is itself a function of the pH and ionic strength of the bulk solution, and (ii) bioavailability when considering mass transport through charged membranes.	The definition is considered relevant as a regulatory requirement [26]; there is no OECD guideline for measurement [27].

(Continued)

Table 1.2 (*Continued*)

Category	Term	Definition (unit of measure [um])	Relevance in (eco)toxicological assessment	REACH requirement or OECD recommendation
Reactivity	Steric hindrance	It refers to the repulsion short-range effects a rising from surface characteristics (e.g., presence of polymers) [51].	It influences agglomeration [52] and protein binding [43] of NMs. It is considered relevant as a property driving toxicity of NMs [53].	The definition is not required in REACH.
	Protein binding	It is the mechanism of chemical interaction with proteins. It can be covalent (irreversible; common with electrophilic toxicants, such as nonionic and cationic electrophiles and radical cations) or noncovalent (apolar interactions or the formation of hydrogen and ionic bonds and typically involved in the interaction of toxicants with targets such as membrane receptors, intracellular receptors, ion channels, and some enzymes) [54].	Protein binding results in increased stability of NMs [55] and affects biodistribution of NMs throughout the body [44].	The definition is not required in REACH.
	Hydrolysis	It refers to the cleavage of chemical bonds by the addition of water.	Uptake is modulated by hydrolysis and hence may influence exposure to a substance [56].	The definition is already required in REACH [5].
	Catalytic activity	This is the ability of some NMs to speed up certain reactions as a catalyst [mol/(g*s); mol/(g*h)].	It causes ionization events, leading to free radical production [57].	The definition is under discussion [25].
	Photocatalytic activity	This is the ability of some NMs to speed up certain photoreactions as a catalyst, eventually in combination with light (sunlight, ultraviolet [UV] light) [mol/(g*s); mol/(g*h)].	It is recognized as a relevant PC property in toxicity testing [52]. Photocatalytic activity may drive oxidative stress under UV light [58].	The definition is under discussion [25, 26].
	Conduction band (valence band, bandgap)	This is the energy required to free an electron from its bond to an atom (eV).	Substances with conduction bands in the range of cellular membrane redox potential (−4.12 to −4.84 eV) possibly cause oxidative stress [59, 60].	The definition is not required in REACH and not considered in OECD recommendations.

Category	Term	Definition (unit of measure [um])	Relevance in (eco)toxicological assessment	REACH requirement or OECD recommendation
	Electrophilicity and nucleophilicity (acidity and basicity)	Electrophilicity is the attraction to electrons; electrophiles tend to accept electrons. Nucleophilicity is the tendency to donate an electron pair to an electrophile in a reaction.	Electrophilicity/nucleophilicity is related to the reactivity of NMs [61, 62].	The definition is not considered in REACH and not considered in OECD recommendations.
	Radical formation potential	This is the ability of a substance to produce reactive oxygen species (ROS), including, for instance, superoxide radical O_2·, hydrogen peroxide H_2O_2, and hydroxyl radical OH·.	These ROS compounds exert severe cellular damage, such as oxidation of DNA, proteins, or lipids, which may cause cell death [112].	The definition is under discussion [25, 26].
	Redox potential	This refers to the measure of the tendency of a chemical species to acquire electrons and thereby be reduced (mV).	Oxidizing substances may interact with cells, thus altering their redox balance and causing oxidative stress [63]. This is related to the conduction band.	The definition is already required in REACH [5].
	Dissociation constant	It is the reversible reaction of a substance resulting in two or more chemical species, which may be ionic; the process may be represented as RX · R – X. The dissociation constant K is expressed as the ratio of concentrations of the species on either side of the equation at equilibrium: $K = \frac{[R][X]}{[RX]}$, where the cation R+ is hydrogen, the substance can be considered an acid, and so this constant becomes an acid dissociation constant (K_a) [28].	It is related to protein binding [64, 65] and may be important for interpreting the agglomeration of NMs [66].	The definition is already required in REACH [5].

There is a distinction between intrinsic and extrinsic properties. Intrinsic properties are characteristics of the material itself that do not account for interactions with other components [25]. Extrinsic properties are characteristics resulting from interactions occurring at the interface (i.e., boundary) and the surrounding medium (which may be an environmental or biological matrix or medium in an experimental test system). This distinction is important in the development of in silico models for PC properties. With regard to this distinction, the PC properties listed in Table 1.2 are to be considered as intrinsic properties that aim at defining the nature of the pristine NMs. Any property that may be modified by the interaction of the NM with the environment can be considered as an extrinsic property and may be then defined according to the environment (e.g., behavior properties may be also considered extrinsic properties). Table 1.2 shows the list of parameters that can be submitted following the voluntary ECHA practical guide and those that are taken into consideration by the OECD WPMN testing program [26, 27]. PC properties that are considered relevant in the literature are also added to the table.

1.4.1 Theories Underlying Environmental and Biological Fate

The OECD published 118 guidelines for the testing of chemicals, and the WPMN evaluated which of these is applicable to NMs [67]. This review concluded that 4 out of 22 TGs for physicochemical properties are applicable to NM, 16 guidelines might be applicable under some circumstances or to some classes of NMs, and 2 guidelines are not applicable to NMs or, if applicable, provide no useful information.

Standard methods are needed in order to fulfil the information requirements listed in Table 1.3. Measurement methods and protocols for providing information listed in the previous table are not yet identified, although several projects are developing measurement protocols. In particular, NanoReg developed several standard operating procedures on particle size distribution [68].

The OECD WPMN also aims at providing test guidance for the determination of PC properties for NMs to support the submission of information on substances within IUCLID.

Table 1.3 List of the existing methods supporting the measurement of NM properties

PC property	OECD guideline[1]	Applicability for NM testing	ISO standards	Other available methods
Dispersion/ agglomeration/ aggregation	None available [32]	-	Centrifugal liquid sedimentation, dynamic light scattering, small- angle X-ray scattering, single-particle ICP-MS, particle-tracking analysis, field-flow fractionation, transmission electron microscopy (TEM), and X-ray diffraction [70].	-
Water solubility	105	Existing guidance needs to be revised [32]	-	Solubility of particles of size below 4 mm considered in CEN [73]
Zeta potential	None available [32]	-	Electrophoretic and electroacoustic methods available [72–74]	-
Composition of nanomaterial	None available [67]	-	ISO documents also available [75–79], one specifically for NMs [80]	Bulk: Elemental analysis, ICP-MS, EDX, EELS, dynamic-SIMS, 3D atom probe [67]
Particle size distribution—dry and in relevant media	110	Might be applicable to NMs under certain circumstances [67]	Many techniques available from ISO, many of them recently revised [81–87]	On TEM analyses [88–90]

(*Continued*)

Table 1.3 *(Continued)*

PC property	OECD guideline[1]	Applicability for NM testing	ISO standards	Other available methods
Basic morphology	None available [67]	-	Scanning electron microscopy (SEM) and TEM [75, 76, 79] Specific guidelines for carbon nanotubes [76]	-
Crystallite type	-	-	-	Crystalline phase guidance available from Japanese Industrial Standards [91]
Crystallite size	-	-	Grain size determined via SEM and TEM [75, 81]; a guideline available for aerosols [92]	Powder XRD, HR-TEM, Raman spectroscopy [67, 93]
Surface chemistry	None available [32]	-	Available guidelines for single-walled carbon nanotubes given by ISO [78, 94], plus other guideline on other methods [95–97]	Fourier-transform infrared spectroscopy, time-of-flight secondary-ion mass spectrometry, matrix-assisted laser desorption time-of-flight mass spectrometry, Auger electron spectroscopy, XPS, electron energy loss spectroscopy, UV-vis, and molecular spectroscopy; ICP-MS, ICP-AAS, LC-MS, GC-MS, and TGA [32]; other guidance [100]
SSA	None available [32]	-	Brunauer Emmett Teller (BET) [83]	BET method and TEM methods [32, 67, 98][4]; small-angle X-ray scattering (SAXS) [99]

PC property	OECD guideline[1]	Applicability for NM testing	ISO standards	Other available methods
Photocatalytic activity	None available [32]	-	-	DCF fluorescence–based analysis for reactive oxygen species (possible NP interference with the assay), colorimetric methods (possible NP interference with the assay), electron spin resonance, quartz crystal microbalance, and gas-phase techniques (such as gas chromatography)
Hydrolysis	111	Might be applicable to NMs under certain circumstances [67]	-	-
Dustiness	-	-	-	Relevant guideline on airborne dust (not specific to NMs) [101, 102]
Pore density	None available [32]	-	BET [32]; methods available from ISO but not specific to NMs [103–105]	-
Octanol-water partition coefficient where relevant	107, 117, 123	Not suitable for NMs (except fullerene)		Other kinetics to be considered for NMs; methods not sufficiently developed to become standardized [32]
Hydrophobicity	116	Applicable to NMs [67]	-	-

[1]ISO methods are not defined specifically for NMs.

Table 1.3 lists the methods for relevant PC and environmental fate properties according to the OECD WPMN discussions [32, 69]. Characteristics that are not reported in the OECD guidelines are not listed here. When OECD guidance is available for the identified PC property, its applicability as addressed in OECD [67] is reported in a dedicated column. When an OECD guideline is not available, other relevant guidelines are reported [32, 67].

1.5 Understanding NMs' Fate and Toxicity

Exposure of humans and the environment to NMs can occur throughout their life cycles, from their production and manufacture (incorporation into products) to their use and disposal. The different phases in its life cycle may affect the NM's properties and hence its interactions with environmental and biological media [106].

Once humans are exposed to NMs, the particles can travel throughout the body and get deposited in target organs, where they may penetrate cell membranes and even enter organelles, such as mitochondria. They can, therefore, affect biology at different levels [107], which may result in injurious responses.

This section is aimed at (i) introducing well-established theories that have been applied in investigating the fate of NMs in environmental and biological systems, (ii) summarizing the available information on the kinetics of NMs, and (iii) reporting evidence on their toxicodynamics (TD).

1.5.1 Theories Underlying Environmental and Biological Fate

The fate of NMs in environmental media and biological systems is partly determined by particle-particle and particle-surface[3] interactions, leading to agglomeration/aggregation processes that may result in colloidal instability and gravitational sedimentation. These processes are relevant in both aqueous and gaseous compartments.

---

[3]A surface is conceptualized as an infinite flat plate and is relevant in studying the deposition of NMs [260]. In some instances a natural colloid interacting with an NM is assumed as a surface because of the big difference in dimension [39].

1.5.1.1 Agglomeration and aggregation kinetics in fluid media

The behavior of NMs in fluids is dependent on particle-specific properties, the chemistry of the surrounding medium (e.g., pH, ionic strength, ionic composition, presence of proteins, or colloids), and hydrodynamic conditions [108–110].

The transport of NMs in fluids follows Brownian diffusion, a random and chaotic motion of particles resulting in collisions between the particles and the fluid molecules surrounding them. The trajectory of a given particle is of a self-similar nature, meaning that any magnified portion of the trajectory would look qualitatively similar to the original one [111].

Depending on the environmental conditions and on intrinsic PC properties, NMs can agglomerate or aggregate in fluid media (see Table 1.2 for definitions). Agglomerates can deagglomerate in the presence of stabilizing agents such as proteins [112]. Changes in particle size resulting from agglomeration or aggregation may influence the interactions of NMs with natural colloids (e.g., humic and fulvic acids) and macromolecules (e.g., proteins and peptides) [113], as well as their reactivity and toxicity. Although the processes of agglomeration and aggregation are relevant in predicting the fate of NMs, there are no theoretical models capable of distinguishing between the two processes.

Ongoing research is focusing on identifying which environmental parameters influence particle coalescence [114, 115] and on the parameters that can be used to model and predict the fate of NMs [116, 117]. In these models, aggregate formation is assumed to be irreversible, and aggregation and agglomeration processes are not distinguished [118, 119].

In the following paragraphs, the main theories considered in modeling aggregation/agglomeration processes are presented.

1.5.1.2 DLVO theory

The classical Derjaguin, Landau, Verwey, and Overbeek (DLVO) theory of colloidal stability has been proposed to address the kinetics of agglomeration processes. DVLO theory combines the opposing effects of the van der Waals attractive force and the electrostatic repulsive force due to the so-called double layer of counterions, that is, zeta potential of NMs. The repulsive force depends on the double-layer potential and thickness, the particle radius, and the dielectric constant of the medium. The potential energy curve as a function of the separation distance between particles has an energy barrier and two minima, as shown in Fig. 1.1.

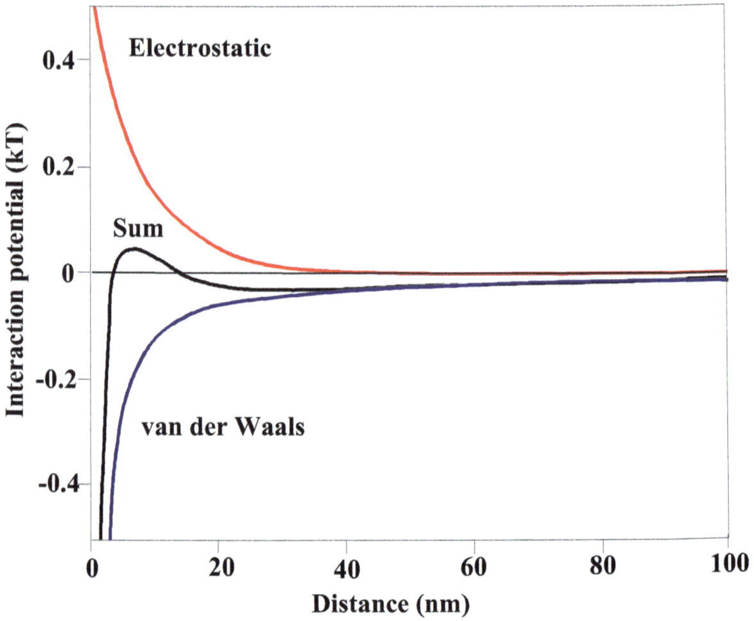

Figure 1.1 Potential energy curve for the interaction between colloidal particles. The interaction depends on a balance between attractive van der Waals forces and repulsive electrostatic forces.

General rules can be used to determine colloidal stability. For example, the thicker the double layer around the particles, the greater the repulsion between them. Repulsive forces between particles decrease when the pH of the medium is close to the isoelectric point value of the particle. An increased concentration of an inert electrolyte causes the electrical double-layer potential to decrease, and the same effect is observed when an organic, water-miscible solvent is present, which causes the dielectric constant of the medium to decrease. Repulsive forces between particles can be caused by adsorption of polymers, proteins, nonionic detergents, and polyelectrolytes. The intentional coating of particles with polymers causes steric stabilization, although partial coverage can lead to agglomeration.

Van der Waals forces are expressed by the Hamaker constant, which is related only to the properties of the interacting bodies and the medium and consists of the summation of intermolecular interactions [49]. In aqueous environments, when particles approach each other (coalescence) or a surface (deposition), the overlap of the diffuse electric double layers results in electrostatic double-layer interactions.

Non-DLVO interactions that are relevant for the agglomeration/aggregation and deposition of NMs in an aqueous environment include steric interactions, magnetic forces (for iron-based NMs), and hydration forces [108]. Expressions describing the extent of steric forces have been derived for particles with adsorbed layers of polymers or surfactants that might lead to steric repulsion. Magnetic materials exhibit dipole behavior that overcomes the particle-particle interactions. Finally, particles that carry hydrophilic material, such as proteins, on their surfaces may attract water molecules and hence have a hydrated surface. Such particles will exert a repulsive interaction, thus influencing NM stability [120]. Parameters for the calculation of non-DLVO interactions are not available, and this is the reason why this theory is not extensively applied [59].

DLVO is taken into consideration in environmental fate models in the absence of measurements for calculating the attachment efficiency α [39]. According to the DLVO theory, the total interaction energy between suspended particles can be evaluated as the sum of the attractive van der Waals and the repulsive electrical double-layer energies. The aggregation efficiency α is defined as:

$$\alpha \approx k_{\text{Debye}} 2 r_{\text{i}} e^{\left(-\frac{V_{\max(i,j)}}{k_{\text{b}} T}\right)}, \tag{1.1}$$

where $V_{\max(i,j)}$ is the total interaction energy between particles i and j, k_{b} is the Boltzmann constant, T is the temperature, k_{Debye} is a parameter related to the length of the double layer [121], and r_i is the radius of particle i. Other authors have proposed different calculations of α, for instance, by taking into consideration the relevance of both interaction energy and Brownian diffusion in the transport of NMs [122, 123]. The DLVO theory is also taken into consideration in biological environments in predicting protein binding to NMs [124], where it is applied in simulating NM-protein interactions [4][4].

1.5.1.3 Smoluchowski–Friedlander theory

Smoluchowski [125] and Friedlander [126] calculated the kinetics of spherical dispersed particles in fluids, and this theory is sometimes considered to represent the kinetics involved in aggregation and coagulation processes such as flocculation and Ostwald ripening. Simplified versions of the Smoluchowski equation are considered, for instance, in the calculation of aggregation constants [117, 127], as a theory at the base of model development [128].

This approach considers three types of kinetic process: Brownian diffusion (perikinetic aggregation, due only to particle diffusion, Eq. 1.2), fluid motion (orthokinetic aggregation, which happens in the presence of flow, Eq. 1.3), and differential settling (Eq. 1.4) [118]:

$$k_{i,j} = \frac{2kT}{3\mu}\frac{(a_i + a_j)^2}{a_i a_j} \tag{1.2}$$

$$k_{i,j} = \frac{4}{3}G(a_i + a_j)^3 \tag{1.3}$$

$$k_{i,j} = \left(\frac{2\pi g}{9\mu}\right)(\rho_s - \rho)(a_i - a_j)(a_i + a_j)^3, \tag{1.4}$$

where k_{ij} are the different collision rate constants of particles (or aggregates) i and j, k is the Boltzmann constant, T is the absolute temperature, a_i and a_j are the particle radii, μ is the viscosity of the fluid, G is the shear rate, g is the acceleration due to gravity, ρ is the density of the fluid, and ρ_s is the density of the particle. The diffusion coefficient of a spherical particle undergoing Brownian motion is expressed by the Einstein Stokes equation (1.5), from which Smoluchowski derived the definition for the perikinetic rate constant (Eq. 1.4):

$$D_i = \frac{kT}{6\pi a_i \mu} \tag{1.5}$$

Assuming each collision is effective in forming an aggregate, it is possible to write the rate of change of concentration of aggregates as Eq. 1.6:

[4]The software is available online at http://ovilanova.github.io/BUBBLES/ and is an output of the NanoTranskinetic FP7 project. Publications are not available yet.

$$\frac{dn_k}{dt} = \frac{1}{2} \cdot \sum_{\substack{i=1 \\ i+j \to k}}^{i=k-1} k_{i,j} n_i n_j - n_k \sum_{k=1}^{k=\infty} k_{i,k} n_i, \tag{1.6}$$

where the first term on the right-hand side represents the rate of formation of k-fold[5] aggregates by collision of any pair of aggregates, i and j, such that $i + j = k$. The second term accounts for the loss of k-fold aggregates by aggregation with any other particles or aggregates. The terms k_{ij} and k_{ik} are the appropriate rate constants (Eqs. 1.2–1.4), n_i and n_j are the number of concentrations of different aggregates (n_i are particles of size i, and n_j are particles of size j).

These equations have been combined in several applications aimed at studying the behavior of NMs in environmental media [108, 117, 128, 129] and in human exposure [130]. For instance, Arvidsson et al. [128] applied Eq. 1.6 in defining an environmental fate kinetic model where the number of particles of k-fold aggregates varies in time as a function of aggregation efficiencies[6] $\alpha_{i,j}$ (see definition in Table 1.4), adding a term for sedimentation and one term for advection in the system. In the field of human exposure this equation was applied in the ENPRA EU FP7 project for the estimation of coagulation kinetics of particles in the aerosol, where particles of any size can coagulate to form aggregates [130].

Considering the low environmental concentrations of NMs and the corresponding lower probability of interaction of two NMs in real fluids, scientists in the field identified heteroagglomeration/aggregation rather than homoagglomeration as a relevant process in the environmental fate and transport of NMs [39]. The inherent challenges in the prediction are still high as there is no clear measurement technique or recognized protocol for the measurement of attachment efficiencies, which is so far only possible though the application of semiempirical models [123].

---

[5]Equations are defined assuming an initial dispersion of identical particles that after a period during which aggregation occurs contains aggregates of various sizes and different concentrations: n_i particles of size i, etc., where n_i refers to the number concentrations of different aggregates and "size" implies the number of primary particles comprising the aggregate; we can speak of 'i-fold' and 'j-fold' aggregates [118].

[6]Arvidsson et al. [128] refer to $\alpha_{i,j}$ as "collision efficiency"; since it is defined as irreversible collision, it corresponds to aggregation efficiency [118].

Table 1.4 Relevant parameters taken into account by multimedia modeling (MMM) approaches

Parameter relevant for environmental fate models	Definition
Collision (or attachment or aggregation) efficiency, also referred to as "sticking probability" or "stickiness coefficient," α.	It is defined as the fraction of collisions effective in agglomeration in Elimelech et al. [118]. It is recalled by Praetorius et al. [38] and by Quik et al. [117] as attachment efficiency. If there is strong repulsion between particles then practically no collision gives an aggregate and $\alpha = 0$. When there is no significant net repulsion or when there is an attraction between particles, then the collision efficiency can approach unity. Zhang et al. [123] report that aggregation kinetics of various NMs have been extensively studied using attachment efficiency (α), commonly determined by normalizing the hydrodynamic size growth rate in initial aggregation curves to the growth rate under the favorable (or fast) aggregation condition in which the ionic strength is equal to or greater than the critical coagulation concentration.
Homo- and heteroagglomeration/aggregation rate constants (also defined as coagulation or flocculation rate constants according to Ref. [118].	It is the rate constant governing agglomeration/aggregation processes in kinetic models [128]. The homoagglomeration rate constant is referred to as the interactions between NMs of the same type and is defined as $k_{\text{hom-agg}} = \alpha_{\text{hom}}\ k_{\text{colli},j}$, where α_{hom} is the attachment efficiency for homoagglomeration and $k_{\text{colli},j}$ is the collision rate constant. Heteroagglomeration/aggregation is the interaction of natural colloids or of suspended particulate matter with NMs. Praetorius et al. [38] define it as $k_{\text{het-agg}} = \alpha_{\text{het}} \cdot k_{\text{colli},j}$, where α_{het} is the attachment (or collision) efficiency for heteroagglomeration.
Collision rate constant for particles of type i and j.	It depends on factors such as particle size and transport mechanism. It is usual to assume that the collision rate is independent of colloid interactions and depends only on particle transport (see Eqs. 1.3–1.5).

Note: The table reports definitions of the identified parameters that are used as input values in MMMs.

Following these approaches, and in order to express the kinetics of interactions of different nature, different authors defined homo- and heteroattachment efficiencies α_{homo} and α_{hetero} (definitions in Table 1.4) [38, 39, 128].

Brownian motion applies also in the gas medium, and diffusion is considered the main mechanism for collision of NM particulates in aerosols. Reflecting this, some authors have defined the polydisperse coagulation coefficient as a function of particle size and diffusivity in air [39]. Polydisperse particles are particles of different sizes (i.e., particles and their agglomerates or aggregates).

Agglomeration kinetics are also taken into account in the kinetics of NMs in in vitro systems, where the process is considered to be affected by NM concentration, surface chemistry, and zeta potential [131], where the concepts of gravitational settling and diffusion are also taken into account.

1.5.1.4 Fractal approaches

The Einstein Stokes equation (Eq. 1.6) is also applied in the analytical technique of dynamic light scattering assuming well-dispersed primary spherical particles [30]. When agglomerates or aggregates are formed, their loose packing can be accounted for by considering the aggregate/agglomerate fractal dimension [132].

The term "fractal" comes from the Latin adjective *fractus*, which means "to break," intended as creating irregular fragments [133]. Fractal geometry is an extension of conventional Euclidean geometry that allows size-dependent measures to change in a noninteger or fractional way when the scale changes. This characteristic can be described by assigning a fractional number—a fractal dimension—to the dimension of the object. The relationship between the mass M of an agglomerate/aggregate and its size L can be expressed in terms of a fractal dimension (Eq. 1.7):

$$M \approx L^{d_f}, \tag{1.7}$$

where M is the mass of the particles, L is a linear measure of size[7], and d_f is the mass fractal dimension [118]. Aggregates and agglomerates are considered fractal objects with fractal dimension $d_f < 3$. When

[7] L is a measure of length of a particle or aggregate/agglomerate. It may be the gyration radius of the aggregate or the largest diameter of an irregular aggregate.

Eq. 1.8 is valid at different scales, then aggregates have a self-similar structure, independent of the scale of observation.

Fractal approaches have been applied so far in a series of studies that have demonstrated that scale invariance is a common characteristic of biological systems, ranging from tissues to cultured cells, nucleus, and chromatin [134, 135]. In colloidal chemistry, fractal distribution is assumed in diffusion-limited agglomeration, leading to fractal morphology characterized by a $d_f \approx 2.1$ [136]. Being fractal, the structure of the colloidal aggregates can be studied in more detail in terms of the relationship of the cluster structure and its aggregation kinetics.

Fractal dimension d_f is mentioned in the ECHA Guidance on Information Requirements and Chemical Safety Assessment regulatory requirements (Appendix R7-1) in relation to the definition of NM shape [9]. It is also considered as a structural descriptor in QSAR applications [137] as well as environmental fate models [38, 128, 138], where it is used as a rough estimate of the density of the aggregates.

1.5.2 Human Kinetics

Human biokinetics/toxicokinetics (TK)/pharmacokinetics (PK) (all synonymous but from here on called TK) is the science that studies the concentration-time course of a chemical substance in blood, tissues, and excreta. It is dependent on the rates and extents of absorption, distribution, metabolism, and excretion (ADME). Distribution, metabolism, and excretion together are sometimes called "disposition," where metabolism and excretion may be lumped with the term "elimination."

1.5.2.1 Preabsorption processes

TK for classical substances starts with absorption. Absorption is usually defined as the passage across an outer lining (membrane), such as the GI tract epithelium, the skin, or the epithelium covering the airways. For soluble chemicals, the rate of absorption is not considered to be limited by the dissolution process. Thus, absorption is considered to occur directly from the mucus layer (produced by mucous membranes) in which the substance is dissolved, from the GI fluid, or from a very thin aqueous layer covering the skin.

For NMs, the situation may be different. Some NMs are relatively soluble, and once they are dissolved, the resulting ions behave like the soluble chemicals. Other NMs are poorly soluble but may still undergo various kinds of transformations, for example, change in speciation and/or attachment of organic molecules, taken from a paper on aquatic and terrestrial organisms in the environment [139]. This may occur in humans following airborne exposure that results in deposition in the aqueous lung lining phase, oral exposure that might result in release from the food matrix in the aqueous contents of the GI tract, or dermal exposure that might result in migration of a consumer product or direct deposition (following airborne exposure or skin contact with contaminated surfaces in occupational settings) toward the thin aqueous layer on the epidermis.

The above implies that "preabsorption" processes can have an influence on the systemic bioavailability of NMs (the amount that becomes available in the systemic circulation relative to the total exposure). The low solubility or its absence implies that there is "time" for other processes to occur. For example, local clearance (i.e., removal) may occur at the same time and leave less NM for absorption. In other words, unlike soluble chemicals, NMs can be cleared to varying extents before absorption.

Inhalation exposure: Considering the inhalation pathway, it is recognized that particles smaller than 10 μm can enter the lung. Particle deposition in the lung depends on particle size, density, and hygroscopicity (ability of a substance to attract and hold water molecules from the surrounding environment) and is influenced by the local anatomy and airflow, as reviewed recently by Braakhuis et al. [140]. They report that NMs with diameters in the range 10–100 nm enter the alveolar areas preferentially. For particles in the mentioned diameter range, the deposition of NMs is mainly governed by diffusion of the NMs in the inhaled air (Brownian motion) and the density is less relevant whereas for particles (or agglomerates) larger than 100 nm, diffusion is less likely but the density increasingly determines the deposition. Probably the shape and the extent of absorption contribute as well. Once deposited in the lung, (partially) soluble NMs dissolve (partially) in the lining fluid (mucus layer) of the epithelium, where inert NMs might form nondissolved but colloidal suspensions. Local clearance from the airways occurs as macrophages transport nondissolved NMs (single and agglomerated but still relatively small NMs) by mucociliary transport up to the laryngopharynx [141]. This can be followed by swallowing, after which absorption in the GI tract may occur via the so-called "mucociliary escalator". This is the transport by macrophages from the lower lung compartments "upward" to the bronchioles, which can be followed by coughing and swallowing. As an example, for agglomerates of 7 and 20 nm gold NPs, macrophage-mediated escalation followed by fecal excretion is the major pathway of clearing the inhaled NPs from the lungs in rat [142].

Local clearance may also result from the solubility of the NM, as suggested by Konduru et al. (2014) in a study on biokinetics and effects of barium sulfate NPs following inhalation and instillation in rats [143]. Pulmonary exposure to instilled $BaSO_4$ NPs caused dose-dependent lung injury and inflammation. Four-week and thirteen-week inhalation exposures to a high concentration (50 mg/m^3) of $BaSO_4$ NPs elicited minimal pulmonary response and no systemic effects. Instilled and inhaled $BaSO_4$ NPs were cleared quickly yet resulted in higher tissue retention than when ingested. Injected $BaSO_4$ NPs were localized in the reticuloendothelial organs and redistributed to the bones over time. $BaSO_4$ NPs exhibited lower toxicity and biopersistence in the lungs compared to other poorly soluble NPs such as CeO_2 and TiO_2. On the basis of these observations, particle dissolution was regarded as a likely mechanism.

Oral exposure: NMs can reach the GI tract via intake of food or water, or accidental ingestion, or following inhalation and clearance by the mucociliary escalator. Once ingested, NMs face different environments while passing through the GI tract. An important factor is the pH value as it affects the net surface charge of the NM [62]. In the stomach, it ranges from 1.5 to 2 in the fasting state and might rise up to 7 after ingestion of a meal. As the motility of the stomach increases, the contents are transported to the small intestine, where pancreatic bicarbonate secretion leads to neutralization of the chyme to a pH ranging from 6.4 to 7.5, depending on the intestinal section. The SA of the small intestine is amplified immensely toward the more distal part, due to the presence of crypts (the crypts of Lieberkühn) and projections (microvilli), thereby facilitating nutrient absorption (altogether, the intestine features an SA of 200–300 m^2). In addition to the wide pH range an NM has to face while passing through the GI tract, proteins such as mucins, lactoferrin, albumin, and other factors, are secreted that are capable of mineral chelation. These proteins have the potential to influence the absorption of an NM by affecting corona formation, solubility, agglomeration, and aggregation.

Dermal exposure: Personal care products such as sunscreens may contain nano-TiO_2 or nano-ZnO to increase their efficiency in filtering UV light and their transparency [144].

1.5.2.2 Absorption

NMs may enter the body via different routes, including inhalation, ingestion, and penetration of the skin.

Inhalation exposure: Soluble NMs that dissolve in the lining fluid of the lung epithelium can be transferred to the blood and distributed to the whole body [107]. Solubility (rate and extent of dissolution) depends on chemical composition, size, coating, stability, and the biological environment [140]. In the respiratory tract, less soluble NMs may be absorbed via cell-mediated active translocation from the site of deposition through the lung epithelium to interstitial sites. From there NMs may be directed to the local lymph nodes, and as lymph nodes are drained by blood, NMs may ultimately reach the systemic blood circulation. Uptake from the site of deposition into systemic blood may also happen directly by the NMs crossing the lung barrier in the alveoli [145]. Once less soluble NMs are deposited in the alveoli, there is a competition between alveolar macrophages and lung endocytes for cellular uptake. Cellular uptake by alveolar macrophages may result in local clearance via mucociliary transport before absorption to the systemic circulation can occur (as mentioned above). Once the NMs are taken up by endocytes, cell-mediated transport across the epithelial membrane results in their absorption.

Oral exposure: The level of NM uptake within the GI tract appears to be relatively low, albeit dependent on properties such as size, surface structure, chemical composition, and charge [43]. Various uptake routes are relevant for various size distributions: endocytosis through "regular" epithelial cells, uptake by M-cells (phagocytizing enterocytes in Peyer's patches), and possibly also persorption (via gaps at the villous tip when enterocytes are lost to the gut lumen) and paracellular uptake [146, 147]. Interspecies differences occur, for example, because there are more M-cells in rodents compared to humans [148].

Dermal exposure: The number of studies on skin penetration of NMs is limited. When Zn was applied topically as labeled ZnO-nano [149], most of healthy and intact skin SA seemed impermeable to the NM. Only the hair follicles and the openings of the sweat glands were available for some particle penetration. Also, the effect of damaged skin needs further investigation [150]. Furthermore, most studies seem to be performed under static conditions whereas flexing of the skin during exposure might enhance penetration [107].

The SCCS has evaluated the dermal penetration potential of several NM ingredients in cosmetics, including TiO_2 (SCCS/1516/13 and SCCS/1539/14), ZnO (SCCS/1489/12 and SCCS/1518/13), and carbon black (SCCS/1515/13). In its opinion on TiO_2 (nano form), the SCCS concluded the following:

> From the limited relevant information provided in the submission, and the information from open literature, the SCCS considers that TiO_2 nanomaterials in a sunscreen formulation are unlikely to lead to: systemic exposure to nanoparticles through human skin to reach viable cells of the epidermis, dermis, or other organs. (SCCS/1539/14)

However, the SCCS also states the following:

> Although there is no conclusive evidence at present to indicate penetration of TiO_2 nanoparticles through the skin to viable cells of the epidermis, a number of studies have shown that they can penetrate into the outer layers of the stratum corneum, and can also enter hair follicles and sweat glands.

For ZnO NPs the SCCS stated that there is no indication for dermal penetration of ZnO NPs according to the available scientific literature. However, a very small proportion of Zn ions that are released from the ZnO NPs may be available for systemic exposure after dermal application (SCCS/1489/12). The SCCS also found an "overall lack of dermal absorption of carbon black nanoparticles" (SCCS/1515/13).

A study funded by the Danish EPA used in vitro (EpiDerm^{TM}) and in vivo (mouse, human skin graft) models to investigate whether the size or surface coating of TiO_2 NPs and ZnO NPs has an effect on the dermal penetration/absorption of the NPs [151]. On the basis of the results, the following was concluded:

> Dermal penetration of TiO_2 and ZnO NPs did not occur at or above the limit of detection of the used experimental methods. Should absorption of TiO_2 and ZnO nanoparticles occur at levels below the detection limit of the assays used herein, the systemic dose would be very small (far lower than the doses used in the studies discussed above) and so highly unlikely to cause systemic toxicity based on toxicological evidence in rodents.

1.5.2.3 Distribution

Once systemically available (following absorption), poorly soluble NMs get typically distributed via the systemic circulation or within phagocytizing cells via lymph drainage to various parts of the body. Distribution via blood is followed by rapid uptake by organs and tissues containing significant phagocytizing capacity. Thus, distribution is mainly to the liver and spleen although distribution to the heart, kidneys, and immune-modulating organs has been reported as well [43, 107, 152]. General characteristics, such as rate of dissolution, surface treatment, chemical composition, shape, agglomeration, and aggregation, all may have an impact on distribution. In whatever tissue (including blood), NMs can bind with different affinities to biological targets, such as peptides, proteins, lipids, and DNA, to form a corona [157]. NMs are reported to be able to get distributed to the central nervous system via sensory neurons in the nose [107]. A recent paper argues that NMs have the right size and shape for interacting with transport proteins, such as apolipoproteins, offering them access to all cells via the low-density lipoprotein receptor [153].

1.5.2.4 Metabolism/Dissolution/Transformation/Bio-nano interaction

Metabolism, defined as the enzyme-mediated conversion from one chemical species to another (a metabolite), is probably not relevant for NMs, not least because they are generally too large to fit into the active site of the well-known biotransformation enzymes. However, some abiotic oxidation may occur at the outer surface of NMs. Two other transformation processes are more relevant for NMs: dissolution into ions and corona formation through bio-nano interactions [153]. These processes will depend on the PC characteristics of the NM as well as those of the environment (e.g., pH and the presence of salts and proteins).

Once the NM is exposed to a protein-rich environment, a protein corona is formed. This is the biomolecule coating that forms around NPs upon their contact with biological molecules. It is formed because of the high surface reactivity of the NMs resulting in adsorption of various molecules (mainly proteins). Desorption also occurs at different time scales depending on the binding energy of the NP-protein interaction. Characteristic binding forces are van der Waals interactions, hydrogen bonds, hydrophobic interactions, electrostatic interactions, and π-π stacking [154]. There is evidence that the corona is stratified into two layers, one of which is "soft" (loosely associated macromolecules, characterized by short exchange times) and the other "hard" (tightly bound, characterized by long exchange times) [155, 156].

The corona is continuously exchanging with the proteins in the environment, and its composition varies in time and depending on the environmental conditions [157]. The corona is stable for a longer period of time than the typical time scale of cellular uptake, thus acting as a cell "mediator" in the interaction of the NM with cell receptors [158]. Wang et al. [158] also found that after entering the cell, the NP and the corona reach the lysosomes, where peptides are degraded by lysosomal enzymes. Subsequently, membrane damage may occur followed by the release of the lysosomal content into the cytosol and subsequent cell death due to apoptosis.

Qualitatively, the composition of the protein corona on a given NM at a given time depends on the concentration and type of proteins in the surrounding medium (e.g., plasma) and their affinity for the NM surface. It is generally assumed that proteins with high concentrations in plasma and high association rates will initially cover the surface of the NMs. Over time, however, these proteins will dissociate and will be replaced by proteins of lower concentration and slower exchange rates but higher affinities [159]. Therefore, as NMs are distributed from the blood to various locations or between different cell compartments, the evolution of the protein corona will change the surface properties of the NMs, changing also their toxicological profiles [62, 153].

The plasma membrane is a highly selective and effective barrier that limits the entry and exit of large macromolecular substances and materials. Nevertheless, NMs are capable of entering cells, either through one of the several endocytic pathways or by passive penetration of the plasma membrane. Following endocytosis, NMs are enclosed within the early endocytic vesicles and are thus not directly carried into the cytosol. In contrast, the NMs internalized via membrane penetration enter the cytoplasm directly. The term "endocytosis" can be broadly divided into pinocytosis ("cell drinking") and phagocytosis ("cell eating"). Pinocytosis is commonly involved in the internalization of fluids and molecules by small vesicles, whereas phagocytosis is the process by which the cells such as monocytes/macrophages, neutrophils, and dendritic cells engulf large particulate matter and form intracellular phagosomes. The mechanisms of cellular uptake by endocytosis have been reviewed elsewhere [160].

1.5.2.5 Excretion

Excretion via the usual routes (biliary, urinary, via mammal glands, via saliva) is generally unknown. The few studies available, however, suggest that excretion is very slow. The longer the half-life, the longer (and more expensive) the study needed for a reliable assessment of the rate of excretion. This is, however, highly relevant information, since the longer the half-life, the higher the risk for bioaccumulation [161].

Generally, a large proportion of ingested NMs in food appears to be excreted directly via the feces; the excretion rate is usually above 90% [43, 152, 162]. Only upon intravenous administration, limited biliary secretion into the feces was found for polystyrene NMs [43].

For very small quantum dots (up to approximately 20 nm, which is dependent on the coating) as well as for carbon nanotubes, urinary excretion seems relevant [43].

1.5.2.6 Elimination (sum of solubilization and excretion)

Once absorbed and phagocytized, insoluble NMs are eliminated (cleared) from the body to a minor extent. For many NMs, the whole-body elimination half-lives are significantly longer than the absorption half-lives, which means that the rate of appearance in the body is larger than the rate of disappearance. Persistency is an early indication of accumulation upon repeated exposure. As far as the authors are aware, however, no studies have been reported that investigate the time course of tissue concentrations of NMs upon repeated exposure. What has been investigated is the time course after a single exposure or after the last exposure in recovery studies. This has been recognized as a data gap and need for further investigation in the EU [161]. The results of a recent preliminary risk assessment based on computer modeling of human intake and clearance of nanosilica showed that the concentration of Si in the liver is expected to keep on increasing for about 2 years (more than 500 days) before a steady state is reached between uptake and elimination [152].

1.5.3 Toxicodynamics

Once an NM reaches an organ or a cell following exposure, it may exert different toxicological effects [55, 163, 164].

NMs have been implicated in a range of cellular interactions that may result in cytotoxicity and other cellular responses [165]. These include:

- Interaction with the plasma membrane that physically disrupts membrane processes (ion transport, signal transduction, etc.) and may lead to cell death
- Biochemical interaction with membrane receptors, leading to activation of signal transduction pathways
- Uptake into the cell via endocytosis, with subsequent effects on intracellular macromolecules and organelle
- Interaction with mitochondria, leading to alterations in metabolism and ROS production and interference with thc antioxidant defense [164, 166]
- DNA binding and damage, leading to the arrest of cell cycle division and protein synthesis

- Interaction with the cytoskeleton, halting vesicular trafficking and causing mechanical instability and cell death
- Interaction with proteins, lipids, and other biomolecules, leading to different types of corona and biological effects

Depending on the NM, oxidative stress results in inflammation, DNA adduct formation, and apoptosis [167, 168].

The reticuloendothelial system (RES) primarily consists of monocytes and macrophages accumulated in lymph nodes, spleen, and liver. Once NMs reach the liver, they may induce systemic DNA damage and mutagenesis because of uptake and subsequent induction of systemic inflammation. Macrophages in the RES of the liver and the spleen are known to take up particles bound with serum proteins, which can trigger autoimmune responses [169]. Once NMs reach the liver and the spleen, they have the potential to induce acute toxicity associated with histological changes in terms of focal infiltration with phagocytic and inflammatory cells, loss of neurons, microgliosis, astrogliosis, and apoptosis [170].

Once taken up by macrophages, NMs can interfere with their physiological function. Autophagy is an active uptake mechanism during which a portion of the cytoplasm or old and damaged organelles is engulfed by double- or multimembrane structures (autophagosomes). These organelles later fuse with lysosomes so that their contents can be digested by lysosomal enzymes. Autophagy can be activated by cells as a prosurvival mechanism in response to cellular stress and to sustain energy production via alternative routes, but it can also be involved in a cell death program, depending on the stimuli and the cell type [158]. This protective mechanism could be disturbed by NM uptake and may lead to the enhancement of NM-induced toxicity. Moreover, the NM may directly interact with the lysosomal membrane of the macrophage, inducing oxidative stress, cell death, and inflammation [171–173].

As a consequence of inflammation, the NLRP-3 inflammasome is activated, a process that has been described for both crystalline and amorphous SiO_2 and also for TiO_2 [174–176]. Particles of low solubility and toxicity (LSLTP) may cause inflammation in proportion to their specific surface area (SSA) and their zeta potential [177, 178], and their adverse effects have been demonstrated to be associated with their intrinsic inflammatory potency [107, 179].

As it has been shown that these LSLTP induce rat lung overload at a certain threshold level, Donaldson and coworkers considered dosimetry, rather than particle number or size, as the driving factor for inflammatory responses in vivo [180]. In fact, the determined threshold value for TiO_2 and $BaSO_4$ was found to be 1 cm^2/cm^2 of the proximal alveolar region (site of high particle retention in the lung), as calculated using previously published results [177], below which the inflammatory response would not be triggered. This was found true not only for NPs but also for their micron-sized counterparts. Equal results could be found in in vitro studies using A549 lung epithelial type II cells and IL-8 as the relevant readout for proinflammatory effects [180]. In a later publication, Donaldson and Poland underpinned the comparability of mechanisms of toxicity for nano- and micron-sized particles and stated that the dose is the main trigger for adverse effects [203]. Taking into account above considerations, SA was recommended as the most appropriate metric in particle toxicity testing.

Inhaled particles are known to play an important role not only in fibrosis or cancer development but also in chronic obstructive pulmonary disease and asthma [145, 181]. An important player in the relationship between inflammation and carcinogenesis is the formation of ROS during inflammatory phagocyte respiratory burst. Also macrophages are known to play a significant role in inflammatory processes induced by NMs in the lungs: they are a key player in the uptake and removal of inhaled NMs [58]. Vandebriel and De Jong [55] report that the toxic effect of ZnO NMs is due to their solubility resulting in increased intracellular concentration of Zn^{2+}. A correlation exists between particle size, surface charge, and pulmonary inflammation, as indicated by neutrophil influx and the induction of proinflammatory mediators: the smaller the particles, the greater the inflammatory response [58, 182].

Chronic inflammation in the liver may occur when biopersistent NMs cannot be removed and inflammatory cells are continuously recruited. In vivo experiments with injected (intraperitoneal or intravenously) TiO_2 have reported increases in general markers for liver damage (such as alanine transaminase and aspartate transaminase) [183, 184], inflammation (increase in proinflammatory cytokines and/or infiltration of inflammatory cells) [185, 186], oxidative stress [187, 188], apoptosis, necrosis, and fibrosis [163, 189]. Upon treatment via oral gavage, similar findings were reported: apoptosis, necrosis or inflammation [185], and general liver damage after treatment with higher concentrations of TiO_2 (starting from 125 mg/kg BW, as detected via above-mentioned markers) [184] were found usually above subacute exposure (60 and 30 days). Also, specific DNA damage has been reported after oral gavage [190] and administration via drinking water [191]. It must be noted, however, that the reported effects were only found at the highest concentration groups (200 mg/kg BW and 500 mg/kg BW). ZnO and SiO_2 NMs are also reported to induce liver damage and inflammation. Vandebriel and De Jong [55] report that intravenous instillation and oral administration of ZnO NPs result in accumulation in the liver (and also in the spleen, lungs, and kidneys); oral exposure to nanoparticulate SiO_2 also leads to accumulation in the liver [192]. Intravenous administration of SiO_2 NPs causes DNA damage in the liver [193]. When lower doses induce an adverse effect, such as influx of inflammatory cells, recovery to normal levels is sometimes reported [194].

In the blood, NMs can mechanically obstruct the vasculature, which leads to congestion in multiple vital organs and subsequent organ failure. For example, there are reports of NP-induced coagulopathies (i.e., coagulation disorders caused by perturbation of the blood coagulation system), dependent on the size and charge of the NM, which determines if an NM has pro- or anticoagulant properties. A consequence can be disseminated intravascular coagulation, which can be either acute or chronic and, if left untreated, may lead to multiple organ failure and death [109, 110].

Moreover, NMs such as TiO_2 have been found to cause an imbalance in hemostasis and thus disturbance of the immune system. For example, in vivo studies in mice reported decreased proliferation of T and B lymphocytes as well as natural killer cells [184]; exposure occurred through intragastric administration. Disturbance of the cell cycle can result from interactions of TiO_2 with proteins, as reported in in vitro studies on human lung epithelial cells [195].

Thus, NMs can cause adverse effects through their interactions with cells in target organs and with proteins in the blood and tissue fluids. An increasing number of studies are investigating these effects and their relationships with the PC properties of NMs.

1.6 Standard Test Guideline Methods for Toxicity Testing

The OECD provides test guidelines (TGs) that contain internationally harmonized test methods for mutual acceptance of data, used by government, industry, and independent and certified laboratories. Originally, they were developed to determine the safety of chemicals and chemical preparations. Their use for safety assessment of NMs has been evaluated and published in the OECD's "Preliminary Review of OECD Test Guidelines for Their Applicability to Manufactured Nanomaterials" [196]. A number of toxicity endpoints are listed, as well as the respective TGs and their possible implementation, for nanotoxicity testing.

More recently, results from expert workshop recommendations have been published, such as on genotoxicity [197] or inhalation toxicity testing [198]. Inhalation toxicity testing has been further discussed by an expert group [199]. This group has considered issues such as the NM size range and the analysis of the bronchoalveolar lavage fluids. The links between the toxicity test results and these parameters are considered essential, as is knowledge on TK, to determine, for example, lung burden and biokinetics. The TG 417 on TK was discussed for its applicability for NM toxicity testing, as discussed at the OECD expert meeting on the TK of manufactured NMs in 2014, where challenges for its application in NM toxicological assessment were identified [200]. TG 417 is still being updated to accommodate NMs [201]. Table 1.5 lists the endpoints, the discussed methods, and the respective OECD TGs and their current applicability status.

Furthermore, the OECD has published a living document on NM sample preparation and dosimetry, with the latest update being made available in 2012 [202]. The issue on the right dosimetry is discussed in previous publications by Donaldson and colleagues [180, 203], according to whom, "The mass metric appears not always to be the most appropriate or relevant one. Indeed for some nanomaterials the results may be better expressed as a function of surface area or particle number because particle size and specific area may play a main role in determining the toxicity of nanomaterials." However, it is also acknowledged that this is currently still a challenging task.

Table 1.5 Endpoints, the discussed method, and the respective OECD Test Guidelines and their current applicability status

Endpoint	Method	OECD Test Guideline	Applicability for nanotoxicity testing
Acute toxicity	Oral exposure	420, 423, 425	Appropriate for initial investigation [196].
Repeated dose toxicity (28 and 90 days) in rodents	Oral exposure to detect neurotoxic and immunotoxic effects and effects on the reproductive and endocrine system	407, 409	Applicable; enhancement of methods to detect NM-specific effects (e.g., cardiovascular effects) needs to be considered [196].
	Inhalation	412, 413	Not specifically intended for the testing of NMs [196].
Skin and eye irritation	In vivo application	404, 405	Applicable [196].
Skin corrosion	**Transcutaneous electrical resistance test (TER), reconstructed human epidermis (Rhe) test method, and in vitro membrane barrier test method**	430, 431, 435	Applicable; however some assays (e.g., MTT assay or other assays using metabolically converted vital dyes) may not be appropriate [196].
Skin sensitization	Local lymph node assay	429	Most appropriate method, permits estimation of the potency of the sensitization reaction [196].

(*Continued*)

Table 1.5 *(Continued)*

Endpoint	**Method**	**OECD Test Guideline**	**Applicability for nanotoxicity testing**
Genotoxicity	**Bacterial reverse mutation assay**	471	Not applicable [197].
	In vitro mammalian cell gene mutation assay	473, 476	Applicable [197].
	In vitro micronucleus assay	487	Applicable after modification [197].
Reproductive and developmental toxicity	Oral exposure	421, 422, 414, 415, 416	Applicable for oral exposure (inhalation exposure would require additional considerations) [196].
Toxicokinetics	Oral exposure, intravenous (IV) administration, and dermal or inhalation exposure (as applicable)	417	Not applicable [196].

Note: This status is under regular review. In vitro test methods are in bold.

Other issues raised in this document concern the storage and stability of NMs in appropriate test media, making the point that it is key to know and report the exact composition of the prepared sample. Furthermore, the document expresses the fundamental importance of proper PC characterization of the materials and lists the relevant properties.

In additional to these TGs, several different working groups are currently dealing with the fate of NMs in aquatic media, as discussed in a rolling work plan. Here, new TGs/guidance documents for determining different PC properties, their ecotoxicological and environmental effects [204], or their aquatic and algal toxicity are being discussed or developed further.

1.7 Alternative Approaches to Animal Testing

The risk assessment of chemicals has traditionally been based on toxicity studies on animals, mainly rodents like rats and guinea pigs, which have served as surrogates for humans. With a growing concern about animal welfare and technological advances, new possibilities to determine the toxic properties of substances that do not require the use of animals are increasingly available. Alternative (nonanimal) methods can be used as exploratory (e.g., to better understand mechanisms of action) or predictive tools (to extrapolate observations to the whole organism level).

In vitro experiments provide an important source of information on possible (toxic) effects of chemicals at the cellular level. However, for many endpoints, they still do not represent a full replacement for in vivo tests, with the exception of some validated and accepted methods (e.g., for eye irritation/corrosion).

Another alternative to in vivo testing is the use of computer-based techniques commonly referred to as in silico tools. When little is known about the activity of substances, in silico tools are usually used to cluster and group those that are "similar," that is, have common structural features, metabolites, or PC properties and are thus expected to have similar biological activity [205–207]. This is the basis of the read-across approach, in which the unknown toxic effects of a chemical of interest are deduced from the known effects of its analogues.

Differently from read-across, QSAR models are in silico tools that correlate specific activities (toxicity endpoint) of a series of substances with structural features. QSARs can thus be used to predict the activities of substances from their own properties, without the need to know the activities of analogues.

Physiologically based toxicokinetic (PBTK) models are another set of tools that at least in theory have in vivo predictive potential based on an understanding of anatomy and physiology and on the application of mathematical modeling to data derived from in silico and in vitro models. In this case, predictions are not of a toxicological endpoint but of the kinetics of absorption, distribution, metabolism, and excretion resulting in an overall time course of a chemical in blood plasma or organs.

Given their reductionist nature and uncertainties in their applicability, alternative methods do not usually represent stand-alone alternatives to classical test methods. However, it is expected that approaches based on the integrated use of multiple alternative methods, commonly called integrated testing strategies or integrated approaches to testing and assessment, will help to minimize or avoid animal testing.

1.7.1 Adverse Outcome Pathways

An adverse outcome pathway (AOP) is a conceptual construction that portrays existing knowledge about the link between a molecular initiating event (MIE) and an adverse outcome (AO) by capturing the sequential chain of causally-linked events at different levels of biological organization [208]. AOPs provide a means of systematically organizing mechanistic knowledge on the biological and toxicological effects of chemicals; they can guide the further development of alternative methods, facilitate the integration and use of newer "nonstandard" data such as high-throughput screening (HTS) and -omics, and support the identification of knowledge gaps to inform intelligent testing strategies.

An AOP program was launched by the OECD in 2012 (http://www.oecd.org/chemicalsafety/testing/adverse-outcome-pathways-molecular-screening-and-toxicogenomics.htm). Specific rules are applied to AOP development. For example, one AOP always consists of a single MIE and a single AO but can have multiple key events (KEs) that are causally linked by the so-called KE relationships. This leads to a simplified and "linear" representation of a mechanism of toxicity. The AO can be defined at various levels: in humans, an AO can relate to the whole population, but also to individual organ damage, for example, liver fibrosis or skin sensitization; in environmental science, the AO relates to the death or reproductive impairment of an individual (e.g., a fish), which ultimately has consequences for the whole population. To promote the harmonized development of

AOPs, an AOP-wiki was launched in September 2014 (https://aopkb.org/).

So far, AOPs have been developed on the basis of chemical-induced MIEs [209]. However, due to the unique properties of NMs, their mechanisms of toxicity might not be directly comparable to those of chemicals in solution. In particular, NM-triggered MIEs could be different, for example, due to the size-dependent and mechanical effects of particles.

1.7.2 In silico Methods

In silico methods comprise a diverse range of techniques that solely use a computer to obtain an outcome. They are used to extract patterns and knowledge from large amounts of data (i.e., perform data mining) and to use this knowledge to build models (i.e., machine learning) that are able to predict the outcome of an experimental test, thereby avoiding the need to carry it out. The terms "data mining" and "machine learning" are often confused as they employ the same methods but with different goals.

1.7.2.1 Supervised and unsupervised methods

Data mining methods may be categorized as either supervised or unsupervised. In unsupervised methods, no target variable is identified as such. They can be used during the early stages of an investigation to detect patterns, find features that can be useful for categorization, and gain insight into the nature or structure of the data. In contrast, supervised methods are mostly applied to reproduce and/or predict patterns found in the data.

The most common unsupervised data mining method is clustering, which is the grouping of objects into meaningful categories. Given a representation of N objects, a clustering exercise will assign the N objects to k clusters with respect to a similarity measure. There are various clustering algorithms. Here we briefly describe two of the most common ones, hierarchical clustering and k-means clustering.

Hierarchical clustering assigns each object (e.g., NM) to a cluster. In step 1, N objects are assigned to N clusters, each containing just one object. The distances (similarities) between the clusters are, therefore, the same as the distances (similarities) between the items they contain. In step 2, the most similar pair of clusters is identified and these are merged into a parent cluster so that there is one cluster less. The distances between this new cluster and all the other clusters are recalculated in step 3. Then, steps 2 and 3 are repeated until the whole data set has been merged into a single cluster.

An example of hierarchical clustering is shown in Fig. 1.2. In the ENPRA project (www.enpra.eu), various NMs (ZnO, TiO_2, Ag, and carbon nanotubes) were tested with different coatings, crystalline structures, and sizes. The NMs were clustered with respect to their lowest observed effect concentrations obtained for a series of cytotoxicity, genotoxicity, inflammation, and oxidative stress tests [210].

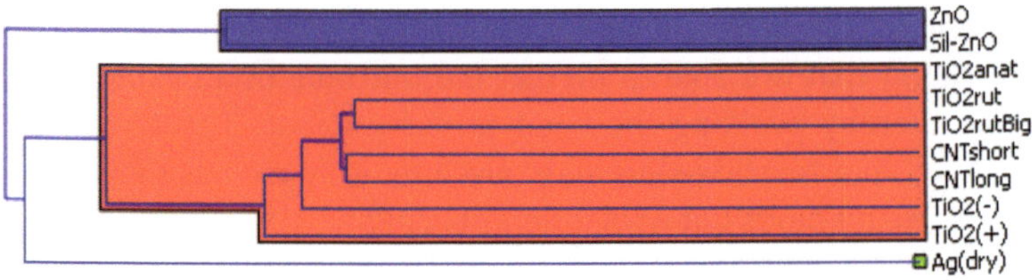

Figure 1.2 Hierarchical clustering of a set of NMs on the basis of their toxicological profiles.

The way in which the distance between the clusters is calculated (single-linkage, complete-linkage, average-linkage, etc.) determines the hierarchy of the cluster. If a predetermined number of clusters is desired, the dendrogram is cut across the appropriate number of branches/links.

K-means clustering is an iterative algorithm that groups data into a preset number of clusters (*K*) that is defined by the user. In step 1, the algorithm randomly guesses the center locations of the clusters. In step 2, each data point is assigned to its closest center. In step 3, each cluster finds the centroid of the points that it is assigned to, that is, each cluster finds its center. Steps 2 and 3 are then repeated until convergence is achieved, that is, there is no variation in the positions of the centroids.

Supervised learning consists of finding a model/rule that relates the data associated with objects (e.g., NM descriptors or properties) to a predefined class (e.g., carcinogen/noncarcinogen, sensitizer/nonsensitizer, etc.). Data sets of objects with known classes are defined as *training sets* and are used to train the model. The trained model will be applied to a *test set* of compounds for its validation. If the training sets are sufficiently diverse (general), the model will be applicable to a wide variety of compounds (large applicability domain). Otherwise, its applicability domain will be specific to a small family of compounds (small applicability domain). For example, given a database that consists of a list of chemicals with their structural features (presence of heteroatoms, presence of rings, number of conjugated bonds, etc.) and their toxic activity (e.g., carcinogen or noncarcinogen), a supervised machine learning algorithm will find a function or pattern that differentiates carcinogens from noncarcinogens. The function or model could be as simple as this: compounds containing aromatic nitro groups and with molecular weight < 1000 Da are indicators of carcinogens [211]. Given the database of carcinogens and noncarcinogens, the model "learns" that aromatic nitro groups are characteristic of carcinogens. There are many types of machine learning algorithms. Decision trees like the one explained previously are probably the most simple and intuitive models. More sophisticated supervised learning approaches include support vector machines [212], Bayesian networks [213], and artificial neural networks [214]. See Table 1.6 for other methods.

Table 1.6 Supervised methods that are used to derive predictive models.

Method	Short description
Linear discriminant analysis (LDA)	A method that uses a linear combination of continuous descriptors to discriminate between two or more categorical variables.
Classification algorithms and regression trees (CART)	A recursive process to partition the data set into different subsets with respect to attributes values. The data are partitioned on the "nodes" of a tree, creating branches (subsets of data). Each branch is recursively partitioned until no added value in predicting a category is observed. Then, a leaf corresponding to a category is created. This approach can use either continuous or categorical data in both, descriptors and category.

(Continued)

Table 1.6 (*Continued*)

k nearest neighbors (*k*-NN)	A local regression or classification method that uses the average of the *k*-NN to the test chemical in order to classify or predict the category or property of interest.
Support vector machines (SVM)	A model that represents the data set as points in a space that maximizes the separation between the different classes of the category. The test chemical is categorized or predicted with respect to the points in the space that are within a given threshold.
Bayesian networks (BN)	A graphical model that encodes probabilistic relationships (arcs) among random variables (nodes). The distribution of these variables with respect to the categories is used to assign a probability of pertinence to each category. The accumulated pertinence probability across all nodes, which are presumed independent, is used for categorization. One of the advantages of this method is that not all descriptors (variables) are needed in order to have a prediction and that the influence of having additional data on the final classification can be tested beforehand.
Artificial neural networks (ANN)	Set of interconnected nodes (neurons) that resemble biological neuronal networks. The neurons can compute values from inputs, and a hidden layer of neurons establishes self-adapted relations that derive in an output. Depending on the input values, different neurons are activated and provide a result.

1.7.2.2 QSAR/QSPR

QSARs are a specific type of in silico method. These methods are based on the assumption that the activity of a substance is related to its structure. The concept is well established (the first QSAR for biological activity can be dated back to 1868 [215], but in the past few decades it has been applied more efficiently and extensively due the availability of computational methods. The term "structure–activity relationship" (SAR) is sometimes used to express a simple qualitative association between an activity and the presence of a structural feature.

Quantitative structure–property relationships (QSPRs) are conceptually the same as QSARs, but they typically relate structure to PC properties (rather than biological activities) of the chemicals.

There are different ways to encode structural information in a QSAR/QSPR. The simplest way is identifying the presence or absence of certain chemical groups, that is, fingerprints. These chemically based encodings are usually not quantitative but discrete, giving rise to an SAR. Alternatively, quantitative representations of chemical structures are used, giving rise to a QSAR. One of the most famous QSARs is the Hansch equation [216], which relates general bioactivity of compounds with their hydrophilicity. The extended Hansch [217] equation (Eq. 1.8) shows the curvilinear relationship between log1/C50 and hydrophobicity normally found in single-dose tests:

$$\log {}^{1}\!/_{C_x} = \text{-}a(\log P)^2 + b\log P + \rho\sigma + \delta Es + K, \qquad (1.8)$$

where C is the molar toxicant concentration producing a standard response (e.g., 50% mortality or effect) in a constant time interval, log$P$ is the water-octanol partition coefficient, σ is the Hammett electronic parameter [218], E_S is Taft's steric factor [219], and ρ and δ are constants related to the sensitivity of the reaction to electron density.

A descriptor is a mathematical representation of the chemical structure [220] and has been defined as follows:

> The molecular descriptor is the final result of a logic and mathematical procedure which transforms chemical information encoded within a symbolic representation of a molecule into a useful number or the result of some standardized experiment.

There are thousands of descriptors that can represent a chemical structure. The main families are presented in Table 1.7.

Table 1.7 List of chemical descriptors usually used in QSARs

Descriptor family	Description
Zero-dimensional (0-D)/ constitutional/information	These are derived from just the structural formula and only describe the composition, for example, the number of *N* atoms and the number of double bonds.
1-D/fingerprints	These describe the composition in terms of structural fragments, for example, the number of benzene rings.
2-D/topological	These include information on the connectivity of atoms and structural fragments. They are obtained from molecular graph theory (2-D representations of the chemical structure) and can differentiate molecules according to their size, degree of branching, flexibility, etc. Examples include atom connectivity indexes [221, 222] that are derived from the bonds present in the molecule.
3-D/geometric	These are properties that depend on a 3-D representation of the chemical structures. They include size-, shape-, surface-, and volume-related descriptors, as well as electronic (quantum chemical) descriptors. They are typically derived from the coordinates of an energy-minimized conformation of the molecule. They may also take conformational flexibility into account, for example, using molecular dynamic simulations [223].

Descriptor family	Description
4-D	These are similar to 3-D descriptors, but they also capture interaction energies. These energies are calculated in terms of the interaction of the molecule with a steric, electronic, or hydrophobic probe (depending on the type of interaction energy). They are derived by various methods, including comparative molecular field analysis (ComFA) [224]. ComFA is based on the rationale that differences in a target property, for example, biological activity, are often closely related to changes in shapes and strengths of noncovalent interaction fields surrounding the molecules. In ComFA, field values are systematically calculated for ligands at each grid point of a regularly sampled 3-D grid box that extends 4 Å beyond the dimension of all molecules in the data set, using an sp^3 carbon atom with +1 charge as the probe molecule. Other 3-D programs are comparative molecular similarity indices analysis[a] [225] and generating optimal linear partial least squares estimations[b] [226].
Physicochemical properties	These can also be used as "descriptors" or, more accurately, as predictor variables. The property may be calculated from chemical structure, or an experimental measurement may be used directly. A common example is log*P*.

[a]Also known as CoMSIA.
[b]Also known as GOLPE.

Descriptors that have been found useful in the QPSR/QSAR modeling of particles (including NMs) are identified in Chapter 3.

1.7.2.3 Validation of QSARs for regulatory purposes

The first condition for using QSARs in regulatory decision-making is the demonstration of model validity. Since the concept of validation is incorporated into legal texts and regulatory guidelines, it is important to clearly define what this means and to describe what the practical validation process might entail. According to the OECD Guidance Document on the Validation and International Acceptance of New or Updated Test Methods for Hazard Assessment [227], the term "validation" is defined as follows:

> The process by which the reliability and relevance of a particular approach, method, process or assessment is established for a defined purpose.

This wide-ranging definition is intended to cover all kinds of traditional and alternative testing methods. In the context of QSARs, this definition is rather abstract and difficult to apply. However, in the case of QSARs, a set of five validation principles has been established by the OECD report [228]. Thus, for a QSAR to be valid it must have:

- A defined endpoint
- An unambiguous algorithm
- A defined domain of applicability
- Appropriate measures of goodness of fit, robustness, and predictivity
- A mechanistic interpretation, if possible

The "appropriate documentation" for demonstrating model validity is recognized in the QSAR Model Reporting Format, which is structured according to the OECD principles for QSAR validation. Information on QSAR model validity, including peer-reviewed documentation, is available from various sources, including the Joint Research Centre (JRC) QSAR Model Database (https://qsardb.jrc.ec.europa.eu).

1.7.2.4 Expert systems

Expert systems have traditionally been associated with tools that capture and reproduce the knowledge of an "expert". One of the best-known expert systems is Derek (Lhasa Ltd.), a knowledge-based expert system that predicts the toxicity of a chemical from its structure. Its predictions are based in part on structural alerts that describe structural features or toxicophores associated with toxicity. These alerts are usually derived by human expert interpretation of toxicity data for known compounds within the same chemical class and an understanding of toxicological mechanisms.

A broader definition of the term "expert system" was recommended during a European Union Reference Laboratory for alternatives to animal testing (EURL-ECVAM) workshop [229]: "Any formalized system, not necessarily computer-based, which enables a user to obtain rational predictions about the toxicity of chemicals." According to this definition, any available software tool for toxicity prediction is considered an expert system, including those based entirely on statistical models (statistically based systems), those based entirely on expert knowledge (knowledge-based systems), and those based on a combination of expert knowledge and statistical induction (hybrid systems). Widely used examples include Toxtree, TIMES-SS, TOPKAT, EPISuite, PASS, and Vega. A review of expert systems and their regulatory applicability (in the cosmetics sector) has been published by the International Cooperation on Cosmetics Regulation (http://www.iccrnet.org/topics).

1.7.2.5 Applicability of QSAR/QSPR approaches to NMs

The challenge of developing QSARs for NMs has been discussed in the literature [213, 230–232]. Due to the novelty of applying the QSAR approach to NMs, the resulting models have sometimes been called QNARs[8] and the underlying descriptors nanodescriptors [212, 233].

However, these terms are misleading since the main modeling challenges are not nano-specific, but rather they relate to the difference between modeling the behavior of chemicals in solution ("classical" or "bulk form") and in particulate form. Traditionally, QSPR and QSAR modeling has been applied to substances in solution, typically undissociated molecules, but in principle, ionized molecules and soluble metal ions can be modeled too. This is sometimes complicated by the fact that the substance of interest can undergo abiotic or biotic (metabolic) conversions (in the case of molecules) or speciation (in the case of metal ions), creating a mixture whose composition can vary in time and space.

[8]Quantitative nanostructure–activity relationships.

The modeling of particles, including NPs, has been a challenge since relatively few theoretical descriptors are available for particles. In some cases, the particles have an amorphous (e.g., certain clays) rather than a crystalline structure, making it difficult to establish theoretical descriptors. Nevertheless, experimental descriptors (Table 1.2) may be available. In the case of NMs, it will be necessary to develop descriptors that are genuinely "nanodescriptors" in the sense that they express the novel and size-dependent characteristics of NMs. A further complication is that a nano-sized (or micro-sized) particle does not form a homogeneous collection of species—it may undergo aggregation/agglomeration processes, may adsorb and desorb macromolecules present in the surrounding medium, and may (partially) dissolve as well, leading to a distribution of masses/sizes/shapes (i.e., polydispersity). Rather than modeling a single species, it may, therefore, be necessary to model a distribution/mixture of species, which is increasingly difficult as the material deviates more and more from monodispersity. Furthermore, as with most "classical" chemicals, the mode of toxicological action is often unknown, making it difficult to identify, a priori, the most relevant and predictive descriptors [234, 235].

As a consequence of these challenges, as well as the currently limited availability of systematic collections of PC, in vitro, and in vivo data [230], there are relatively few QSARs for NMs in the scientific literature.

In principle, it should be easier to develop QSPRs [236] for NMs, since PC (including magnetic and optical) properties are more closely linked to nanostructure than biological activity, and the property data are more likely to be generated in a systematic way (using standardized protocols). A few reviews of QSPRs and QSARs for NMs have been published elsewhere [45, 230, 237–239].

1.7.2.6 Physiologically based kinetic modeling

Kinetics (sometimes referred to as toxicokinetics [TK] or pharmacokinetics [PK]) is the science that studies the fate of a substance (including NMs) in the body, whereas dynamics (sometimes referred to as toxicodynamics [TD] or pharmacodynamics) studies what a substance does to the body once it comes in contact with the body (local toxicity) or enters the body (systemic toxicity).

The usual OECD Test Guidelines for investigating TK are OECD TG 417 (TK), TG 427 (in vivo skin absorption), and TG 428 (in vitro skin absorption). When TK is studied using TG 417, separate studies are usually carried out and reported to study some ADME parameters in isolation: one to determine relative (%) absorption and excretion (feces, urine, and sometimes exhaled air), one to establish organ distribution, and one to establish metabolism. They are usually performed at two dose levels and with single and repeated (usually once per day for seven days) dosing. Indications of accumulation might be obtained when comparing tissue levels after seven days of exposure compared to a single exposure. Increasingly, concrete TK parameters like AUC, C_{max}, T_{max}, and half-life ($t_{1/2}$) are also being investigated. To establish half-life ($t_{1/2}$), follow-up TK (tissue depletion) during a wash-out (no exposure) period following single or repeated dosing is needed.

Irrespective of whether a test substance is an NM or not, the TK is studied separately from the TD. It is, therefore, a challenge to evaluate whether data gathered in TK studies is relevant to the hazard characterization information obtained from TD studies. One way of addressing this is to include some preliminary TK sampling in the TD studies. This is currently under discussion by the OECD Expert Group on the inhalation TGs 412 and 413 (28 and 90 days inhalation).

Traditionally, TK has been studied empirically in living organisms. In this approach, whole-body parameters, like absorption rate constant (k_a or k_{in}) and elimination rate constant (k_e or k_{out}) can be obtained using classical toxicokinetic (CTK) computer approaches to fit the empirical data. CTK modeling can be used for interpolation between dose levels if measurements at these dose levels exist. However, it can be used for extrapolation only to a limited extent, and care should be taken due to the possibility of nonlinear relationships. Once k_a and k_e are established, CTK modeling can be used to assess the potential for $k_a > k_e$ and to determine the level of accumulation at given exposure levels.

A more recent approach, PBTK modeling, is based on physiologically relevant compartments and processes (including bioavailability in the relevant organs and the systemic blood flow). In contrast to CTK, PBTK allows various kinds of extrapolation, for example, interspecies extrapolation, high-to-low dose extrapolation, and route-to-route extrapolation, and can take interindividual variation into account. This has been used for non-NM substances. In addition, PBTK modeling has been applied to simulate TK profiles by integrating independently obtained information from in silico (QSAR) and in vitro methods for ADME. In that respect, it has more characteristics of a predictive modeling than the CTK modeling approach.

Although CTK modeling can be used to perform interpolation and assessment of accumulation on the basis of at least some experimental in vivo data, it is generally considered an empirical rather than predictive approach. Multicompartmental classical empirical modeling (i.e., CTK) has provided some initial insights into the NM-specific (Ag and SiO_2) rates of absorption from the GI tract into the systemic circulation and into the distribution from the systemic circulation to specific organs [152, 240, 241].

PBTK modeling approaches have greater potential to simulate the concentration-time profile. This is especially the case for relatively simple chemicals for which the TK is (mainly) driven by passive diffusion across biological membranes. Incorporation of active processes, such as active transport across external and internal membranes [242], is more demanding. Although for very small NMs, like quantum dots (<10 nm), diffusion might still be relevant, for NMs in general most processes that drive the absorption, tissue distribution, and excretion (if any) are probably active processes, such as macrophage uptake. These processes have not been mimicked in vitro or modeled in silico in any detail.

An issue that is relevant not only for kinetic modeling but also for assessing the suitability of analogues for read-across (e.g., between bulk particles and nanoforms) is solubility (the amount that dissolves and the rate of dissolution) under physiologically relevant conditions. Typically, a considerable proportion of an NM does not dissolve[9] to a large extent (the NM is "durable"). With respect to kinetic modeling, it is much easier to model the behavior of an NM

[9]Dissociation means to go from salt to dissociated ions (e.g., from AgCl to Ag^+ and Cl^-). To dissolve is to change from a solid to a "dissolved in solution" (solute) status. For some chemicals, dissolving requires dissociation at the same time (from NaCl crystal into Na^+ and Cl^-). For other chemicals, dissociation does not occur when they are going into solution (e.g., from glucose crystals to glucose monomers in solution).

that readily dissolves, that is, becomes a "classical readily soluble chemical" for which many of the usual assumptions hold. However, when it comes to insoluble particles, uptake by phagocytizing cells needs to be considered, which is not trivial from a modeling point of view.

The computational modeling of this type of biodistribution is more demanding than simulating simple diffusion using differential equations. As such, the development and use of PBTK modeling has not kept pace with experimental TKs and TDs. The number of papers covering PBTK modeling of NMs is currently very low, that is, 10 ≈ 20. These are all hybrid PBTK models that simulate whole-body kinetics of NMs but rely heavily on calibration with experimental concentration–time course data, that is, they have a strong empirical grounding [152, 240, 241, 243]. This means that the physiological parameters, such as blood flow and organ weights, are measured data, but chemical-dependent parameters, such as absorption rate, are not known and are set by fitting the data to existing blood concentration–time course data.

Nevertheless, interesting results have been found from PBTK studies of NMs. For example, on dietary exposure to TiO_2 NMs between 15 and 150 nm, the size and crystalline structure of the particles had a minor influence on the biodistribution and at high internal exposure the particles agglomerate in vivo and are taken up by macrophages [66, 243]. Agglomeration and deagglomeration are very relevant processes for the exposure assessment as well as hazard characterization of NMs. High concentrations in an animal study might cause agglomeration, exposure to larger (maybe micron-sized) particles, and a blurred dose response.

The future power of applying PBTK modeling to NMs lies in the possibility of simulating TK profiles a priori rather than fitting them a posteriori. In other words, the parameterization of the equations in the model (e.g., giving values to absorption rates and transfer rates from blood to tissues) can be done by using QSPR predictions or in vitro measurements. This will require the further development of QSPR models and in vitro models for kinetic parameters. For example, the development of in vitro assays to measure active uptake by different types of phagocytizing cells in different tissues may greatly enhance PBTK modeling approaches.

1.7.3 In vitro Methods

In 2009, Stone et al. provided the first critical overview of in vitro test systems in nanotoxicology [244]. They identified some general limitations of in vitro test systems, one of them being the fact that these test systems are limited to one or only a few different cell types and thus cannot fully represent the biological responses in a whole organism. Furthermore, these cell systems are usually derived from cancer or other long-lived cell lines, which can result in different outcomes when compared to in vivo tests [245]. Single exposures are typically used in in vitro work, which usually lasts from a few minutes up to a few days, depending on the tested endpoint. Therefore, chronic NM exposure cannot be tested sufficiently. Defining a suitable dose range is a challenge for both in vitro and in vivo tests, as often unrealistically high doses are chosen in order to observe the effect of interest. In vitro studies often aim to acquire mainly mechanistic information and are difficult or impossible to conduct in a whole-body system.

The in vitro testing of NMs poses some specific technical challenges. The exposure medium plays an important role as it can affect the agglomeration or aggregation state of an NM and subsequently its uptake and toxicity [245]. NMs can also interfere with the test system. For example, when using colorimetric assays such as the MTT assay (Table 1.8), one has to consider that an NM may generate an absorbance at the same wavelength as that used to quantify the colored product, which will lead to an overestimation of the cell viability [244]. Also, the NM might adsorb the colored reagent due to its high specific surface area (SSA). When using fluorescent endpoints, one needs to consider the interference of the NM by, for example, physical blocking of the light emitted (e.g., carbon), reflection of the excitation light (e.g., TiO_2), and autofluorescence of the particle. Therefore, thorough care needs to be taken to choose the right systems, conduct appropriate pretests, and include suitable controls [244].

Despite these technical challenges, in vitro models offer a number of advantages for nanotoxicity testing [246]. They can be used in mechanistic studies to investigate the potential of NMs to cross barriers; to enter cells; to influence cellular morphology and viability; to cause genotoxicity; and to trigger cell signaling, gene expression, and protein production. Table 1.8 lists in vitro systems that have been found useful for detecting cytotoxicity or cellular membrane damage. These assays can support the derivation of values for regulatory toxicity (e.g., LC_{50} or no observable adverse effect level) and also to determine sublethal concentrations for the assessment of mechanistic endpoints such as apoptosis and formation of ROS and oxidative stress, markers of proinflammatory enzymes, and general gene-regulation assays.

Table 1.8 Overview of the most commonly used in vitro test methods to detect NP-induced adverse effects in vitro

General effect	Read-out	Assay
Cell viability	Mitochondrial function is determined by measuring the activity of mitochondrial enzymes; quantified by light absorbance.	3-(4,5-Dimethylthiazol-2-yl)-2,5-diphenyltetrazolium bromide (MTT) assay 3-(4,5-Dimethylthiazol-2-yl)-5-(3-carboxymethoxyphenyl)-2-(4-sulfophenyl)-2H-tetrazolium (MTS) assay 2,3-Bis-(2-methoxy-4-nitro-5-sulfophenyl)-2H-tetrazolium-5-carboxanilide (XTT) assay Water-soluble tetrazolium salts (WST) assay
	Adenosine triphosphate (ATP) levels are an indicator of metabolically active, viable cells; quantified by luminescence.	Cellular ATP content
Cell death via necrosis	There is an increase in the leakiness of the plasma membrane; quantified by light absorbance.	Lactate dehydrogenase (LDH) assay

(*Continued*)

Table 1.8 (*Continued*)

General effect	Read-out	Assay
	Cells with an intact cell membrane are able to prevent uptake of the dye; compromised cells, however, are colored blue within seconds of exposure; quantified by manual counting.	Trypan blue exclusion Propidium iodide (PI) staining
Apoptosis	There is staining of cleaved caspase 3/7; analyzed by immunofluorescence staining or luminescence.	Caspase 3/7 staining
	Annexin V binds to phosphatidyl serine on the surface of apoptotic cells; quantified by flow cytometry.	Annexin V–FITC (fluorescein isothiocyanate) staining
ROS production	Cell-free or in the presence of cells, the presence of ROS converts DCFH to 2,7-dichlorofluorescein; analysis by fluorimetry or flow-cytometry.	DCF(D)H (2,7-dichlorofluorescin) assay
	Specific spin traps or probes allow for the quantification and specific identification of free radical species generated.	Electroparamagnetic resonance (EPR) spectrometry
	Unwinding and linearization of a coiled bacterial DNA plasmid is used to estimate free radical and/or ROS exposure.	Plasmid assay

General effect	Read-out	Assay
	To detect a superoxide, it exhibits blue fluorescence in the cytosol until it's oxidized, where it intercalates within the cell's DNA, staining its nucleus a bright fluorescent red.	DHE (dihydroethidium) assay
Oxidative stress	Its reduced form acts as an antioxidant by reacting directly with ROS to neutralize it; the oxidized glutathione (GSSG) is formed and accumulated; protein contents are analyzed.	GSH (glutathione) assay
	TBARS are formed as a byproduct of lipid peroxidation; colorimetrical quantification.	Lipid peroxidation (e.g., thiobarbituric acid reactive substances [TBARS] assay)
	mRNA expression changes of oxidative stress–dependent genes are quantified.	Heme oxygenase-1 (HO-1), etc.
Inflammation	Cytokine and/or chemokine protein production is detected; colorimetrical quantification.	Enzyme-linked immunosorbent assay (ELISA)
	mRNA expression changes of cytokine regulating genes are quantified.	Tumor growth factor beta (TGFβ), interleukin 8 (IL-8), etc.

(*Continued*)

Table 1.8 (*Continued*)

General effect	Read-out	Assay
Genotoxicity	Clastogenic (chromosome breaking) effects are evaluated; microscopical analysis.	Alkaline comet assay Micronucleus assay
	Specific comet assay: There is specific quantification of DNA double-strand lesions.	Neutral comet assay
	Specific comet assay: Oxidative DNA adducts are quantified.	8-OHdG (8-hydroxydeoxyguanosine) formation
	Salmonella reverse mutation assay; quantification by counting of the mutants.	Ames test[a]

[a]The Ames test has recently been described as not recommended for NMs by an OECD workshop [197].

High-content-imaging systems, such as Cellomics, provide a means to upgrade traditional in vitro analyses to HTS and offer a way to prevent the interference of NM with colorimetric products, as in-cell fluorescence is analyzed via an inverted microscope. Another advantage is the fact that it allows for live- and fixed-cell imaging and multiple endpoints can be analyzed in parallel in the whole-cell system. HTS is becoming an increasingly important means of detecting mechanistically relevant endpoints that can then serve for in vivo toxicity prediction [247].

At present, multiple endeavors are on to develop standardized in vitro methods for NM testing, but as yet, none have been formally validated by validation bodies such as the EURL-ECVAM, which is hosted by the JRC. In the meantime, available and potentially useful methods have been summarized by the SCCS in their Guidance on the Safety Assessment of Nanomaterials in Cosmetics (SCCS 1484/12: SCCS 2012) and in the SCCS Notes of Guidance for the Testing of Cosmetic Ingredients and Their Safety Evaluation 9th Revision (SCCS 1564/15; SCCS 2015).

1.7.4 Grouping and Read-Across

The concept of chemical category has been defined within the REACH technical guidance [207] and the OECD grouping guidance [248, 249] as follows:

> A chemical category is a group of chemicals whose physicochemical and human health and/or environmental toxicological properties and/or environmental fate properties are likely to be similar or follow a regular pattern as a result of structural similarity.

In practical terms, this involves treating a group of similar substances as a category. Missing data on endpoints or properties within a category are predicted by read-across from data-rich analogues within the category.

The way in which similarity is defined within a group is essential to read-across. However, there is not a single way to measure/define it, and it can actually be endpoint dependent. According to REACH and OECD guidance, similarities may be based on:

- Common functional groups, for example, aldehyde
- Common constituents or chemical classes and similar carbon range numbers, for example, substances of unknown or variable composition, complex reaction products, or biological (UVCB) materials
- An incremental and constant change across the category, for example, a chain-length category for boiling point range
- The likelihood of common precursors and/or breakdown products, via physical or biological processes, which result in structurally similar chemicals

The terms "category approach" and "analogue approach" are used to describe techniques for grouping chemicals. An analogue approach is often used when a chemical grouping is based on a very limited number of chemicals, typically two substances. A chemical category is used to describe a grouping of three or more chemicals. The term "read-across" applies to both the category and analogue approaches.

One of the key considerations in grouping and read-across is to demonstrate that the grouping method is not invalidated by "extrinsic" factors such as bioavailability, metabolism, or reactivity.

This is especially relevant for NMs. Although the development of chemical categories and QSARs is underpinned by the same principles of chemical similarity, there is no formal process for validating a category, that is, there is no official body, like the EURL-ECVAM, to determine the validity of chemical categories. Nevertheless, a robust justification to demonstrate the scientific robustness of the data gap–filling approach needs to be presented. In other words, there is a need to justify the read-across argument. This should be done using the standard document known as the Reporting Formats for the Analogue and Category Approaches (Section 6.2.6 in Ref. [207]).

There is a preference for the use of interpolation within grouping approaches under REACH, presumably because this gives rise to less uncertainty than extrapolation. Extrapolation is, therefore, considered less reliable due to this higher level of uncertainty associated with predictions. The exception to this is where extrapolation from one substance to another leads to an equally severe or more severe hazard assessment for the target substance. Although it may seem logical to assume that interpolation is subject to less uncertainty than extrapolation, in reality the degree of uncertainty is not due to the interpolation or extrapolation of data but rather due to the strength of the relationship forming the basis of the category/analogue approach itself. This, in turn, is dependent on the size of the category and the amount and quality of the experimental data for the category members themselves. If the relationship underpinning the category is poorly defined, then interpolation or extrapolation can result in significant uncertainty.

In addition to the available regulatory guidance, practical guiding principles and considerations for developing analogue and category approaches are described in several papers [250–252].

It is important to bear in mind that there are no officially accepted chemical categories in the EU, so read-across arguments are presented (by the registrant) and evaluated (by the regulatory assessor) on a case-by-case basis. To ensure consistency in the evaluations of different assessors, ECHA is developing a read-across assessment framework (RAAF) for both human health and ecotoxicological effects. This is an attempt to identify the different kinds of arguments/hypotheses that can be made and the importance of confounding factors (e.g., differences in metabolism or mode of action between source and target chemicals). It also applies a scoring scheme to determine the validity of the read-across argument.

While there is still considerable experience to be gained in the application of read-across to NMs, one can anticipate that the RAAF will need to be adapted at some point in order to better define and apply the concept of particle similarity.

Read-across and categories of nanomaterials

In principle the development of categories of NMs should provide a valuable means for filling data gaps for the hazardous effects of NMs, especially in the absence of reliable QSARs. For the prediction of NM properties, three main kinds of read-across can be foreseen (even though strictly speaking the REACH legal text only refers to read-across between different substances): (i) from bulk to all nanoforms, (ii) from bulk to specific nanoforms, and (iii) from one or more nanoforms to one or more other nanoforms (nanoforms of different chemical identities or the same chemical identity but with differences in PC characteristics).

As mentioned above, any grouping and read-across approach must be based on a robust scientific justification and the use of toxicologically relevant properties. It has been shown that the drivers for the toxicological effects of NMs include parameters such as composition (including the presence of impurities, coatings, and surface treatment), shape and size, charge, surface reactivity, solubility, biological persistence, dispersibility, and the ability to translocate across biological barriers [203].

1.7.5 Weight of Evidence

According to the REACH legal text, weight of evidence (WoE) is one of the options for meeting the information requirements while arguing that testing is not scientifically necessary. In particular, according to REACH Annex XI:

> There may be sufficient weight of evidence from several independent sources of information leading to the assumption/conclusion that a substance has or has not a particular dangerous property, while the information from each single source alone is regarded insufficient to support this notion.

> There may be sufficient weight of evidence from the use of newly developed test methods, not yet included in the test methods referred to in Article 13(3) or from an international test method recognised by the Commission or the Agency as being equivalent, leading to the conclusion that a substance has or has not a particular dangerous property.
> Where sufficient weight of evidence for the presence or absence of a particular dangerous property is available:
>
> — further testing on vertebrate animals for that property shall be omitted,
>
> — further testing not involving vertebrate animals may be omitted.
>
> In all cases adequate and reliable documentation shall be provided.

While WoE is a widely used scientific concept, the way in which it is applied is very heterogeneous. According to a review carried out by Linkov et al. [253], the term "WoE" can be found in the scientific literature with a variety of meanings, ranging from the purely colloquial use of the word to structured approaches to data integration and interpretation. Focusing on the structured approaches, and building on previous work [254, 255], Linkov et al. proposed a taxonomy of WoE methods, ranging from mainly qualitative approaches to increasingly more quantitative ones (Table 1.9).

Table 1.9 Taxonomy of WoE methods

Method	Nature	Description
Listing evidence	Qualitative	Presentation of individual lines of evidence without attempt to integrate them
Best professional judgment	Qualitative	Qualitative integration of multiple lines of evidence, usually invoking a professional opinion
Causal criteria	Semiquantitative	Use of a consistent set of criteria for determining cause and effect
Logic	Semiquantitative	Use of a systematic framework for evaluating multiple lines of evidence

Method	Nature	Description
Scoring	Quantitative	Quantitative integration of multiple lines of evidence by applying (different) weights and ranking them
Indexing	Quantitative	Weighting of multiple lines of evidence and integration into a single score (numerical value)
Quantification	Quantitative	Use of formal decision analytic methods, such as Bayesian statistics or multicriteria decision analysis (MCDA)

Adapted from Ref. [253].
Note: The methods are listed in order of increasing quantitative rigor.

Guidance on how to apply WoE in REACH is given in an ECHA practical guide [256]. The guide recommends that WoE arguments take into account the reliability, relevance, adequacy, and quantity of the individual lines of evidence. While the ECHA guidance does not make reference to the Linkov taxonomy, from the examples presented, it would appear to fall under the best professional judgment category, although more quantitative approaches are not excluded.

The EFSA has also developed "The Use of the Weight of Evidence Approach in Scientific Assessments" guidance [257].

1.7.6 Integrated Approaches to Testing and Assessment

Integrated assessment approaches are based on data generated by multiple alternative methods combined with other types of information (e.g., exposure considerations). These approaches have been developed to effectively and efficiently support decision-making while reducing the reliance on animal testing. An integrated approach to testing and assessment (IATA) may include both formalized (e.g., in vitro prediction models and QSARs) and nonformalized methods (e.g., WoE and read-across).

The OECD [258] defines an IATA as follows:

> An approach based on multiple information sources used for the hazard identification, hazard characterization and/or safety assessment of chemicals. An IATA integrates and weights all relevant existing evidence and guides the targeted generation of new data, where required, to inform regulatory decision-making regarding potential hazard and/or risk. Within an IATA, data from various information sources (i.e. physicochemical properties, in silico models, grouping and read-across approaches, *in vitro* methods, *in vivo* tests and human data) are evaluated and integrated to draw conclusions on the hazard and/or risk of chemicals.

Formalized methods within an IATA are referred to as "defined approaches" [258], which consist of a defined set of information sources and data interpretation procedure that converts the results of the information sources into a prediction of the property of interest.

One of the first published IATAs corresponds to skin corrosion and irritation [259] and consists of three steps:

1. Existing information (in vivo and in vitro data; PC properties; and nontesting methods, like QSAR and read-across)
2. WoE analysis
3. Additional testing

The WoE analysis (Step 2) is carried out with the data obtained in Step 1, which will mainly be obtained from literature, databases, and other reliable sources of data. If the WoE is found conclusive, no more testing is needed. However, if it is found inconclusive, Step 3 will be used in order to obtain further data and re-evaluate Step 2.

Essentially the same sequential strategy for skin irritation and/or corrosion is proposed in the ECHA Guidance on Information Requirements and Chemical Safety Assessment [28], which also includes IATA for other human health and environmental effects.

More recently, with the advent of the adverse outcome pathway (AOP) approach, there have been discussions on the development and evaluation of AOP-informed IATA [200]. It has been concluded that the degree to which an IATA needs to be populated by a full complement of methods addressing each of the key events (KEs) will be dependent on the ultimate purpose it is being used for. For chemical categorization purposes, for example, to facilitate read-across, it is conceivable that using approaches to address the first KE, the molecular initiating event might suffice. On the other hand, if a risk assessment decision is being made where uncertainty needs to be minimized as far as possible, generating information to address a number of other KEs and their quantitative relationship with the adverse outcome as well as information on the expected exposure may be necessary. Thus, flexibility is foreseen in the choice of the various information sources depending on the purpose of the IATA and the substance under investigation.

1.8 Concluding Remarks

In this chapter, we have summarized the regulatory information requirements for NMs (mainly in REACH; but other regulatory frameworks have also been considered) with reference to the state of play in terms of possible amendments to the annexes to the REACH regulation and respective guidance. In addition, we have given an overview of the scientific basis for understanding the behavior (fate and effects) of NMs, as well the different types of nonanimal methods and approaches that are under development to determine the properties of NMs.

References

1. European Commission (2012). Commission staff working paper: types and uses of nanomaterials, including safety aspects accompanying the commission communication on the second regulatory review on nanomaterials.
2. European Commission (2011). Commission recommendation of 18 October 2011 on the definition of nanomaterial, L 275/38.
3. Rauscher, H., Rasmussen, K. and Sokull-Klüttgen, B. (2017). Regulatory aspects of nanomaterials in the EU, *Chem. Ing. Tech.*, **89**(3), 224–231.
4. European Union (2007). Corrigendum to Regulation (EC) No 1907/2006 of the European Parliament and of the Council of 18 December 2006 concerning the Registration, Evaluation, Authorisation and Restriction of Chemicals (REACH), establishing a European

Chemicals Agency, amending, *Official Journal of the European Union*, L 396/7, p. 278.

5. European Parliament and Council (2006). Regulation (EC) No 1907/2006 of the European Parliament and of the Council of 18 December 2006 concerning the Registration, Evaluation, Authorisation and Restriction of Chemicals (REACH), establishing a European Chemicals Agency, amending Directive 1999/4, *Official Journal of the European Union*, L 396/1, p. 849.
6. European Commission (2008). Nanomaterials in REACH. Follow-up to the 6th meeting of the REACH competent authorities for the implementation of regulation (EC) 1907/2006 (REACH), European Commission, Brussels, p. 21.
7. EC (2012). Second regulatory review on nanomaterials: communication from the commission to the European Parliament, the council and the European economic and social committee, COM(2012)572.
8. Hankin, S. M., Peters, S. A. K., Poland, C. A., Foss Hansen, S., Holmqvist, J., Ross, B. L., Varet, J. and Aitken, R. J. (2011). Specific advice on fulfilling information requirements for nanomaterials under REACH (RIP-oN 2): final project report, RNC/RIP-oN2/FPR/1/FINAL.
9. European Chemicals Agency (2012). Guidance on information requirements and chemical safety assessment: Appendix R7-1 Recommendations for nanomaterials applicable to Chapter R7a Endpoint specific guidance, Helsinki, ECHA-12-G-03-EN.
10. European Chemicals Agency (2012). Guidance on information requirements and chemical safety assessment: Appendix R7-1 recommendations for nanomaterials applicable to Chapter R7b endpoint specific guidance.
11. European Chemicals Agency (2012). Guidance on information requirements and chemical safety assessment: Appendix R7-2 recommendations for nanomaterials applicable to Chapter R7c endpoint specific guidance, pp. 1–7
12. European Parliament and Council (2009). Regulation (EC) No 1223/2009 of the European Parliament and of the Council of 30 November 2009 on cosmetic products, *Official Journal of the European Union*, L 342, pp. 59–209.
13. Scientific Committee on Consumer Safety (2012). Guidance on the safety assessment of nanomaterials in cosmetics, Scientific Committee on Consumer Safety, Brussels.
14. SCCS (2015). Notes of guidance for the testing of cosmetic ingredients and their safety evaluation, SCCS/1564/15.

15. SCCS (2014). Memorandum on relevance, adequacy and quality of data in safety dossiers on nanomaterials, SCCS/1524/13.
16. European Parliament and Council (2012). Regulation (EU) No 528/2012 of the European Parliament and of the Council of 22 May 2012 concerning the making available on the market and use of biocidal products, *Official Journal of the European Union*, L 167, pp. 1–123.
17. European Parliament and Council (2009). Regulation (EC) No 1107/2009 of the European Parliament and of the Council of 21 October 2009 concerning the placing of plant protection products on the market and repealing Council Directives 79/117/EEC and 91/414/EEC, *Official Journal of the European Union*, Luxembourg, L 309/1, pp. 1–50.
18. Union and Council of the European Union (1991). Council directive of 15 July 1991 concerning the placing of plant protection products on the market (91/414/EEC), *Official Journal of the European Union*, L 230, p. 290.
19. EFSA (2011). Guidance on the risk assessment of the application of nanoscience and nanotechnologies in the food and feed chain, *Parma*, **9**(5), 2140.
20. European Parliament and Council (1997). Regulation (EC) No. 258/97 of the European Parliament and of the Council of 27 January 1997 concerning novel foods and novel food ingredients, L 43, pp. 1–6.
21. European Commission (2013). Proposal for a regulation of the European Parliament and of the Council on novel foods, COM(2013) 894 final.
22. ECHA (2011). Guidance on information requirements and chemical safety assessment; Chapter R.5: adaptation of information requirements, Helsinki, ECHA-2011-G-15-EN.
23. SCENIHR (2009). Risk assessment of products of nanotechnologies, Scientific Committee on Emerging and Newly Identified Health Risks, Brussels.
24. EFSA Scientific Committee (2009). The potential risks arising from nanoscience and nanotechnologies on food and feed safety- scientific opinion of the scientific committee, *EFSA J.*, **958**, 1–39.
25. OECD (2010). Guidance manual for the testing of manufactured nanomaterials: OECD's sponsorship programme; first revision, ENV/JM/MONO(2009)20/REV.
26. Rasmussen, K., Rauscher, H., Mech, A., Riego Sintes, J., Gilliland, D., González, M., Kearns, P., Moss, K., Visser, M., Groenewold, M. and

Bleeker, E. A. J. (2018). Physico-chemical properties of manufactured nanomaterials - characterisation and relevant methods. An outlook based on the OECD Testing Programme, *Regul. Toxicol. Pharmacol.*, **92**, 8–28.

27. OECD (2016a). ENV/JM/MONO(2016)2 joint meeting of the chemicals committee and the working party on chemicals, pesticides and biotechnology physical-chemical parameters: measurements and methods relevant for the regulation of nanomaterials OECD workshop report series on.
28. ECHA (2014). Guidance on information requirements and Chemical safety assessment Chapter R.7a: endpoint specific guidance, Helsinki.
29. Aitken, R., Bassan, A., Friedrichs, S., Hankin, S. M., Hansen, S. F., Holmqvist, J., Peters, S. A. K., Poland, C. A. and Tran, C. L. (2011). Specific advice on exposure assessment and hazard/risk characterisation for nanomaterials under REACH (RIP-oN 3) final project report.
30. Landsiedel, R., Ma-Hock, L., Kroll, A., Hahn, D., Schnekenburger, J., Wiench, K. and Wohlleben, W. (2010). Testing metal-oxide nanomaterials for human safety, *Adv. Mater.*, **22**(24), 2601–2627.
31. ECHA (2017a). Appendix R7-1 recommendations for nanomaterials applicable to Chapter R7a endpoint specific guidance and Appendix R7-2 recommendations for nanomaterials applicable to Chapter R7c endpoint specific guidance, Helsinki.
32. OECD (2014). Report of the OECD expert meeting on the physical chemical properties of manufactured nanomaterials and test guidelines. Series on the Safety of Manufactured Nanomaterials No. 41, Paris, ENV/JM/MONO(2014)15.
33. ECHA (2017b). How to prepare registration dossiers that cover nanoforms: best practices, ED-02-17-415-EN-N.
34. Rivolo, P. (2012). *Encyclopedia of Nanotechnology* (Springer Netherlands, Dordrecht).
35. European Commission, EC, and IHCP JRC (2011). REACH implementation project substance identification of nanomaterials (RIP-oN 1).
36. ISO (2012). ISO 9276-6:2008 - representation of results of particle size analysis. Part 6: Descriptive and quantitative representation of particle shape and morphology.
37. ECHA (2017c). Appendix R.6-1: recommendations for nanomaterials applicable to the guidance on QSARs and grouping, Helsinki.
38. Praetorius, A., Scheringer, M. and Hungerbühler, K. (2012). Development

of environmental fate models for engineered nanoparticles-a case study of TiO_2 nanoparticles in the Rhine river, *Environ. Sci. Technol.*, **46**(12), 6705–6713.

39. Meesters, J., Koelmans, A., Quik, J., Hendriks, J. and van de Meent, D. (2014). Multimedia modeling of engineered nanoparticles with SimpleBox4nano: model definition and evaluation, *Environ. Sci. Technol.*, **48**(10), 5726–5736.
40. Kreyling, W. G., Semmler-Behnke, M. and Chaudhry, Q. (2010). A complementary definition of nanomaterial, *Nano Today*, **5**(3), 165–168.
41. Utembe, W., Potgieter, K., Stefaniak, A. B. and Gulumian, M. (2015). Dissolution and biodurability: important parameters needed for risk assessment of nanomaterials, *Part. Fibre Toxicol.*, **12**, 11.
42. Jensen, K. A., Kembouche, Y. and Nielsen, S. H. (2013). Nanogenotox deliverable 4.7: hydrochemical reactivity, solubility, and biodurability of NANOGENOTOX nanomaterials, Copenhagen.
43. Landsiedel, R., Fabian, E., Ma-Hock, L., van Ravenzwaay, B., Wohlleben, W., Wiench, K. and Oesch, F. (2012). Toxico-/biokinetics of nanomaterials, *Arch. Toxicol.*, **86**(7), 1021–1060.
44. Aggarwal, P., Hall, J. B., McLeland, C. B., Dobrovolskaia, M. and McNeil, S. E. (2009). Nanoparticle interaction with plasma proteins as it relates to particle biodistribution, biocompatibility and therapeutic efficacy, *Adv. Drug Delivery Rev.*, **61**(6), 428–437.
45. Gallegos, A., Burello, E. and Worth, A. (2009). Review of computational approaches for predicting the physicochemical and biological properties of nanoparticles, Ispra.
46. Epa, V. C., Burden, F. R., Tassa, C., Weissleder, R., Shaw, S. and Winkler, D. A. (2012). Modeling biological activities of nanoparticles, *Nano Lett.*, **12**(11), 5808–5812.
47. Lidén, G. (2006). Dustiness testing of materials handled at workplaces, *Ann. Occup. Hyg.*, **50**(5), 437–439.
48. Fermin, D. and Riley, J. (2010). Charge in colloidal systems, in *Colloid Science: Principles, Methods and Applications*, 2nd ed., Cosgrove, T., ed. (Wiley), pp. 23–44.
49. Hamaker, H. C. (1937). The London—van der Waals attraction between spherical particles, *Physica*, **4**(10), 1058–1072.
50. Sellers, K., Deleebeek, N., Messiaen, M., Jackson, M., Bleeker, E., Sijm, D. and Broekhuizen, F. (2015). Grouping nanomaterials: a strategy towards grouping and read-across, 2015-0061.

51. Elimelech, M., Gregory, J., Jia, X. and Williams, R. A. (1995). Surface interaction potentials, in *Particle Deposition and Aggregation: Measurement, Modelling and Simulation* (Butterworth-Heinemann Ltd.), pp. 33–68.

52. SCENIHR (2010). Risks scientific basis for the definition of the term 'nanomaterial,' Brussels.

53. Puzyn, T., Rasulev, B., Gajewicz, A., Hu, X., Dasari, T. P., Michalkova, A., Hwang, H.-M., Toropov, A., Leszczynska, D. and Leszczynski, J. (2011). Using nano-QSAR to predict the cytotoxicity of metal oxide nanoparticles, *Nat. Nanotechnol.*, **6**(3), 175–178.

54. Casarett, L. and Doull, J. (2008). *Toxicology*, 7th ed. (McGraw-Hill, Kansas City).

55. Vandebriel, R. J. and De Jong, W. H. (2012). A review of mammalian toxicity of ZnO nanoparticles, *Nanotechnol. Sci. Appl.*, **5**, 61–71.

56. Patlewicz, G., Ball, N., Booth, E. D., Hulzebos, E., Zvinavashe, E. and Hennes, C. (2013). Use of category approaches, read-across and (Q) SAR: general considerations, *Regul. Toxicol. Pharmacol.*, **67**(1), 1–12.

57. WHO (2012). Nanotechnology and human health: scientific evidence and risk governance, Bonn, Germany.

58. Donaldson, K., Schinwald, A., Murphy, F., Cho, W.-S., Duffin, R., Tran, L. and Poland, C. (2013). The biologically effective dose in inhalation nanotoxicology, *Acc. Chem. Res.*, **46**(3), 723–732.

59. Zhang, H., Ji, Z., Xia, T., Meng, H., Low-Kam, C., Liu, R., Pokhrel, S., Lin, S., Wang, X., Liao, Y.-P., Wang, M., Li, L., Rallo, R., Damoiseaux, R., Telesca, D., Mädler, L., Cohen, Y., Zink, J. I. and Nel, A. E. (2012). Use of metal oxide nanoparticle band gap to develop a predictive paradigm for oxidative stress and acute pulmonary inflammation, *ACS Nano*, **6**(5), 4349–4368.

60. Burello, E. and Worth, A. P. (2011). A theoretical framework for predicting the oxidative stress potential of oxide nanoparticles, *Nanotoxicology*, **5**(2), 228–235.

61. Poater, A., Gallegos Saliner, A., Solà, M., Cavallo, L. and Worth, A. P. (2010). Computational methods to predict the reactivity of nanoparticles through structure-property relationships, *Expert Opin. Drug Deliv.*, **7**(3), 295–305.

62. Burello, E. (2013). Profiling the biological activity of oxide nanomaterials with mechanistic models, *Comput. Sci. Discov.*, **6**(1), 014009.

63. Burello, E. and Worth, A. P. (2013). A rule for designing safer nanomaterials: do not interfere with the cellular redox equilibrium, *Nanotoxicology*, 1–2.
64. Yang, S.-T., Liu, Y., Wang, Y.-W. and Cao, A. (2013). Biosafety and bioapplication of nanomaterials by designing protein-nanoparticle interactions, *Small*, **9**(9–10), 1635–1653.
65. Vilaseca, P., Dawson, K. A. and Franzese, G. (2013). Understanding and modulating the competitive surface-adsorption of proteins through coarse-grained molecular dynamics simulations, *Soft Matter*, **9**(29), 6978.
66. Bruinink, A., Wang, J. and Wick, P. (2015). Effect of particle agglomeration in nanotoxicology, *Arch. Toxicol.*, **89**(5), 659–675.
67. OECD (2009). Preliminary review of OECD test guidelines for their applicability to manufactured nanomaterials, Paris.
68. Mast, J. and De Temmerman, P.-J. (2016). Protocol(s) for size-distribution analysis of primary NM particles in air, powders, and liquids, D2.10.
69. OECD (2014). Ecotoxicology and environmental fate of manufactured nanomaterials: test guidelines. Expert meeting report. Series on the Safety of Manufactured Nanomaterials No. 40, Paris, ENV/JM/MONO(2014)1.
70. ISO (2013). Guidelines for the characterization of dispersion stability, ISO/TR 13097:2013.
71. CEN (2002). CEN/TC 292 - Characterization of waste, 12457-1:2002.
72. ISO (2012). Colloidal systems – methods for zeta-potential determination. Part 1: Electroacoustic and electrokinetic phenomena, ISO 13099-1:2012.
73. ISO (2012). Colloidal systems – methods for zeta-potential determination. Part 2: Optical methods, ISO 13099-2:2012.
74. ISO (2013). Colloidal systems – methods for zeta potential determination. Part 3: Acoustic methods, ISO 13099-3:2014.
75. ISO (2012). Nanotechnologies – characterization of single-wall carbon nanotubes using transmission electron microscopy, ISO/TS 10797:2012.
76. ISO (2012). Nanotechnologies – characterization of multiwall carbon nanotube (MWCNT) samples, ISO/TR 10929:2012.
77. ISO (2011). Microbeam analysis – quantitative analysis using energy-dispersive spectrometry (EDS) for elements with an atomic number of 11 (Na) or above, ISO 22309:2011.

78. ISO (2011). Nanotechnologies - characterization of single-wall carbon nanotubes using ultraviolet-visible-near infrared (UV-Vis-NIR) absorption spectroscopy, ISO/TS 10868:2011.
79. ISO (2011). Nanotechnologies – charaterization of single-wall carbon nanotubes using scanning electron microscopy and energy dispersive X-ray spectrometry analysis, ISO/TS 10798:2011.
80. ISO (2011). Nanotechnologies – characterization of single-wall carbon nanotubes using thermogravimetric analysis, ISO/TS 11308:2011.
81. ISO (2004). Microbeam analysis - scanning electron microscopy - guidelines for calibrating image magnification, 16700:2004.
82. ISO (2009). Particle size analysis – laser diffraction methods, 13320-1:2009.
83. ISO (1996). Determination of the specific surface area of solids by gas adsorption – BET method, ISO 9277:2010.
84. ISO (1996). Particle size analysis - photon correlation spectroscopy, 13321:1996.
85. ISO (2007). Determination of particle size distribution – single particle light interaction methods. Part 2: light scattering liquid-borne particle counter, 21501-2:2007.
86. ISO (2007). Determination of particle size distribution – single particle light interaction methods. Part 3: light extinction liquid-borne particle counter, 21501-3:2007.
87. ISO (2007). Determination of particle size distribution – single particle light interaction methods. Part 4: light scattering airborne particle counter for clean spaces, 21501-4:2007.
88. ASTM (2010). Standard test method for wipe sampling of surfaces, indirect preparation, and analysis for asbestos structure number concentration by transmission electron microscopy, D6480-05(2010).
89. ASTM (2009). Standard test method for airborne asbestos concentration in ambient and indoor atmospheres as determined by transmission electron microscopy direct transfer (TEM), D6281-09.
90. ASTM (2009). Standard test method for microvacuum sampling and indirect analysis of dust by transmission electron microscopy for asbestos structure number surface loading, D5755-09(2014)e1.
91. JIS (2007). Measurement of lattice parameters and crystallite sizes of carbon materials, 7651:2007.
92. ISO (2009). Determination of particle size distribution - differential electrical mobility analysis for aerosol particles, 15900:2009.

93. JIS (2005). Method for particle size determination in metal catalysts by electron microscope, JIS H 7804:2005.

94. ISO (2010). Nanotechnologies – characterization of single-wall carbon nanotubes using near infrared photoluminescence spectroscopy, ISO/TS 10867:2010.

95. ISO (2002). Surface chemical analysis – high-resolution Auger electron spectrometers – calibration of energy scales for elemental and chemical-state analysis, 17974:2002.

96. ISO (2010). Surface chemical analysis – X-ray photoelectron spectrometers – calibration of energy scales, 15472:2010.

97. ISO (2006). Surface chemical analysis – Auger electron spectroscopy – derivation of chemical information, 18394:2006.

98. ISO (2003). Fine ceramics (advanced ceramics, advanced technical ceramics) – determination of specific surface area of ceramic powders by gas adsorption using the BET method, ISO 18757:2003.

99. American Society for Testing and Materials (2010). Standard test method for metal powder specific surface area by physical adsorption, ASTM B922-10.

100. Stefaniak, A., Hoover, M., Day, G., Ekechukwu, A., Whitney, G., Brink, C. and Scripsick, R. (2005). Characteristics of beryllium oxide and beryllium metal powders for use as reference materials, *J. ASTM Int.*, **2**, 15.

101. CEN (2006). Workplace atmospheres - measurement of the dustiness of bulk materials - requirements and reference test methods, Brussels, EN 15051:2006.

102. ASTM (1990). Evaluation of airborne dust measurement methods for agricultural chemical carriers, STP25370S.

103. ISO (2007). Pore size distribution and porosity of solid materials by mercury porosimetry and gas adsorption. Part 3: analysis of micropores by gas adsorption, ISO 15901-3:2007.

104. ISO (2005). Pore size distribution and porosity of solid materials by mercury porosimetry and gas adsorption. Part 1: mercury porosimetry, ISO 15901-1:2005.

105. ISO (2006). Pore size distribution and porosity of solid materials by mercury porosimetry and gas adsorption. Part 2: Analysis of mesopores and macropores by gas adsorption, ISO 15901-2:2006.

106. Sun, T. Y., Gottschalk, F., Hungerbühler, K. and Nowack, B. (2014). Comprehensive probabilistic modelling of environmental emissions of engineered nanomaterials, *Environ. Pollut.*, **185**, 69–76

107. Oberdörster, G., Oberdörster, E. and Oberdörster, J. (2005). Nanotoxicology: an emerging discipline evolving from studies of ultrafine particles, *Environ. Health Perspect.*, **113**(7), 823–839.

108. Petosa, A. R., Jaisi, D. P. E. B. P., Quevedo, I. R., Elimelech, M., Tufenkji, N. and Haven, N. (2010). Aggregation and deposition of engineered nanomaterials in aquatic environments: role of physicochemical interactions, *Environ. Sci. Technol.*, **44**(17), 6532–6549.

109. Ilinskaya, A. and Dobrovolskaia, M. (2013). Nanoparticles and the blood coagulation system. Benefits of nanotechnology, *Nanomedicine*, **8**(5), 773–784.

1110. Ilinskaya, A. and Dobrovolskaia, M. (2013). Nanoparticles and the blood coagulation system. Part II: Safety concerns, *Nanomedicine (Lond).*, **8**(6), 969–981.

111. Elimelech, M., Gregory, J., Jia, X. and Williams, R. A. (1995a). Colloidal hydrodynamics and transport, in *Particle Deposition and Aggregation: Measurement, Modelling and Simulation*, Vol. 1 (Butterworth-Heinemann Ltd.), pp. 68–109.

112. Rivolo, P., Rose, J., Auffan, M., Proux, O., Niviere, V. and Bottero, J. Y. (2012). *Encyclopedia of Nanotechnology* (Springer Netherlands, Dordrecht).

113. Klaine, S. J., Alvarez, P. J. J., Batley, G. E., Fernandes, T. F., Handy, R. D., Lyon, D. Y., Mahendra, S., McLaughlin, M. J. and Lead, J. R. (2008). Nanomaterials in the environment: behavior, fate, bioavailability, and effects, *Environ. Toxicol. Chem.*, **27**(9), 1825.

114. Li, Y., Yang, C., Guo, X., Dang, Z., Li, X. and Zhang, Q. (2015). Effects of humic acids on the aggregation and sorption of nano-TiO_2, *Chemosphere*, **119**, pp. 171–176.

115. Quik, J. T. K., Stuart, M. C., Wouterse, M., Peijnenburg, W., Hendriks, A. J. and van de Meent, D. (2012). Natural colloids are the dominant factor in the sedimentation of nanoparticles, *Environ. Toxicol. Chem.*, **31**(5), 1019–1022.

116. Cornelis, G., Pang, L., Doolette, C., Kirby, J. K. and McLaughlin, M. J. (2013). Transport of silver nanoparticles in saturated columns of natural soils, *Sci. Total Environ.*, **463–464**, 120–130.

117. Quik, J. T. K., Velzeboer, I., Wouterse, M., Koelmans, A. A. and van de Meent, D. (2014). Heteroaggregation and sedimentation rates for nanomaterials in natural waters, *Water Res.*, **48**, 269–279.

118. Elimelech, M., Gregory, J., Jia, X. and Williams, R. A. R. A. (1995b). Modelling of aggregation processes, in *Particle Deposition and*

Aggregation: Measurement, Modelling and Simulation (Butterworth-Heinemann, Woburn), pp. 157–202.

119. Pippa, N., Dokoumetzidis, A., Demetzos, C. and Macheras, P. (2013). On the ubiquitous presence of fractals and fractal concepts in pharmaceutical sciences: a review, *Int. J. Pharm.*, **456**(2), 340–352.

120. Healy, T. W., Homola, A., James, R. O. and Hunter, R. J. (1978). Coagulation of amphoteric latex colloids: reversibility and specific ion effects, *Faraday Discuss. Chem. Soc.*, **65**, 156.

121. Pashley, R. and Karaman, M. (2005). *Applied Colloid and Surface Chemistry* (John Wiley & Sons).

122. Zhang, W., Rittmann, B. and Chen, Y. (2011). Size effects on adsorption of hematite nanoparticles on E. coli cells, *Environ. Sci. Technol.*, **45**(6), 2172–2178.

123. Zhang, W., Crittenden, J., Li, K. and Chen, Y. (2012). Attachment efficiency of nanoparticle aggregation in aqueous dispersions: modeling and experimental validation, *Environ. Sci. Technol.*, **46**(13), 7054–7062.

124. Vilanova, O., Franzese, G. and Barnabei, M. (2014). BUBBLES user-friendly bundle for bio-nano large-scale efficient simulations.

125. Smoluchowski, M. V. (1917). Versuch einer mathematischen Theorie der Koagulationskinetik, *Z. Phys. Chem.*, **92**, 129–168.

126. Friedlander, S. K. (1977). *Smoke, Dust, and Haze: Sheldon K. - Oxford University Press*, 2nd ed. (Oxford University Press, New York).

127. Jiang, J., Oberdörster, G. and Biswas, P. (2008). Characterization of size, surface charge, and agglomeration state of nanoparticle dispersions for toxicological studies, *J. Nanopart. Res.*, **11**(1), 77–89.

128. Arvidsson, R., Molander, S., Sandén, B. A. and Hassellöv, M. (2011). Challenges in exposure modeling of nanoparticles in aquatic environments rickard, *Hum. Ecol. Risk Assess.*, **17**(1), 245–262.

129. Buffle, J., Wilkinson, K. J., Stoll, S., Filella, M. and Zhang, J. (1998). A generalized description of aquatic colloidal interactions: the three-colloidal component approach, *Environ. Sci. Technol.*, **32**(19), 2887–2899.

130. EC (2012). D6.4 – Optimised PBPK/PD and exposure model, NMP4-SL-2009-228789.

131. Teeguarden, J. G., Hinderliter, P. M., Orr, G., Thrall, B. D. and Pounds, J. G. (2007). Particokinetics in vitro: dosimetry considerations for in vitro nanoparticle toxicology assessments, *Toxicol. Sci.*, **97**(2), 614–614.

132. Wohlleben, W. (2012). Validity range of centrifuges for the regulation of nanomaterials: from classification to as-tested coronas, *J. Nanopart. Res.*, **14**(12), 1300.

133. Mandelbrot, B. (1982). *The Fractal Geometry of Nature* (Henry & Company, Inc., Holt).

134. Metze, K. (2013). Fractal dimension of chromatin: potential molecular diagnostic applications for cancer prognosis, *Expert Rev. Mol. Diagn.*, **13**(7), 719–735.

135. Moreno, P. A., Vélez, P. E., Martínez, E., Garreta, L. E., Díaz, N., Amador, S., Tischer, I., Gutiérrez, J. M., Naik, A. K., Tobar, F. and García, F. (2011). The human genome: a multifractal analysis, *BMC Genomics*, **12**(1), 506.

136. Lin, M. Y., Lindsay, H., Weitz, D. A., Ball, R., Klein, R. and Meakin, P. (1990). Universal reaction-limited colloid aggregation, *Phys. Rev. A*, **41**(4), 2005–2020.

137. Le, T., Epa, V. C., Burden, F. R. and Winkler, D. A. (2012). Quantitative structure-property relationship modeling of diverse materials properties, *Chem. Rev.*, **112**(5), 2889–2919.

138. Liu, H. H. and Cohen, Y. (2014). Multimedia environmental distribution of engineered nanomaterials, *Environ. Sci. Technol.*, **48**(6), 3281–3292.

139. Schultz, C., Powell, K., Crossley, A., Jurkschat, K., Kille, P., Morgan, A. J., Read, D., Tyne, W., Lahive, E., Svendsen, C. and Spurgeon, D. J. (2014). Analytical approaches to support current understanding of exposure, uptake and distributions of engineered nanoparticles by aquatic and terrestrial organisms, *Ecotoxicology*, **24**(2), 239–261.

140. Braakhuis, H. M., Park, M. V. D. Z., Gosens, I., De Jong, W. H. and Cassee, F. R. (2014). Physicochemical characteristics of nanomaterials that affect pulmonary inflammation, *Part. Fibre Toxicol.*, **11**(1), 18.

141. Yang, W., Peters, J. I. and Williams, R. O. (2008). Inhaled nanoparticles–a current review, *Int. J. Pharm.*, **356**(1–2), 239–247.

142. Balasubramanian, S. K., Poh, K.-W., Ong, C.-N., Kreyling, W. G., Ong, W.-Y., Yua, L. E. and Yu, L. E. (2013). The effect of primary particle size on biodistribution of inhaled gold nano-agglomerates, *Biomaterials*, **34**(22), 5439–5452.

143. Konduru, N., Keller, J., Ma-Hock, L., Gröters, S., Landsiedel, R., Donaghey, T. C., Brain, J. D., Wohlleben, W. and Molina, R. M. (2014). Biokinetics and effects of barium sulfate nanoparticles, *Part. Fibre Toxicol.*, **11**(1), 55.

144. Nohynek, G. J. and Dufour, E. K. (2012). Nano-sized cosmetic formulations or solid nanoparticles in sunscreens: a risk to human health? *Arch. Toxicol.*, **86**(7), 1063–1075.

145. Borm, P. J. A., Robbins, D., Haubold, S., Kuhlbusch, T., Fissan, H., Donaldson, K., Schins, R., Stone, V., Kreyling, W., Lademann, J., Krutmann, J., Warheit, D. and Oberdörster, E. (2006). The potential risks of nanomaterials: a review carried out for ECETOC, *Part. Fibre Toxicol.*, **3**, 11.

146. Stern, S. T. and McNeil, S. E. (2008). Nanotechnology safety concerns revisited, *Toxicol. Sci.*, **101**(1), 4–21.

147. Powell, J. J., Faria, N., Thomas-McKay, E. and Pele, L. C. (2010). Origin and fate of dietary nanoparticles and microparticles in the gastrointestinal tract, *J. Autoimmun.*, **34**(3), J226– J233.

148. Hagens, W. I., Oomen, A. G., de Jong, W. H., Cassee, F. R. and Sips, A. J. A. M. (2007). What do we (need to) know about the kinetic properties of nanoparticles in the body? *Regul. Toxicol. Pharmacol.*, **49**(3), 217–229.

149. Osmond-McLeod, M. J., Oytam, Y., Kirby, J. K., Gomez-Fernandez, L., Baxter, B. and McCall, M. J. (2014). Dermal absorption and short-term biological impact in hairless mice from sunscreens containing zinc oxide nano- or larger particles, *Nanotoxicology*, **8**(Suppl 1), 72–84.

150. EASAC and JRC (2011). Impact of engineered nanomaterials on health, Luxembourg.

151. DK EPA (2015). Dermal absorption of nanomaterials titanium dioxide and zinc oxide based sunscreen. ISBN 978-87-93352-53-7. Available from: http://www2.mst.dk/Udgiv/publications/2015/08/978-87-93352-53-7.pdf.

152. van Kesteren, P. C. E., Cubadda, F., Bouwmeester, H., van Eijkeren, J. C. H., Dekkers, S., de Jong, W. H. and Oomen, A. G. (2014). Novel insights into the risk assessment of the nanomaterial synthetic amorphous silica, additive E551, in food, *Nanotoxicology*, 1–10.

153. Lynch, I., Feitshans, I. L. and Kendall, M. (2015). 'Bio-nano interactions: new tools, insights and impacts': summary of the Royal Society discussion meeting, *Philos. Trans. R. Soc. Lond. B: Biol. Sci.*, **370**(1661), 20140162.

154. Wolfram, J., Yang, Y., Shen, J., Moten, A., Chen, C., Shen, H., Ferrari, M. and Zhao, Y. (2014). The nano-plasma interface: implications of the protein corona, *Colloids Surf. B: Biointerfaces*, **124**, 17–24.

155. Monopoli, M. P., Walczyk, D., Campbell, A., Elia, G., Lynch, I., Bombelli, F. B. and Dawson, K. A. (2011). Physical-chemical aspects of protein

corona: relevance to in vitro and in vivo biological impacts of nanoparticles, *J. Am. Chem. Soc.*, **133**(8), 2525–2534.

156. Monopoli, M. P., Aberg, C., Salvati, A. and Dawson, K. A. (2012). Biomolecular coronas provide the biological identity of nanosized materials, *Nat. Nanotechnol.*, **7**(12), 779–786.

157. Cedervall, T., Lynch, I., Lindman, S., Berggård, T., Thulin, E., Nilsson, H., Dawson, K. A. and Linse, S. (2007). Understanding the nanoparticle-protein corona using methods to quantify exchange rates and affinities of proteins for nanoparticles, *Proc. Natl. Acad. Sci. U. S. A.*, **104**(7), 2050–2055.

158. Wang, F., Yu, L., Monopoli, M. P., Sandin, P., Mahon, E., Salvati, A. and Dawson, K. A. (2013). The biomolecular corona is retained during nanoparticle uptake and protects the cells from the damage induced by cationic nanoparticles until degraded in the lysosomes, *Nanomedicine*, **9**(8), 1159–1168.

159. Vroman, L. and Adams, A. L. (1969). Findings with the recording ellipsometer suggesting rapid exchange of specific plasma proteins at liquid/solid interfaces, *Surf. Sci.*, **16**, 438–446.

160. Salatin, S. and Yari Khosroushahi, A. (2017). Overviews on the cellular uptake mechanism of polysaccharide colloidal nanoparticles, *J. Cell. Mol. Med.*, doi:10.1111/jcmm.13110

161. Savolainen, K., Backman, U., Brouwer, D., Fadeel, B., Fernandes, T., Kuhlbusch, T., Landsiedel, R., Lynch, I. and Pylkkänen, L. (2013). Nanosafety in Europe 2015–2025: Towards safe and sustainable nanomaterials and nanotechnology innovations nanosafety in europe towards safe and sustainable nanomaterials and nanotechnology innovations, EDITA, Helsinki.

162. Hillery, A. M., Jan, P. U. and Florence, A. T. (1994). Comparative, quantitative study of lymphoid and non-lymphoid uptake of 60 nm polystyrene particles, *J. Drug Test.*, **2**(1992), 151–156.

163. Alarifi, S., Ali, D., Al-Doaiss, A. A., Ali, B. A., Ahmed, M. and Al-Khedhairy, A. A. (2013). Histologic and apoptotic changes induced by titanium dioxide nanoparticles in the livers of rats, *Int. J. Nanomed.*, **8**, 3937–3943.

164. Park, E.-J., Yi, J., Chung, K.-H., Ryu, D.-Y., Choi, J. and Park, K. (2008). Oxidative stress and apoptosis induced by titanium dioxide nanoparticles in cultured BEAS-2B cells, *Toxicol. Lett.*, **180**(3), 222–229.

165. Lai, D. Y. (2012). Toward toxicity testing of nanomaterials in the 21st century: a paradigm for moving forward, *Wiley Interdiscip. Rev. Nanomed. Nanobiotechnol.*, **4**(1), 1–15.

166. Xia, X.-R., Monteiro-Riviere, N. A. and Riviere, J. E. (2010). An index for characterization of nanomaterials in biological systems, *Nat. Nanotechnol.*, **5**(9), 671–675.

167. Donaldson, K. and Poland, C. A. (2012). Inhaled nanoparticles and lung cancer - what we can learn from conventional particle toxicology, *Swiss Med. Wkly.*, **142**, w13547.

168. Cheng, L.-C., Jiang, X., Wang, J., Chen, C. and Liu, R.-S. (2013). Nano-bio effects: interaction of nanomaterials with cells, *Nanoscale*, **5**(9), 3547–3569.

169. Nel, A., Xia, T., Mädler, L. and Li, N. (2006). Toxic potential of materials at the nanolevel, *Science*, **311**(5761), 622–627.

170. Knudsen, K. B., Northeved, H., Ek, P. K., Permin, A., Andresen, T. L., Larsen, S., Wegener, K. M., Lam, H. R. and Lykkesfeldt, J. (2014). Differential toxicological response to positively and negatively charged nanoparticles in the rat brain, *Nanotoxicology*, **8**(7), 764–774.

171. Ho, M., Wu, K.-Y., Chein, H.-M., Chen, L.-C. and Cheng, T.-J. (2011). Pulmonary toxicity of inhaled nanoscale and fine zinc oxide particles: mass and surface area as an exposure metric, *Inhalation Toxicol.*, **23**(14), 947–956.

172. Koch, W. and Stöber, W. (2008). A simple pulmonary retention model accounting for dissolution and macrophage-mediated removal of deposited polydisperse particles, *Inhalation Toxicol.*, **13**(2), 129–148.

173. Kreyling, W. G. (1992). Intracellular particle dissolution in alveolar macrophages, *Environ. Health Perspect.*, **97**, 121–126.

174. Winter, M., Beer, H.-D., Hornung, V., Krämer, U., Schins, R. P. F. and Förster, I. (2011). Activation of the inflammasome by amorphous silica and TiO_2 nanoparticles in murine dendritic cells, *Nanotoxicology*, **5**(3), 326–340.

175. Hornung, V., Bauernfeind, F., Halle, A., Samstad, E. O., Kono, H., Rock, K. L., Fitzgerald, K. A. and Latz, E. (2008). Silica crystals and aluminum salts activate the NALP3 inflammasome through phagosomal destabilization, *Nat. Immunol.*, **9**(8), 847–856.

176. Tschopp, J. and Schroder, K. (2010). NLRP3 inflammasome activation: the convergence of multiple signalling pathways on ROS production? *Nat. Rev. Immunol.*, **10**(3), 210–215.

177. Tran, C. L., Buchanan, D., Cullen, R. T., Searl, A. and Jones, A. D. (2000). Inhalation of poorly soluble particles. II Influence of particle surface area on inflammation and clearence, *Inhalation Toxicol.*, **12**, 1113–1126.

178. Duffin, R., Tran, L., Brown, D., Stone, V. and Donaldson, K. (2007). Proinflammogenic effects of low-toxicity and metal nanoparticles in vivo and in vitro: highlighting the role of particle surface area and surface reactivity, *Inhalation Toxicol.*, **19**, 849–856.

179. Bakand, S., Hayes, A. and Dechsakulthorn, F. (2012). Nanoparticles: a review of particle toxicology following inhalation exposure, *Inhalation Toxicol.*, **24**(2), 125–135.

180. Donaldson, K., Borm, P. J. A., Oberdörster, G., Pinkerton, K. E., Stone, V. and Tran, C. L. (2008). Concordance between in vitro and in vivo dosimetry in the proinflammatory effects of low-toxicity, low-solubility particles: the key role of the proximal alveolar region, *Inhalation Toxicol.*, **20**, 53–62.

181. Geiser, M., Wigge, C., Conrad, M. L., Eigeldinger-Berthou, S., Künzi, L., Garn, H., Renz, H. and Mall, M. A. (2014). Nanoparticle uptake by airway phagocytes after fungal spore challenge in murine allergic asthma and chronic bronchitis, *BMC Pulm. Med.*, **14**, 116.

182. Mura, S., Hillaireau, H., Nicolas, J., Le Droumaguet, B., Gueutin, C., Zanna, S., Tsapis, N. and Fattal, E. (2011). Influence of surface charge on the potential toxicity of PLGA nanoparticles towards Calu-3 cells, *Int. J. Nanomed.*, **6**, 2591–2605.

183. Liu, H., Ma, L., Zhao, J., Liu, J., Yan, J., Ruan, J. and Hong, F. (2009). Biochemical toxicity of nano-anatase TiO_2 particles in mice, *Biol. Trace Elem. Res.*, **129**(1–3), 170–180.

184. Duan, Y., Liu, J., Ma, L., Li, N., Liu, H., Wang, J., Zheng, L., Liu, C., Wang, X., Zhao, X., Yan, J., Wang, S., Wang, H., Zhang, X. and Hong, F. (2010). Toxicological characteristics of nanoparticulate anatase titanium dioxide in mice, *Biomaterials*, **31**(5), 894–899.

185. Cui, Y., Liu, H., Zhou, M., Duan, Y., Li, N., Gong, X., Hu, R., Hong, M. and Hong, F. (2011). Signaling pathway of inflammatory responses in the mouse liver caused by TiO_2 nanoparticles, *J. Biomed. Mater. Res. A*, **96**(1), 221–229.

186. Ma, L., Zhao, J., Wang, J., Liu, J., Duan, Y., Liu, H., Li, N., Yan, J., Ruan, J., Wang, H. and Hong, F. (2009). The acute liver injury in mice caused by nano-anatase TiO_2, *Nanoscale Res. Lett.*, **4**(11), 1275–1285.

187. Liu, H., Ma, L., Liu, J., Zhao, J., Yan, J. and Hong, F. (2010). Toxicity of nano-anatase TiO_2 to mice: liver injury, oxidative stress, *Toxicol. Environ. Chem.*, **92**(1), 175–186.

188. Soliman, M. M., Attia, H. F., Hussein, M. M., Mohamed, E. H. and Ismail, T. A. (2013). Protective effect of N-Acetylcystiene aganst titanium dioxide nanoparticles modulate immune responses in male albino rats, *Am. J. Immunol.*, **9**(4), 148–158.

189. Chen, J., Dong, X., Zhao, J. and Tang, G. (2009). In vivo acute toxicity of titanium dioxide nanoparticles to mice after intraperitioneal injection, *J. Appl. Toxicol.*, **29**(4), 330–337.

190. Sycheva, L. P., Zhurkov, V. S., Iurchenko, V. V., Daugel-Dauge, N. O., Kovalenko, M. A., Krivtsova, E. K. and Durnev, A. D. (2011). Investigation of genotoxic and cytotoxic effects of micro- and nanosized titanium dioxide in six organs of mice in vivo, *Mutat. Res.*, **726**(1), 8–14.

191. Trouiller, B., Reliene, R., Westbrook, A., Solaimani, P. and Schiestl, R. H. (2009). Titanium dioxide nanoparticles induce DNA damage and genetic instability in vivo in mice, *Cancer Res.*, **69**(22), 8784–8789.

192. van der Zande, M., Vandebriel, R. J., Groot, M. J., Kramer, E., Herrera Rivera, Z. E., Rasmussen, K., Ossenkoppele, J. S., Tromp, P., Gremmer, E. R., Peters, R. J. B., Hendriksen, P. J., Marvin, H. J. P., Hoogenboom, R. L. A. P., Peijnenburg, A. A. C. M. and Bouwmeester, H. (2014). Sub-chronic toxicity study in rats orally exposed to nanostructured silica, *Part. Fibre Toxicol.*, **11**(1), 8.

193. Downs, T. R., Crosby, M. E., Hu, T., Kumar, S., Sullivan, A., Sarlo, K., Reeder, B., Lynch, M., Wagner, M., Mills, T. and Pfuhler, S. (2012). Silica nanoparticles administered at the maximum tolerated dose induce genotoxic effects through an inflammatory reaction while gold nanoparticles do not, *Mutat. Res.*, **745**(1–2), 38–50.

194. Kermanizadeh, A. (2013). Engineered nanomaterial impact in the liver following exposure via an intravenous route–the role of polymorphonuclear leukocytes and gene expression in the organ, *J. Nanomed. Nanotechnol.*, **4**(157), 1–7.

195. Prasad, R. Y., Wallace, K., Daniel, K. M., Tennant, A. H., Zucker, R. M., Strickland, J., Dreher, K., Kligerman, A. D., Blackman, C. F. and Demarini, D. M. (2013). Effect of treatment media of titanium dioxide nanoparticles: impact on genotoxicity, cellular interaction, and cell cycle, *ACS Nano*, **7**(3), 1929–1942.

196. OECD (2009). Guidance manual for the testing of manufactured nanomaterials. OECD's sponsorship programme, Paris, ENV/JM/MONO(2009)20.

197. OECD (2014). Genotoxicity of manufactured nanomaterials: report of the OECD expert meeting. Series on the Safety of Manufactured Nanomaterials No. 43, ENV/JM/WRPR(2014)34.

198. OECD (2011). Report of the WPMN expert meeting on inhalation toxicity testing for nanomaterials, ENV/CHEM/NANO(2911)23/REV1.

199. OECD (2012). Inhalation toxicity testing: expert meeting on potential revisions to OECD test guidelines and guidance document. Series on the Safety of Manufactured Nanomaterials No. 35, Paris, ENV/JM/MONO(2012)14.

200. OECD (2016b). Toxicokinetic of manufactured nanomaterials: report from the OECD expert meeting, Paris, 72.

201. Rasmussen, K., Rauscher, H., Kearns, P., González, M. and Riego Sintes, J. (2019). Developing OECD test guidelines for regulatory testing of nanomaterials to ensure mutual acceptance of test data, *Regul. Toxicol. Pharmacol.*, **104**, 74–83.

202. OECD (2012). Guidance on sample preparation and dosimetry for the safety testing of manufactured nanomaterials, Paris, 36.

203. Donaldson, K. and Poland, C. (2013). Nanotoxicity: challenging the myth of nano-specific toxicity, *Curr. Opin. Biotechnol.*, **24**(4), 724–734.

204. OECD (2014). Ecotoxicology and environmental fate of manufactured nanomaterials: test guidelines expert meeting report, Paris.

205. OECD (2012). The adverse outcome pathway for skin sensitisation initiated by covalent binding to proteins. Part 2: Use of the AOP to develop chemical categories and integrated assessment and testing approaches. Series on Testing and Assessment No. 168, Paris, ENV/JM/MONO(2012)10/PART2.

206. OECD (2012). The adverse outcome pathway for skin sensitisation initiated by covalent binding to proteins. Part 1: Scientific evidence. Series on Testing and Assessment No. 168, ENV/JM/MONO(2012)10/PART1.

207. ECHA (2008). Guidance on information requirements and chemical safety assessment Chapter R.6: QSARs and grouping of chemicals, Helsinki.

208. Ankley, G. T., Bennett, R. S., Erickson, R. J., Hoff, D. J., Hornung, M. W., Johnson, R. D., Mount, D. R., Nichols, J. W., Russom, C. L., Schmieder, P. K., Serrrano, J. A., Tietge, J. E. and Villeneuve, D. L. (2010). Adverse outcome pathways: a conceptual framework to support ecotoxicology research and risk assessment, *Environ. Toxicol. Chem.*, **29**(3), 730–741.

209. Gerloff, K., Landesmann, B., Worth, A., Munn, S., Palosaari, T. and Whelan, M. (2017). The adverse outcome pathway approach in nanotoxicology, *Comput. Toxicol.*, **1**, 3–11.

210. Asturiol, D. and Worth, A. (2012). D6.5 - Report on the development and applications of structure-activity models for toxicity prediction of ENP: analysis of ENPRA dataset, NMP4-SL-2009-228789.

211. Benigni, R. and Bossa, C. (2006). Structural alerts of mutagens and carcinogens, *Curr. Comput. Aided-Drug Des.*, **2**(2), 169–176.

212. Fourches, D., Pu, D., Tassa, C., Weissleder, R., Shaw, S. Y., Mumper, R. J. and Tropsha, A. (2010). Quantitative nanostructure-activity relationship modeling, *ACS Nano*, **4**(10), 5703–5712.

213. Liu, R., Zhang, H. Y., Ji, Z. X., Rallo, R., Xia, T., Chang, C. H., Nel, A. and Cohen, Y. (2013). Development of structure-activity relationship for metal oxide nanoparticles, *Nanoscale*, **5**(12), 5644–5653.

214. Ghorbanzadeh, M., Fatemi, M. H. and Karimpour, M. (2012). Modeling the cellular uptake of magneto fluorescent nanoparticles in pancreatic cancer cells: a quantitative structure activity relationship study, *Ind. Eng. Chem. Res.*, **51**(32), 10712–10718.

215. Fraser, T. R. and Crum-Brown, A. (1868). Action of the salts of the ammonium, *J. Anat. Physiol.*, **2**(2), 224–242.

216. Hansch, C. and Fujita, T. (1964). Additions and corrections - ρ-σ-π analysis. A method for the correlation of biological activity and chemical structure, *J. Am. Chem. Soc.*, **86**(24), 1616–1626.

217. Hansch, C. (1969). Quantitative approach to biochemical structure-activity relationships, *Acc. Chem. Res.*, **2**(8), 232–239.

218. Hammett, L. P. (1937). The effect of structure upon the reactions of organic compounds. Benzene derivatives, *J. Am. Chem. Soc.*, **59**(1), 96–103.

219. Taft, W. (1956). *Steric Effects in Organic Chemistry* (John Wiley & Sons Inc., New York).

220. Todeschini, R. and Consonni, V. (2009). *Molecular Descriptors for Chemoinformatics*, 2nd ed. (Wiley).

221. Randic, M. (1975). Characterization of molecular branching, *J. Am. Chem. Soc.*, **97**(23), 6609–6615.

222. Hall, L. and Kier, L. (1977). Structure-activity studies using valence molecular connectivity, *J. Pharm. Sci.*, **66**, 642–644.

223. Hopfinger, A. J., Wang, S., Tokarski, J. S., Jin, B., Albuquerque, M., Madhav, P. J. and Duraiswami, C. (1997). Construction of 3D-QSAR models using

the 4D-QSAR analysis formalism, *J. Am. Chem. Soc.*, **119**(43), 10509–10524.

224. Cramer, R. D., Patterson, D. E. and Bunce, J. D. (1988). Comparative molecular field analysis (CoMFA). 1. Effect of shape on binding of steroids to carrier proteins, *J. Am. Chem. Soc.*, **110**(18), 5959–5967.

225. Klebe, G., Abraham, U. and Mietzner, T. (1994). Molecular similarity indices in a comparative analysis (CoMSIA) of drug molecules to correlate and predict their biological activity, *J. Med. Chem.*, **37**(24), 4130–4146.

226. Cruciani, G. and Watson, K. A. (1994). Comparative molecular field analysis using GRID force-field and GOLPE variable selection methods in a study of inhibitors of glycogen phosphorylase b, *J. Med. Chem.*, **37**(16), 2589–2601.

227. OECD (2007). Guidance document on the validation of (quantitative) structure-activity relationship [(Q) SAR] models, Paris, ENV/JM/MONO(2007)2.

228. OECD (2004). The report from the expert group on (quantitative) structure-activity relationships [(Q)SARs] on the principles for the validation of (Q)SARs, Paris, ENV/JM/MONO(2004)24.

229. Dearden, J. C., Barratt, M. D., Benigni, R., Douglas, W., Combes, R. D., Cronin, M. T. D., Judson, P. N., Payne, M. P., Richard, A. M., Tichy, M., Worth, A. P. and Yourick, J. J. (1997). The development and validation of expert systems for predicting toxicity, *Altern. Lab. Anim.*, **25**, 223–252.

230. Tantra, R., Oksel, C., Puzyn, T., Wang, J., Robinson, K. N., Wang, X. Z., Ma, C. Y. and Wilkins, T. (2014). Nano(Q)SAR: challenges, pitfalls and perspectives, *Nanotoxicology*, 1–7.

231. Zhou, H., Mu, Q., Gao, N., Liu, A., Xing, Y., Gao, S., Zhang, Q., Qu, G., Chen, Y., Liu, G., Zhang, B. and Yan, B. (2008). A nano-combinatorial library strategy for the discovery of nanotubes with reduced protein-binding, cytotoxicity, and immune response, *Nano Lett.*, **8**(3), 859–865.

232. Hristozov, D. R., Zabeo, A., Foran, C., Isigonis, P., Critto, A., Marcomini, A. and Linkov, I. (2014). A weight of evidence approach for hazard screening of engineered nanomaterials, *Nanotoxicology*, **8**(1), 72–87.

233. Puzyn, T., Leszczynska, D. and Leszczynski, J. (2009). Toward the development of "nano-QSARs": advances and challenges, *Small*, **5**(22), 2494–2509.

234. Shao, C.-Y., Chen, S.-Z., Su, B.-H., Tseng, Y. J., Esposito, E. X. and Hopfinger, A. J. (2013). Dependence of QSAR models on the selection

of trial descriptor sets: a demonstration using nanotoxicity endpoints of decorated nanotubes, *J. Chem. Inf. Model.*, **53**(1), 142–158.

235. Rallo, R., France, B., Liu, R., Nair, S., George, S., Damoiseaux, R., Giralt, F., Nel, A., Bradley, K. and Cohen, Y. (2011). Self-organizing map analysis of toxicity-related cell signaling pathways for metal and metal oxide nanoparticles, *Environ. Sci. Technol.*, **45**(4), 1695–1702.

236. Martin, D., Maran, U., Sild, S. and Karelson, M. (2007). QSPR modeling of solubility of polyaromatic hydrocarbons and fullerene in 1-octanol and n-heptane, *J. Phys. Chem. B*, **111**(33), 9853–9857.

237. Gajewicz, A., Rasulev, B., Dinadayalane, T. C., Urbaszek, P., Puzyn, T., Leszczynska, D. and Leszczynski, J. (2012). Advancing risk assessment of engineered nanomaterials: application of computational approaches, *Adv. Drug Delivery Rev.*, **64**(15), 1663–1693.

238. Puzyn, T., Gajewicz, A., Leszczynska, D. and Leszczynski, J. (2010). Nanomaterials - the next great challenge for QSAR modelers, in *Recent Advances in QSAR Studies*, Puzyn, T., Leszczynski, J. and Cronin, M. T. D., eds. (Springer, New York), pp. 383–403.

239. Burello, E. and Worth, A. P. (2011). QSAR modeling of nanomaterials, *Wiley Interdiscip. Rev. Nanomed. Nanobiotechnol.*, **3**(3), 298–306.

240. Bachler, G., von Goetz, N. and Hungerbuhler, K. (2014). Using physiologically based pharmacokinetic (PBPK) modeling for dietary risk assessment of titanium dioxide (TiO_2) nanoparticles, *Nanotoxicology*, 1–8.

241. Li, D., Johanson, G., Emond, C., Carlander, U., Philbert, M. and Jolliet, O. (2014). Physiologically based pharmacokinetic modeling of polyethylene glycol-coated polyacrylamide nanoparticles in rats, *Nanotoxicology*, **8**(Suppl 1), 128–137.

242. Hagenbuch, B. and Meier, P. J. (2004). Organic anion transporting polypeptides of the OATP/ SLC21 family: phylogenetic classification as OATP/ SLCO superfamily, new nomenclature and molecular/ functional properties, *Pflugers Arch.*, **447**(5), 653–665.

243. Bachler, G., von Goetz, N. and Hungerbühler, K. (2013). A physiologically based pharmacokinetic model for ionic silver and silver nanoparticles, *Int. J. Nanomed.*, **8**, 3365–3382.

244. Stone, V., Johnston, H. and Schins, R. P. F. (2009). Development of in vitro systems for nanotoxicology: methodological considerations, *Crit. Rev. Toxicol.*, **39**(7), 613–626.

245. Joris, F., Manshian, B. B., Peynshaert, K., De Smedt, S. C., Braeckmans, K. and Soenen, S. J. (2013). Assessing nanoparticle toxicity in cell-based

assays: influence of cell culture parameters and optimized models for bridging the in vitro-in vivo gap, *Chem. Soc. Rev.*, **42**(21), 8339–8359.

246. Hirsch, C., Roesslein, M., Krug, H. F. and Wick, P. (2011). Nanomaterial cell interactions: are current in vitro tests reliable? *Nanomedicine (Lond).*, **6**(5), 837–847.

247. Nel, A., Xia, T., Meng, H., Wang, X., Lin, S., Ji, Z. and Zhang, H. (2013). Nanomaterial toxicity testing in the 21st century: use of a predictive toxicological approach and high-throughput screening, *Acc. Chem. Res.*, **46**(3), 607–621.

248. OECD (2007). Guidance on grouping of chemicals. Series on Testing and Assessment No. 80, Paris, ENV/JM/MONO(2007)28.

249. OECD (2014). Guidance on grouping of chemicals. Second edition. Series on Testing and Assessment No. 194, Paris, ENV/JM/MONO(2014)4.

250. ECETOC (2012). Category approaches, read-across, (Q)SAR, Brussels, 116.

251. Patlewicz, G., Roberts, D. W., Aptula, A., Blackburn, K. and Hubesch, B. (2013). Workshop: use of "read-across" for chemical safety assessment under REACH, *Regul. Toxicol. Pharmacol.*, **65**(2), 226–228.

252. Wu, S., Blackburn, K., Amburgey, J., Jaworska, J. and Federle, T. (2010). A framework for using structural, reactivity, metabolic and physicochemical similarity to evaluate the suitability of analogs for SAR-based toxicological assessments, *Regul. Toxicol. Pharmacol.*, **56**(1), 67–81.

253. Linkov, I., Welle, P., Loney, D., Tkachuk, A., Canis, L., Kim, J. B. and Bridges, T. (2011). Use of multicriteria decision analysis to support weight of evidence evaluation, *Risk Anal.*, **31**(8), 1211–1225.

254. Weed, D. L. (2005). Weight of evidence: a review of concept and methods, *Risk Anal.*, **25**(6), 1545–1557.

255. Chapman, P. M., McDonald, B. G. and Lawrence, G. S. (2010). Weight-of-evidence issues and frameworks for sediment quality (and other) assessments, *Hum. Ecol. Risk Assess.*, **8**(7), 1489–1515.

256. ECHA (2010). Practical guide 2: how to report weight of evidence, Helsinki.

257. Hardy, A., Benford, D., Halldorsson, T., Jeger, M. J., Knutsen, H. K., More, S., Naegeli, H., Noteborn, H., Ockleford, C., Ricci, A., Rychen, G., Schlatter, J. R., Silano, V., Solecki, R., Turck, D., Benfenati, E., Chaudhry, Q. M., Craig, P., Frampton, G., Greiner, M., Hart, A., Hogstrand, C., Lambre, C., Luttik, R., Makowski, D., Siani, A., Wahlstroem, H., Aguilera, J., Dorne, J., Fernandez Dumont, A., Hempen, M., Valtueña Martínez, S., Martino, L.,

Smeraldi, C., Terron, A., Georgiadis, N. and Younes, M. (2017). Guidance on the use of the weight of evidence approach in scientific assessments, *EFSA J.*, **15**(8), 4971.

258. OECD (2016c). Guidance document on the reporting of defined approaches to be used within integrated approaches to testing and assessment, ENV/JM/MONO(2016)28.

259. OECD (2014). New guidance document on the integrated approach on an integrated approach on testing and assessment (IATA) for skin corrosion and irritation. Series on Testing and Assessment No. 203, Paris, ENV/JM/MONO(2014)19.

260. Lin, S. and Wiesner, M. R. (2012). Theoretical investigation on the interaction between a soft particle and a rigid surface, *Chem. Eng. J.*, **191**, 297–305.

Chapter 2

Current Developments and Recommendations in Computational Nanotoxicology in View of Regulatory Applications

Andrea-Nicole Richarz, Lara Lamon, David Asturiol, and Andrew P. Worth
European Commission, Joint Research Centre, Ispra, Italy
andrew.worth@ec.europa.eu

This chapter gives an overview of the EU project research landscape in computational nanotoxicology, highlighting the challenges for development and use of computational approaches in the safety assessment of nanomaterials and analyzing recent progress against these needs.

Although considerable progress has been made and is ongoing in view of overcoming these shortcomings, there currently still are gaps

Computational Nanotoxicology: Challenges and Perspectives
Edited by Agnieszka Gajewicz and Tomasz Puzyn

ISBN 978-981-4800-64-8 (Hardcover), 978-0-429-34137-3 (eBook)
www.jennystanford.com

in the knowledge of nanomaterial behavior; there is fragmentation in the scientific results and need for more harmonization and standardization. A lack of public access to the results and tools is preventing uptake and use of the models in regulatory decision-making. More guidance is needed and, in particular, more coordination is required to direct the model development to elucidating relevant gaps of knowledge and addressing questions and endpoints relevant for regulatory applications.

Overall, a number of recommendations are made in view of increasing the availability and uptake of computational methods for regulatory nanosafety assessment. They concern inherent scientific uncertainties in the understanding of nanomaterial behavior; data quality and variability; standardization and harmonization; the availability of data and predictive models; and the accessibility and visibility of models, including the need for infrastructures.

2.1 Introduction

Although there are no explicit requirements for nanomaterials (NMs) under the European Union (EU) Regulation for Registration, Evaluation, Authorisation and Restriction of Chemicals (REACH) [1] or the Classification, Labelling and Packaging Regulation (CLP) [2], NMs meet the regulations' substance definition and thus their provisions apply [3]. Manufacturers, importers, and downstream users need to ensure safe use of each substance, whatever its form under REACH, while reducing the use of in vivo testing through alternative methods with respective adaptations to information requirements according to REACH Annex XI, using, for example, computational approaches such as qualitative or quantitative structure–activity relationships (QSARs) or grouping and read-across [4]. Nano-specific appendices to the REACH guidance on information requirements and chemicals safety assessment have been provided by the European Chemicals Agency (ECHA) [5–8].

The regulatory acceptance of computational models to support the safety evaluation of chemicals is still being discussed, and especially in the case of NMs more experience is needed. The European Commission Joint Research Centre (JRC), through the Nanocomput project [9], evaluated the availability of computational methods, such as QSARs, grouping and read-across, potentially useful for predicting the properties of engineered NMs, and assessed their applicability in order to provide advice on the use of these approaches for the purposes of the REACH regulation. In particular, the EU research project landscape was reviewed in view of the status and potential applicability of computational methods for predicting NM properties/toxicity in regulatory context.

This chapter gives an overview of the European predictive nanotoxicity modeling landscape in research projects, on the basis of the results of the JRC Nanocomput project, and summarizes recommendations for further research and development in view of increasing the availability and uptake of computational methods for the safety assessment of NMs.

2.2 Computational Nanotoxicology Research Project Landscape

2.2.1 European Nanosafety Research Activities

The NanoSafety Cluster (NSC)[1] is the overarching umbrella for European research projects addressing the safety of materials and technologies using NMs, aiming at fostering exchange and synergies and avoiding duplication of work. Topics of the projects include safety assessment, assessment of toxicity to human health and the environment, exposure, and mechanism of interactions as well as definitions, standardization, regulatory aspects, and risk management.

The NSC issues a yearly compendium[2] describing the status of current EU projects related to nanosafety research and giving an

[1]http://www.nanosafetycluster.eu

[2]https://www.nanosafetycluster.eu/home/european-nanosafety-cluster-compendium.html

update on the NSC working groups, for example [10, 11]. The focus is increasingly shifting toward safety-by-design considerations, predictive toxicology, and high throughput/toxicology in the twenty-first-century-type approaches. A biannual newsletter gives updates on the projects. Furthermore, the NSC has published a strategic research agenda "Nanosafety in Europe 2015-2025" [12] and an update on research priorities in 2017 [13]. The latter builds on proposals made by the ITS Nano project in its final report, which stressed the major topics of physicochemical characterization, exposure identification, hazard identification, and modeling approaches for the risk assessment of NMs [14].

Cross-NSC working groups bring together researchers from the different projects to contribute to the discussion on specific topics and tackle issues common to the different projects. The original working groups were dedicated to:

- Materials
- Hazard
- Exposure
- Database
- Risk
- Modeling
- Dissemination
- Systems biology
- Safe-by-design and industrial innovation

With the NSC Action Plan 2017 [15] a reorganization of the working groups took place to match the planned transition of the NSC to an innovation governance platform. The new working groups are focusing on the following topics:

- Communication, training, and education
- Materials and standards
- Exposure and hazard assessment
- Models and tools for risk assessment
- Innovation and safer by design
- Data management

- Regulations and risk governance
- NSC dissemination group

For the scope of computational approaches for nanosafety assessment, the work and outputs of the modeling and data-related working groups are specifically relevant, since they discuss building of and guidance for models for NM characterization and toxicity prediction, as well as databases and ontologies related to NMs. The database working group was particularly active and had fruitful exchanges with the US-EU Communities of Research on Databases[3] and the US Nano Working Group[4] (see Section 2.2.2). They set up initiatives to provide a list of nanosafety-related databases in Europe and worldwide [16], a definition of data quality, and recommendations for a minimal reporting standard, in particular guiding on which experimental details need to be recorded. Members have also contributed to an international effort for assessing data completeness and quality [17]. While data completeness is not explicitly defined in the paper, it generally "assesses the extent to which experimental details are described and associated experimental results are reported" [17]. This can be achieved, for example, by checking against a list of minimum information requirements that would need to be defined in the context of specific applications. The new data management working group carries on discussing issues related to databases and ontologies, in order to be able to ensure data interoperability and a common language for databases in view of the longevity of data and their accessibility and uptake.

Table 2.1 lists the individual EU-funded research projects from the Seventh Framework Programme (FP7)[5], Horizon 2020 (H2020)[6], and a LIFE+[7] project that have been identified as related to computational modeling and grouping approaches for NMs.

More details on the projects can be found in Tables 2.2 and 2.3.

---

[3]https://us-eu.org/communities-of-research
[4]https://nciphub.org/groups/nanowg
[5]https://ec.europa.eu/research/fp7
[6]https://ec.europa.eu/programmes/horizon2020/
[7]https://ec.europa.eu/easme/en/life

Table 2.1 EU-funded projects related to computational modeling and grouping approaches for nanomaterials

Seventh Framework Programme	
eNanomapper	NanoFATE
ENPRA	NanoMILE
FutureNanoNeeds	NanoPUZZLES
ITS-NANO (support action)	NANoREG
MARINA	NanoReTox
MembraneNanoPart	NANOSOLUTIONS
MODENA (COST initiative)	NanoTEST
MOD-ENP-TOX	NanoToxClass (ERA-NET SIINN)
MODERN	NanoTrans-kinetics
ModNanoTox	PreNanoTox
NanoBRIDGES (Marie Curie Action)	
LIFE+	
REACHnano (LIFE+)	
Horizon 2020	
caLIBRAte	NANOGENTOOLS
GRACIOUS	NanoReg2
HISENTS	SmartNanoTox

In FP7, a specific cooperation was set up: five projects dedicated to computational modeling, starting at the same time, were grouped with two other, already running, modeling projects to form the NSC Modeling Cluster (Table 2.4). The projects agreed in a memorandum of understanding to collaborate, share expertise, and use common synergies. They held regular harmonization meetings. eNanomapper joined the cluster as well.

Table 2.2 Seventh Framework Programme EU projects that include aspects of modeling and/or grouping

Project acronym	Project title	Framework	Coordinating partner	Project website	Topics related to modeling and grouping/read-across	Duration
eNanomapper	A Database and Ontology Framework for Nanomaterials Design and Safety Assessment	FP7	Douglas Connect GmbH, Switzerland	http://www.enanomapper.net https://cordis.europa.eu/project/rcn/110961/factsheet/en	Creation of a platform that consists of an ontology addressing all standardization and regulatory requirements supporting the exploration of structure–activity relationships and of a data warehouse for NMs.	02/14–01/17
ENPRA	Risk Assessment of Engineered Nanoparticles	FP7	Institute of Occupational Medicine, UK	https://cordis.europa.eu/project/rcn/91280/factsheet/en	Identification of physicochemical properties that are useful as descriptors for building QSAR models for toxicologically relevant properties and effects.	05/09–10/12

(*Continued*)

Table 2.2 (*Continued*)

Project acronym	Project title	Framework	Coordinating partner	Project website	Topics related to modeling and grouping/read-across	Duration
FutureNanoNeeds	Framework to Respond to Regulatory Needs of Future Nanomaterials and Markets	FP7	University College Dublin, Ireland	http://www.futurenanoneeds.eu https://cordis.europa.eu/project/rcn/110964/factsheet/en	Development of a novel framework to enable naming, classification, and hazard and environmental impact assessment of the next-generation NMs.	01/14–12/17
ITS-NANO	Intelligent Testing Strategies for Engineered Nanomaterials	FP7 (Support Action)	Herriot Watts University, UK	https://www.safenano.org/research/its-nano https://cordis.europa.eu/project/rcn/102987/factsheet/en	Definition of testing strategies on the basis of protocols to support grouping and ranking of NMs.	03/12–05/13
MARINA	Managing Risks of Nanoparticles	FP7	Institute of Occupational Medicine, UK	http://www.marina-fp7.eu https://cordis.europa.eu/project/rcn/101207/factsheet/en	Building of a harmonized database database on reference NM fate-determining parameters in the environment and biota.	11/11–10/15

Project acronym	Project title	Framework	Coordinating partner	Project website	Topics related to modeling and grouping/read-across	Duration
MembraneNanoPart	Modelling the Mechanisms of Nanoparticle-Lipid Interactions and Nanoparticle Effects on Cell Membrane Structure and Function	FP7	University College Dublin, Ireland	https://cordis.europa.eu/project/rcn/106302/factsheet/en	Development of physically justified models and computational tools to quantitatively describe and understand the molecular mechanisms of nanoparticle (NP)/cell membrane interactions, which are crucial in modeling of NP toxicity.	01/13–12/15

(*Continued*)

Table 2.2 (*Continued*)

Project acronym	Project title	Framework	Coordinating partner	Project website	Topics related to modeling and grouping/read-across	Duration
MODENA	Modelling Nanomaterial Toxicity	FP7 (COST Initiative)	Institute of Occupational Medicine, UK	https://www.cost.eu/actions/TD1204	Promotion and coordination of interdisciplinary collaborations aiming at producing quantitative nanostructure–toxicity relationship (QNTR) models for engineered NMs, creating transparent, validated, and rigorous QNTR tools for regulatory purposes in the field of nanotoxicology according to OECD principles.	01/14–12/16

Project acronym	Project title	Framework	Coordinating partner	Project website	Topics related to modeling and grouping/read-across	Duration
MOD-ENP-TOX	Modelling Assays Platform "MAP" for Hazard Ranking of Engineered Nanoparticles	FP7	KU Leuven	http://fys.kuleuven.be/apps/modenptox https://cordis.europa.eu/project/rcn/106207/factsheet/en	Focus on metal NPs. Combining a novel computational modeling package (CMP) based on structural, mechanistic, and kinetic modeling tools and an innovative high-content-screening (HCS) strategy that allows performing multiplexed streamlined assays for calibration, refinement, and validation of the computed models.	01/13–12/15

(Continued)

Table 2.2 (*Continued*)

Project acronym	Project title	Framework	Coordinating partner	Project website	Topics related to modeling and grouping/read-across	Duration
MODERN	**MOD**elling the **E**nvi**R**onmental and Human Health Effects of **N**anomaterials	FP7	Universitat Rovira i Virgili, Spain	https://cordis.europa.eu/project/rcn/106195/factsheet/en	Establishment of new modeling approaches suitable for relating nanotoxicity with the intrinsic molecular and physicochemical properties of NMs. Definition of a categorization and hazard ranking protocol for NMs on the basis of structural similarity principles and on the analysis of their toxicological profiles.	01/13–12/15

Project acronym	Project title	Framework	Coordinating partner	Project website	Topics related to modeling and grouping/read-across	Duration
ModNanoTox	Modelling Nanoparticle Toxicity: Principles, Methods, Novel Approaches	FP7	University of Birmingham, UK	http://www.birmingham.ac.uk/generic/modnanotox/index.aspx https://cordis.europa.eu/project/rcn/101045/factsheet/en	Development of QSAR for NM toxicity on the basis of available data. Project database format to be compatible with emerging standards (ISA-TAB Nano), improving the accessibility and sustainability.	11/11–10/13
NanoBRIDGES	Building Bridges between Specialists on Computational and Empirical Risk Assessment of Engineered Nanomaterials	FP7 (Marie Curie Action)	University of Gdansk, Poland	http://nanobridges.eu https://cordis.europa.eu/project/rcn/102228/factsheet/en	Development of new tools for computational risk assessment of engineered NPs.	01/12–12/14
NanoFATE	Nanoparticle Fate Assessment and Toxicity in the Environment	FP7	Natural Environment Research Council, UK	https://nanofate.eu https://cordis.europa.eu/project/rcn/94229/factsheet/en	Nanoparticle ecotoxicology and bioavailability; environmental fate data and models.	04/10–03/14

(*Continued*)

Table 2.2 (*Continued*)

Project acronym	Project title	Framework	Coordinating partner	Project website	Topics related to modeling and grouping/read-across	Duration
NanoMILE	Engineered Nanomaterial Mechanisms of Interactions with Living Systems and the Environment: A Universal Framework for Safe Nanotechnology	FP7	University of Birmingham, UK	http://nanomile.eu-vri.eu https://cordis.europa.eu/project/rcn/106943/factsheet/en	Development of quantitative structure (property)–activity relationships (QS(P)ARs) enabling predictive work to evolve and feed into risk assessment.	03/13–02/17
NanoPUZZLES	Modelling Properties, Interactions, Toxicity and Environmental Behaviour of Engineered Nanoparticles	FP7	University of Gdansk, Poland	http://www.nanopuzzles.eu https://cordis.europa.eu/project/rcn/106199/factsheet/en	Development of new computational methods for the comprehensive modeling of the relationships between the structure, properties, molecular interactions, and toxicity of selected classes of engineered NPs, grouping, and read-across.	01/13–12/15

Project acronym	Project title	Framework	Coordinating partner	Project website	Topics related to modeling and grouping/read-across	Duration
NANoREG	A Common European Approach to the Regulatory Testing of Nanomaterials	FP7	Ministry of Infrastructure and the Environment, Netherlands	http://www.nanoreg.eu https://cordis.europa.eu/project/rcn/107159/factsheet/en	Development of a proposal for grouping of NMs in categories with similar biological, ecological, and/or toxicological effects.	03/13–02/17
NanoReTox	The Reactivity and Toxicity of Engineered Nanoparticles: Risks to the Environment and Human Health	FP7	Natural History Museum, UK	https://cordis.europa.eu/project/rcn/89919/factsheet/en	Study of molecular and cellular reactivity of selected metal NMs in aquatic species and in cells and cell lines originating from different human target organs.	12/08–11/12

(*Continued*)

Table 2.2 (*Continued*)

Project acronym	Project title	Framework	Coordinating partner	Project website	Topics related to modeling and grouping/read-across	Duration
NANOSOLUTIONS	Biological Foundation for the Safety Classification of Engineered Nanomaterials (ENM): Systems Biology Approaches to Understand Interactions of ENM with Living Organisms and the Environment	FP7	Finnish Institute of Occupational Health Finland	http://nanosolutionsfp7.com https://cordis.europa.eu/project/rcn/106911/factsheet/en	Development of an approach to predicting the safety of NMs, the "ENM Safety Classifier," based on the NM characteristics and an understanding of their interactions at molecular, cellular, and organism levels, including a novel computational method for feature selection and prioritization based on fuzzy logic and random forests. Creation of a data repository, with data specified according to ISA-TAB Nano.	04/13–03/17

Project acronym	Project title	Framework	Coordinating partner	Project website	Topics related to modeling and grouping/read-across	Duration
NanoTEST	Development of Methodology for Alternative Testing Strategies for the Assessment of the Toxicological Profile of Nanoparticles Used in Medical Diagnostics	FP7	Norwegian Institute for Air Research, Norway	http://www.nanotest-fp7.eu https://cordis.europa.eu/project/rcn/86691/factsheet/en	Development of a theoretical model for predicting the reactivity of metal oxide NPs as well as their ability to cause oxidative stress through the generation of reactive oxygen species (ROS).	04/08–03/12

(*Continued*)

Table 2.2 (*Continued*)

Project acronym	Project title	Framework	Coordinating partner	Project website	Topics related to modeling and grouping/read-across	Duration
NanoToxClass	Establishing NM Grouping/ Classification Strategies According to Toxicity and Biological Effects for Supporting Risk Assessment	FP7 ERA-NET SIINN (Safe Implementation of Innovative Nanoscience and Nanotechnology)	Federal Institute for Risk Assessment (BfR), Germany	www.nanotoxclass.eu	Development of grouping principles for NMs on the basis of the physicochemical properties related to their toxicity and toxicity mechanisms. Establishment of criteria for grouping on the basis of well-established toxicity endpoints and including novel approaches, such as transcriptomics, proteomics, and metabolomics.	12/15–11/18

Project acronym	Project title	Framework	Coordinating partner	Project website	Topics related to modeling and grouping/read-across	Duration
NanoTrans-kinetics	Modelling Basis and Kinetics of Nanoparticle Interaction with Membranes, Uptake into Cells, and Sub-cellular and Inter-compartmental Transport	FP7	University College Dublin, Ireland	https://cordis.europa.eu/project/rcn/101046/factsheet/en	Establishment of techniques for modeling relationships between NP properties and toxicity, including interactions with biological systems; understanding of the mechanisms of NP uptake into and subcellular transport within cells and through biological barriers.	11/11–10/14

(*Continued*)

Table 2.2 (*Continued*)

Project acronym	Project title	Framework	Coordinating partner	Project website	Topics related to modeling and grouping/read-across	Duration
PreNanoTox	Predictive Toxicology of Engineered Nanoparticles	FP7	Tel-Aviv University, Israel	http://prenanotox.tau.ac.il https://cordis.europa.eu/project/rcn/106196/factsheet/en	Development of a database of in vitro and in vivo studies; development, adaptation, and validation of novel methodologies for establishing quantitative nanostructure–activity-relationship (QNAR) modeling as a tool for predicting biological effects of selected NMs.	01/13–12/15
REACHnano	Development of a Web Based REACH Help Desk to Support the Chemical Safety Assessment of Nanomaterials	LIFE+	ITENE, Spain	http://ec.europa.eu/environment/life/project/Projects/index.cfm?fuseaction=search.dspPage&n_proj_id=4249	Provide the industry and stakeholders with easy-to-use tools to support the risk assessment of NMs along their life cycles.	10/12–09/15

Table 2.3 H2020 EU projects that include aspects of modeling and/or grouping

Project acronym	Project title	**Frame-work**	**Coordinating partner**	**Project website**	**Topics related to modeling and grouping/read-across**	**Duration**
caLIBRAte	Performance Testing Calibration and Implementation of a Next Generation System-of-Systems Risk Governance Framework for Nanomaterials	H2020	The National Research Center for Work Environment, Denmark	http://www.nanocalibrate.eu https://cordis.europa.eu/project/rcn/203256/factsheet/en	Develop qualitative and fully integrated predictive quantitative human and environmental risk assessment. Further develop of innovative hazard grouping and predictive hazard assessment methods (e.g., physicochemical modeling, in vitro, high-throughput screening [HTS], -omics technologies, bioinformatics, and physiologically based kinetic [PBK] modeling).	05/16–10/19

(*Continued*)

Table 2.3 (*Continued*)

Project acronym	Project title	**Frame-work**	**Coordinating partner**	**Project website**	**Topics related to modeling and grouping/read-across**	**Duration**
GRACIOUS	**G**rouping, **R**ead-**A**cross and **C**lass**I**ficati**O**n Framework for Reg**U**latory Risk Assessment of Manufactured Nanomaterials and **S**afer Design of Nano-enabled Products	H2020	Hariot-Watt University, UK	https://www.h2020gracious.eu https://cordis.europa.eu/project/rcn/212339/factsheet/en	Develop a framework for grouping and read-across on the basis of physicochemical, release, exposure, environmental fate, toxicokinetic, and toxicological information. A series of hypothesis-based intelligent (non-)testing strategies (ITSs) to be integrated into the framework, promoting the use of modeling (e.g., in silico, fate, and exposure), in vitro, and cell-free tests.	01/18–06/21
HISENTS	**H**igh Level **I**ntegrated **SE**nsor for **N**ano**T**oxicity **S**creening	H2020	University of Leeds, UK	https://hisents.eu https://cordis.europa.eu/project/rcn/200813/factsheet/en	Simulate pathway and interaction of NMs in biological organisms using a PBK model: a combination of multimodular screening (from molecular to cell to organ level) with PBK modeling.	04/16–03/19
NanoFASE	Nanomaterial **FA**te and **S**peciation in the **E**nvironment	H2020	Natural Environment Research Council, UK	http://www.nanofase.eu https://cordis.europa.eu/project/rcn/197194/factsheet/en	Deliver an integrated exposure assessment framework of models and characterization protocols comprising validated SOPs. Multimedia fate modeling.	09/15–08/19

Project acronym	Project title	Frame-work	Coordinating partner	Project website	Topics related to modeling and grouping/read-across	Duration
NANOGENTOOLS	Developing and Implementation of a New Generation of Nanosafety Assessment Tools	H2020	Universidad de Burgos-ICCRAM, Spain	http://www3.ubu.es/nanogentools/ https://cordis.europa.eu/project/rcn/198840/factsheet/en	Propose HTS and -omics tools for predicting toxicological properties of NMs. Develop new bioinformatics methodologies for analyzing -omics data and create an open database in collaboration with the EU NSC.	01/16–12/19
NanoReg2	Development and Implementation of Grouping and Safe-by-Design Approaches within **R**egulatory Frameworks	H2020	INERIS, France	http://www.nanoreg2.eu https://cordis.europa.eu/project/rcn/197088/factsheet/en	Develop tools for safe by design, based on regulatory orientated grouping approaches, placed in a general ITS.	09/15–08/18
PATROLS	**P**hysiologically **A**nchored **T**ools for **R**ealistic Nan**O**materia**L** Hazard **AS**sessment	H2020	Swansea University, UK (Research and Innovation Action)	https://www.patrols-h2020.eu https://cordis.europa.eu/project/rcn/215979/factsheet/en	Create robust computational methods for NM exposure and dose modeling, as well as hazard prediction. Provide an innovative and effective set of experimental techniques and computational tools to more reliably predict potential human and environmental hazards resulting from exposure to engineered NM.	01/18–06/21

(*Continued*)

Table 2.3 (*Continued*)

Project acronym	Project title	**Framework**	**Coordinating partner**	**Project website**	**Topics related to modeling and grouping/read-across**	**Duration**
SmartNanoTox	Smart Tools for Gauging Nano Hazards	H2020	University College Dublin, Ireland	http://www.smartnanotox.eu https://cordis.europa.eu/project/rcn/200824/factsheet/en	Establish relationships between NM physicochemical properties and key events; suggest descriptors for grouping of NMs according to their toxicological mode of action. Create a database of bio-nano interactions enabling development of read-across and QSAR tools for toxicity assessment. Develop a smart screening approach with predictions of NM toxicity made on the basis of purely computational or limited in vitro screening tests focused on crucial bio-nano interactions.	03/16–02/20

Table 2.4 Projects forming the NSC Modeling Cluster

MembraneNanoPart	Modeling the mechanisms of nanoparticle-lipid interactions and nanoparticle effects on cell membrane structure and function
MOD-ENP-TOX	Modeling a platform for hazard ranking of engineered nanoparticles (metal-based nanoparticles)
MODERN	Modeling the environmental and human health effects of nanomaterials
NanoPUZZLES	Modeling properties, interactions, toxicity, and environmental behavior of engineered nanoparticles
PreNanoTox	Predictive toxicology of engineered nanoparticles
ModNanoTox	Modeling nanoparticle toxicity: principles, methods, and novel approaches
NanoTrans-kinetics	Modeling basis and kinetics of nanoparticle interaction with membranes, uptake into cells, and subcellular and intercompartmental transport
eNanomapper	A database and ontology framework for nanomaterials design and safety assessment

2.2.2 Related International Activities

The European research projects are in close contact with US colleagues through the Communities of Research (CoRs). Each CoR is dedicated to a specific topic and serves as a discussion forum for researchers, interested stakeholders from academia, government, industry, and nongovernmental organizations (NGOs) from both sides of the Atlantic in this US–Europe dialogue[8] bridging Nano Environmental Health and Safety (NanoEHS) research efforts.

The CoRs are focused on:

- Characterization
- Databases and computational modeling for NanoEHS
- Ecotoxicity
- Human toxicity
- Exposure through product life

[8]http://us-eu.org

- Risk assessment
- Risk management and control

On the US side, the Nanotechnology Working Group (NanoWG) is a working group of the National Cancer Informatics Program (NCIP)[9] and is very active in the community. It includes the Nanomaterial Data Curation Initiative [18], of high relevance for the data-related issues regarding modeling. The NanoWG has close bonds with the NSC database working group.

Furthermore, the National Center for Environmental Research[10] funds projects including aspects of modeling and grouping for NMs. Projects, for example, aimed at providing structure-function relationships for nanoparticle toxicology[11] or carrying out a structure-function analysis for genomics-based determination of nanoparticle toxicity[12], investigating the mechanisms by which manufactured NMs cause cytotoxicity in realistic environments of exposure.

At a global level, the Organisation for Economic Co-operation and Development (OECD) Working Party on Manufactured Nanomaterials (WPMN)[13] promotes international cooperation in human health and environmental safety aspects of manufactured NMs, to develop NM risk assessment approaches and strengthen regulatory risk assessment capacities. The WPMN launched a sponsorship program for testing manufactured NMs (Testing Programme) in November 2007. This testing program verifies the testing methods applied for manufactured NMs. The results of the NM identification, measured physicochemical properties, material characterization, environmental fate, (eco-) toxicological effects, and material safety can also provide information on "intrinsic properties" of NMs, that is, specific to the nanoscale dimension, and their relation with NM toxicity.

Figure 2.1 summarizes the relations between nanosafety-related projects and working groups at the European level and interactions with international activities.

---

[9]https://cbiit.nci.nih.gov/ncip/ncip-home

[10]https://www.federallabs.org/labs/national-center-for-environmental-research-ncer

[11]US EPA Grant no. R832536, 12/05-11/08

[12]US EPA Grant no. R833325, 05/07-05/09

[13]http://www.oecd.org/chemicalsafety/nanosafety/

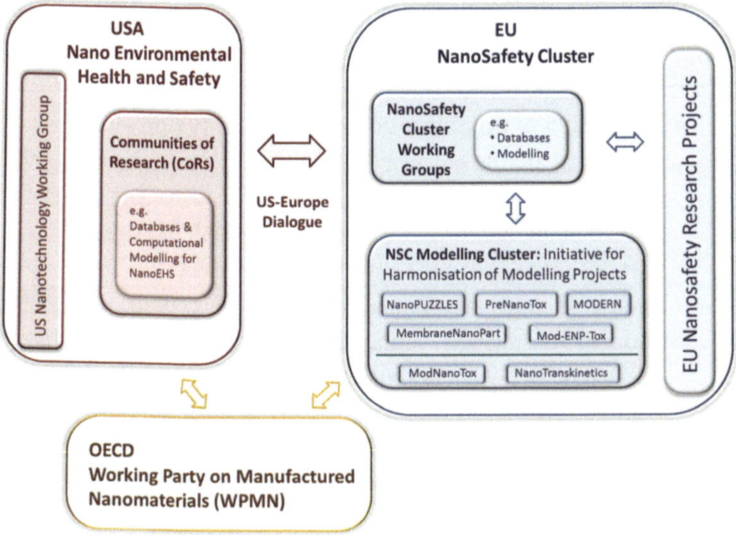

Figure 2.1 Schematic representation of the nanosafety-related projects and working groups at the European level and their interactions with international initiatives.

2.3 Challenges and Needs for the Development and Use of Computational Methods

There is consensus in the literature that computational models for prediction of nanotoxicity will be able to make an important contribution to nanosafety assessment, since it is not feasible to experimentally test all possible nanoforms, which differ in particle size, shape, coating, and other characteristics. This is also in line with general efforts in applying alternative methods to traditional animal testing in chemical risk assessment. However, several limitations and barriers to the development and use of computational prediction models for nanoparticles were identified.

The Nanocomput project [9] has performed a comprehensive review of the model landscape in the scientific literature, including quantitative structure–property relationship (QSPR) and QSAR models for NMs, as well as compartment-based models, including toxicokinetic and toxicodynamic models, in vitro and in vivo dosimetry models, and environmental fate models [19, 20].

From the analysis of available models, limitations on the availability and utility of computational models for NMs for regulatory purposes were found to be related to the following issues:

- Limitation of existing models/applicability domains:
  Models available cover a limited number of endpoints and in some cases, several have been derived from the same data sets. Most models were derived from small data sets.
- Limitation of use in the regulatory context:
  There is an insufficient number of QSPR and QSAR models for directly filling data gaps for REACH endpoints (physicochemical properties, ecotoxicity endpoints, and human health endpoints). However, available models can be used for screening or support grouping and read-across. On the other hand, the applicability and availability of environmental fate models is more promising.

In terms of challenges and knowledge gaps for the development of models, the overarching problems for model development can be summarized as follows, as described in the literature [21–23, 17]. They are mostly linked to the underlying data and thus to experimental issues, apart from chemoinformatics problems and the general uncertainties of linking effects mechanistically to NMs:

- Lack of data (quantity)—large data sets needed for the development of robust models
- Data completeness and data quality insufficiency
- Lack of comparability—limiting creation of larger data sets by pooling data from different sources (e.g., comparability between laboratories and between different types of NMs—variability, different protocols for characterization, etc.)
- Uncertain validity for nanotoxicity (reliability, relevance)—link of NM properties with effect

The knowledge gaps are detailed in Table 2.5.

The quality and robustness of a computational prediction model depend on the quality of the data used for the model development. Therefore, the consideration and evaluation of the data and steps to improve data availability and quality are crucial for modeling. The key concepts of data completeness, minimum information checklist, and data quality are highlighted, for example, in Ref.

[17]. An important point is the aim to allow comparability of data from different sources and provided by different laboratories in order to create a more robust data basis, ideally in an overarching and comprehensive data repository. A robust data basis is indeed needed for analysis of property-effect relationships, for building of predictive models, and for grouping and read-across approaches [24].

Table 2.5 Limitations hampering the development of computational models for NMs and the specific related knowledge gaps

Limitation	Specific related knowledge gaps
Data quantity: Large data sets needed	o Standardized reporting format to enable pooling and comparison of data o Reference materials to allow comparison between studies [25] o Existing data easily and freely accessible o Central data repository
Data quality	o Harmonized protocols (standard operating procedures [SOPs]) for NM characterization and toxicity measurements, also allowing for comparability of data measured in different laboratories o Polydispersity and heterogeneity of NMs o Batch-to-batch variability of nominal identical NMs [17]
Data completeness	o Minimum requirements for parameters to measure and report
Validity for nanotoxicity prediction models	o Finding the appropriate (nano-specific) descriptors to link the complex, nonuniform NM properties to toxicity [22] o Applicability of toxicity test guidelines to NMs, that is, validity of the results for nanosafety assessment; for example, artifacts affecting the reliability of NM biological assessment [26, 27]
Technical issues of modeling	o Transformation of NM structures into computer-usable representation [22]

The Nanocomput project also carried out two case studies on grouping and read-across to explore the practical process of grouping and reading across toxicological properties between different physical forms (analogues) of the same substance as well as the applicability of the ECHA Read-Across Assessment Framework (RAAF) [28] to NMs for evaluating uncertainties in the read-across argumentation. The case studies focused on the genotoxicity of TiO_2 [29, 30] and multiwalled carbon nanotubes (MWCNT) [31]. The uncertainties identified in both case studies point at similar issues as discussed for computational modeling for NMs, that is, uncertainty related to the nanoform identification, high variability in the experimental measurements, uncertainty in applying existing test protocols and relevance for human health endpoints, and lack of knowledge of the mechanism of toxic action.

As a way forward, recommendations have been given in the past, in the literature and, for example, as a result of OECD workshops [32, 33]:

- Setting of minimum requirements on experimental parameters to record
- Standardization of the data format, templates for data exchange, and harmonization
- Creation of a database framework for public availability and access of data
- Creation of ontologies for a harmonized terminology
- Implementation of standard operating procedures (SOPs) for experimental measurements, both for NM characterization and toxicity measurements
- Clarification on which test guidelines are applicable to NMs, establishment of a set of caveats, and adaptation of test guidelines to nano-specific characteristics [34]
- Elucidation of mechanisms underlying nanotoxicity and link to NM properties

2.4 Progress against the Challenges and Needs

2.4.1 Results from EU FP7-Funded Research Projects

The research initiatives described in Section 2.2 were working toward solving the issues hampering model development and use. For example:

- The NSC Modeling Cluster aimed at harmonizing data collection and recording between the projects by using a common standard (ISA-TAB-Nano)[14] [35] in view of bringing all data together in one common database. While the envisaged cross-project data collection was not realized, some projects of the Modeling Cluster did use the ISA-TAB-Nano format and made the collected and formatted data publicly available separately, for example, NanoPUZZLES [36, 37] and MOD-ENP-TOX [38].
- The cross-NSC database working group worked on the definition of minimum information requirements, on the basis of the MINChar Initiative parameters list [39], and discussed data quality and completeness together with US groups, for example, the NCIP Nanomaterial Data Curation Initiative [18].
- The eNanomapper project set out to create a bespoke infrastructure to store data from EU research projects and also share predictive models for NMs. eNanomapper has indeed provided the technical framework to upload and store data from research projects in different formats [40]. So far only a few data have been uploaded: for example, all data from NANoREG; data from MARINA (6 substances), MODENA (59 substances), NanoWiki, and for protein corona fingerprinting predicting the cellular interaction of gold and silver nanoparticles (121 substances); and some lists of materials. The infrastructure can be used for further efforts to create a centralized data repository. The eNanomapper database has been included in the European Union Observatory for Nanomaterials (EUON)[15]. Two research projects (NanoReg2 and caLIBRAte) are using the infrastructure for internal data sharing. With regard to models, eNanomapper has made cheminformatics tools available to create own models, including a nano-read-across application. No models from EU projects are included directly.
- A dialogue between modelers and experimentalists was seen as essential, as stressed, for example, in the NanoBRIDGES project.

---

[14]https://wiki.nci.nih.gov/display/ICR/ISA-TAB-Nano
[15]https://euon.echa.europa.eu/enanomapper

An analysis by Nanocomput of the results from European research projects was carried out in order to assess the (future) availability of more models, tools, or knowledge contributing to close the gaps. This review mainly comprised 20 projects related to modeling approaches under FP7, including a support action, a Marie Curie action, and a COST initiative, as well as a LIFE+ project (Tables 2.1 and 2.2), and was based on project results available by the beginning of 2017. Tasks, work packages, and deliverables in the projects relevant for the scope of NM computational modeling and grouping approaches were identified by a keyword search of the project Descriptions of Work, with the keywords "group," "grouping," "category," "read-across," "QSAR," "QSPR," "in silico," "fate model," "multimedia," "kinetic," and "behavior." Relevant content was identified regarding new models, tools, and other knowledge possibly allowing progress toward the challenges and gaps.

It should be noted that it was difficult in some cases to get access to the results from the projects and there were, for example, concerns about intellectual property. Thus, the review was also a lesson in how difficult it can be to assess research results that should ideally be public and be used to advance public knowledge and science.

Overall the review of the EU project research into nanosafety assessment (see Ref. [9]) showed that at the moment of analysis:

- New models were developed in several projects, as well as grouping approaches (e.g., in MARINA [41] and NANoREG [42]). The new models published in the scientific literature were included in the Nanocomput model landscape review and evaluation of applicability in a regulatory context. Models still under development, not fully described, or not made public at that time might add to the landscape in the future.
- New tools were developed contributing to individual tasks related to computational modeling or designed as comprehensive modeling tools, for example, cheminformatics tools accessible from the eNanomapper website.
- Nano-specific descriptors were studied.
- Progress was made toward data format standardization, ontologies, and data sharing, for example, ISA-TAB-Nano-based

templates for data recording developed within NANoREG [43, 44] and NanoPUZZLES [45] and the eNanomapper database technology to upload and store project data, including the eNanomapper NanoEHS ontology. NANoREG published a harmonized terminology, including terms of relevance in the specific REACH and broader international regulatory context, for example, at the OECD [46].

- SOPs were defined for experimental measurements. For example, in the MARINA project eight OECD test guidelines were adapted on the basis of the testing of at least one ion-releasing NM (Ag) and two inert NMs (TiO_2) [47]. Protocols have been developed by NANoREG[16] and the NanoValid project[17]. NanoDefine[18] has also contributed to the development and/or optimization of key methods required for NM classification.
- New data were generated and/or collected from the literature. Data from, for example, NANoREG, MARINA, and NanoPUZZLES [37] were made publicly available.

Altogether, this contributes toward reducing the knowledge gaps and overcoming current limitations. Examples are summarized in Table 2.6.

The NANoREG project has summarized available tools in the NANoREG Toolbox for the Safety Assessment of Nanomaterials [48, 49], which refers to over 500 tools, for example, test methods, data sets, models, guidance documents, and decision trees, developed in Europe and beyond that can support NM safety assessment specifically in the context of REACH.

Members of the NSC Modeling Cluster have discussed the nano-specific challenges for QSAR models and how they are reflected in the OECD QSAR validation principles [50, 51].

---

[16]https://www.rivm.nl/en/about-rivm/mission-and-strategy/international-affairs/international-projects/nanoreg/work-package

[17]http://www.nanovalid.eu, Development of reference methods for hazard identification, risk assessment and LCA of engineered nanomaterials, 11/11-10/15

[18]www.nanodefine.eu, Development of an integrated approach based on validated and standardized methods to support the implementation of the EC recommendation for a definition of nanomaterial, 11/13-10/17

Table 2.6 Mapping FP7 research project results against challenges

Limitation	New results filling specific related knowledge gaps
Data quantity: Large data sets needed	o Standardized reporting format based on ISA-TAB-Nano developed in NanoPUZZLES [36] and NANoREG [43] o New data created o eNanomapper public database to serve as a repository, compatible with different data formats
Data quality	o SOPs developed, for example, in NANoREG; all measurements in the project carried out according to these SOPs
Data completeness	o Contribution to discussion about data quality, completeness, and minimum requirements [17]
Validity for nanotoxicity prediction models	o Descriptors studied and derived in several projects o New insights into mechanisms, NM behavior and kinetics
Technical issues of modeling	o Representation of NM structures discussed

2.4.2 Horizon 2020 Research Projects

In the framework of H2020, several research projects, including work related to computational approaches, are ongoing (see Table 2.3) that will provide further inputs on advance methodologies and applicability of in silico models as well as grouping and read-across approaches for nanosafety assessment. Furthermore, other projects not focused on modeling also provide further input on and insights into closing the identified knowledge gaps hampering the development and uptake of computational approaches. Examples are listed in Table 2.7.

Table 2.7 H2020 research projects addressing the identified limitations for the development and uptake of computational approaches for nanosafety assessment

Project	Aims
• ACEnano (01/17-12/20) Analytical and Characterisation Excellence in Nanomaterial Risk Assessment: A Tiered Approach http://www.acenano-project.eu	**NM characterization** • Development of a holistic analytical framework for reproducible NM characterization • Robust NM physicochemical characterization for use in reliable read-across framework • Embedded in an operational, linked-up ontology and data framework
• EC4SafeNano (11/16-10/19) European Centre for Risk Management and Safe Innovation in Nanomaterials Nanotechnologies www.EC4SafeNano.eu	**Infrastructure, SOPs** • Development of a distributed Center of European organizations for risk management and safe innovation for NMs and nanotechnologies • Production and promotion of guidance documents on available tools, SOPs, and best practices and an inventory of infrastructures
• NanoCommons (01/18-12/21) The European Nanotechnology Community Informatics Platform: Bridging Data and Disciplinary Gaps for Industry and Regulators https://www.nanocommons.eu/	**Infrastructure** • Facilitating the pooling and harmonizing of methods and data for modeling, safe-by-design product development, and regulatory approval purposes, driving best practice and ensuring maximum access to data and tools • Integrating across EU and US approaches to nanosafety data management

(*Continued*)

Table 2.7 (*Continued*)

Project	Aims
• npSCOPE (01/17-12/20) The Nanoparticle Scope: A New Integrated Instrument for Accurate and Reproducible Physicochemical Characterisation of Nanoparticles http://www.npscope.eu/	**NM characterization** • Development of a new integrated instrument optimized for providing a complete physicochemical NP characterization, using correlative data processing methodologies and algorithms based on statistical methods
• ProSafe (02/15-04/17) Promoting the Implementation of Safe by Design (Coordination and Support Action) https://cordis.europa.eu/project/rcn/194431/factsheet/en)	**Infrastructure** • ProSafe roadmap for reviewing data, protocols, report, and guidance notes for regulatory relevance • EU policy-oriented White Paper • Streamlining data acquisition, collection, and data management

Ongoing efforts relate to the advancement of methods for NM characterization as well as standardized procedures for NM-specific toxicity testing. They concern the development of SOPs and guidance, as well as the building of infrastructure for nanosafety assessment.

For example, the projects PATROLS and GRACIOUS held a joint NSC event on harmonization of SOPs in June 2018, to improve the usability of data and their reliability and comparability for regulatory risk assessment.

An overarching and supporting infrastructure, connecting all other ongoing efforts, is a major need. The EC4SafeNano project is contributing toward this goal by developing a distributed Center of European organizations for risk management and safe innovation for NMs and nanotechnologies. The NanoCommons project aims to integrate EU and US approaches to nanosafety data management and ensure sustainability of the nanosafety knowledge infrastructure. This developed NanoCommons infrastructure is planned to be eventually integrated in EUON (see Section 2.4.3).

ProSafe has published an EU policy-oriented White Paper [52, 53] with recommendations toward a more effective and cost-efficient NM assessment. It is based on the outcomes and analyses from the ProSafe and NANoREG projects, in particular the NANoREG Framework for the Safety Assessment of Nanomaterials [54], which aimed at a common understanding in the field of EHS NM assessment in the current European regulatory context. The ProSafe White Paper refers to limitations due to "science-oriented" and not "regulatory-oriented" nanosafety research. Its recommendations include actions related to the improvement of data quality and data management as well as innovation in risk assessment, such as the determination of mode-of-actions and adverse outcome pathways for representative groups of NMs. Within the ProSafe project, computational models for NM assessment have been reviewed, for example, in Refs. [55–57].

2.4.3 Other Activities

The European Union Observatory for Nanomaterials (EUON)[19], funded by the European Commission and hosted by ECHA, was launched in June 2017. It provides information about NMs on the EU market and addresses all stakeholders, including the industry, NGOs, and consumers. It also aims to collect and make available NM information, including access to the nanotechnology knowledgebase NanoData and the eNanomapper database.

The NSC has published the "Research Regulatory Roadmap" [13], in view of an integrated risk analysis framework, as well as the "Closer to the Market Roadmap" [58], which focuses on a framework to deliver nano-enabled products to the market, including risk assessment. Most recently, the EU US Nanoinformatics Roadmap 2030 has been compiled [59], drawing from previous commentaries, workshop reports, and reviews [60–64]. It addresses specifically considerations on databases and metadata in order to foster the use of computational models. The roadmap also gives an overview of current NM-related database projects, as compiled on the basis of a survey carried out by the NSC Working Group on Databases in collaboration with the caLIBRAte project in December 2016. An

[19]https://euon.echa.europa.eu

example is the NanoDatabank[20] (nanoinformatics group of the UCLA Center for Environmental Implications of Nanotechnology [CEIN]), a centralized web-based database management system for NMs and integral component of the nanoinformatics platform nanoinfo.org.

The US-EU NanoEHS CoR are still actively discussing common issues, such as at the 2018 US-EU Bridging NanoEHS Research Efforts Joint Workshop in October 2018.[21]

In October 2017, the OECD published the first test guidelines developed specifically for NMs: the new Test Guideline 318 on the dispersion stability of NMs in simulated environmental media [65] and the revised Test Guidelines 412 on the 28-day subacute inhalation toxicity study [66] and 413 on the 90-day subchronic inhalation toxicity study [67]. The latter two have been updated to allow testing and hazard identification of inhaled NMs.

2.5 Conclusions from the Research Landscape Review

2.5.1 Conclusions on the Needs Addressed

The review of the EU project research into nanosafety assessment (with a focus on FP7 research project results) showed that considerable progress has been made toward the challenges of modeling NMs. However, not all needs were fully addressed yet. It was concluded that:

- There was fragmentation of work. For example, similar developments were conducted in parallel in several projects, for example, the development of ISA-TAB-Nano-based templates in NANoREG [43] and NanoPUZZLES [36]. Overall, harmonization is fostered, however, by cross-cutting working groups and continued dialogue across the nanosafety communities within Europe and with the US.
- Scientific results, models, and tools are not always made publicly available. While some results were made available through the project websites or public repositories, others

[20]http://nanoinfo.org/#!/nanodatabank/

[21]https://scipol.org/workshop-2018-us-eu-bridging-nanoehs-research-efforts

are not easily accessible. Some new models have been disseminated as journal publications; others are only described in project deliverables and not always in a usable form.

- Even though data templates and SOPs have been developed, they have not gained widespread use within projects yet, thereby creating a barrier to data sharing and dissemination.
- There are significant issues in terms of data quality and reproducibility of experiments. While new and future projects can be expected to produce higher-quality data by following more standardized SOPs and harmonized reporting standards, this does not solve the problem of how to use existing data in a regulatory context. A strategy must be devised on how to interpret and integrate these data, while recognizing their limitations.

2.5.2 Recommendations for Nanosafety Research

To address these still existing shortcomings, the following recommendations should be considered with a view to increasing the availability and utility of computational methods:

- Continued efforts should be made to address the already identified and still existing knowledge gaps and technical challenges. The NSC organization and structure could be exploited even more to support the scientific coordination across projects.
- Within the context of EU research programs, quality guidelines should be implemented to improve the overall consistency in reporting of data, models, and tools. For example, the use of standardized SOPs and reporting formats for data and models could be made a requirement in upcoming projects.
- In addition, the public dissemination of research and development outputs should be a contractual obligation to be monitored closely.
- There is also a need to raise awareness in the research community that model development should be carried out with concrete regulatory applications in mind.
- Models should be made available in a usable form. At the very least, they should be documented appropriately in a public

repository so they can be reproduced, and ideally they should be implemented in the form of a software tool, to increase practical usability.

- A one-stop overarching hub with all necessary information, linking to the respective tools and information sources, will be very valuable. It should include existing infrastructure technologies and refer to available repositories.

2.6 Overall Conclusions on the Availability and Applicability of Computational Approaches for Nanosafety Assessment

The following considerations are drawn from the overall experience in the Nanocomput project, including the analysis of the nanosafety research landscape as well as the literature review of the modeling landscape and read-across case studies.

Figure 2.2 summarizes the major barriers to the development and regulatory use of predictive models in nanotoxicology. These play a role at different levels in the creation and application of knowledge:

- Experimental methods for NM physicochemical and toxicity characterization, creating the knowledge needed for model development, and existing computational models
- Standardization of the obtained information and tools for comparability
- Sharing and accessibility of both data and models
- Dissemination and awareness
- Actual use of data and models for application in regulatory decision-making

Some of these aspects are discussed in more detail in the following. Furthermore, a number of recommendations are given in view of overcoming current shortcomings in the knowledge of NM behavior, the availability of tools (such as databases and predictive models) and practical guidance to use such tools in the regulatory assessment of NMs, in order to improve the uptake and use of computational approaches in regulatory decision-making.

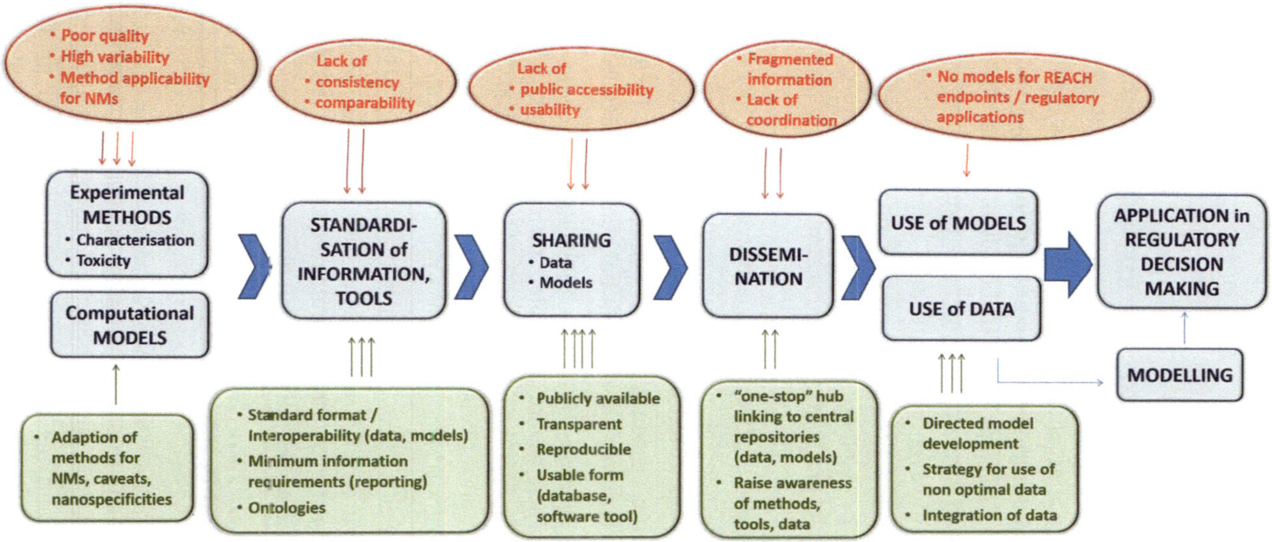

Figure 2.2 Summary of the issues hampering the successful development, uptake, and use of prediction models to support regulatory decision-making on NMs. The problems identified (red) impact computational nanotoxicology at different levels (blue): (A) experimental methods for NM physicochemical and toxicity characterization, existing models, (B) standardization of information and tools for comparability, (C) sharing and accessibility, (D) dissemination and awareness, and (E) use of data and models for application in regulatory decision-making. Possible measures to improve the situation are indicated (green).

2.6.1 Inherent Scientific Uncertainties

There are still significant knowledge gaps regarding the physicochemical characteristics of NMs and their influence on determining physicochemical, biokinetic, and toxicodynamic properties of the materials. Even if no nano-specific mechanisms of toxicity have been identified [54], it is considered that the particle properties of NMs affect their kinetics/fate, as well as the nature of any molecular initiating events in their toxicological pathways [68].

These knowledge gaps need to be filled. European research projects were and are generating new data and/or collecting literature data. However, these are data needed for the specific aims of the respective projects; they are not a targeted approach to fill data gaps for regulatory assessments. Furthermore, there is some difficulty in actually finding and accessing data from research projects.

In developing and applying computational models, it should also be borne in mind that NM behavior depends on many parameters and this dependence could be very sensitive to initial conditions. Therefore, in some circumstances, it could even be theoretically impossible to accurately predict the behavior of NMs, especially over longer time periods. Therefore, modeling efforts should avoid false precision, such as, for example, the exact time-dependent composition of the protein corona. Despite this theoretical challenge, even relatively simple models can be useful aids for decision-making [69].

- ➢ Dedicated and focused research efforts are needed to elucidate the behavior (toxicity and fate) of NMs of regulatory interest and to investigate which fundamental characteristics and properties are relevant for predictive modeling.

2.6.2 Data Quality and Variability

One of the major issues hampering the modeling of NMs is the lack of good-quality data and the high variability of available data. Reliable and relevant data are essential for deriving the underlying principles of NM behavior and for correlating physicochemical properties and activities/toxicity as the basis for building prediction models, grouping similar substances/materials, and performing read-across.

One factor contributing to the inhomogeneity of available data is the lack of standardized and validated methods for physicochemical characterization and toxicity assessment. The other factor is the low reproducibility of some measurements, such as of the zeta potential, due for example to dispersibility problems.

This inhomogeneity in experimental measurements needs to be addressed in practice to make the data sufficiently homogeneous for modeling or read-across. Extensive data analysis and treatment will thus continue to be a significant step before modeling, as long as reliable and comparable methods for physicochemical characterization and toxicological testing are not available and existing data are not recorded in standardized formats, with agreed minimum data inclusion criteria. A lack of comparability between data from different studies means that individual data sets cannot always be merged and will, therefore, tend to be small, limiting their usefulness.

Data treatment such as averaging values from highly variable measurements—for example, for crystallite size values obtained averaging different measures provided by different laboratories—has to be used with some caveats. The use of average values from highly variable measurements could result, for example, in different choices in the read-across process and potentially different conclusions. Therefore, clear guidelines on how to select and pretreat data should be developed for a transparent and consistent approach.

- Data collection should be a directed and systematic community effort to allow for targeted filling of data and knowledge gaps regarding physicochemical, biokinetic, and toxicological NM properties. This process should take into account the reliability and relevance of available experimental methods and apply suitable methods to NMs of regulatory interest.
- More detailed practical guidance could be included, for example, in the ECHA grouping and read-across guidance [5], for example, on which experimental and modeling data to use (ideally based on standardized procedures), on the avoidance of potential bias in analogue and data selection, on data treatment protocols, and on the transparent recording of the results and conclusions.

2.6.3 Model Landscape and Regulatory Relevance

A number of QSPR, QSAR, physiologically based kinetic (PBK), and environmental fate models are available for selected groups of NMs (see Refs. [9, 19, 20]). However, these models cover a limited number of endpoints and have usually been derived from a limited number of small data sets. There are very few QSPR and QSAR models (physicochemical properties, ecotoxicity endpoints, and human health endpoints) with regulatory relevance, for example, which could directly be used to fill data gaps for REACH. For example, many models (76 out of 152) have been developed for cytotoxicity, which is not a regulatory endpoint. Furthermore, there are no models covering acute toxicity, repeated dose toxicity, skin and respiratory sensitization, carcinogenicity, and reproductive toxicity. The only endpoint of REACH relevance is the QSAR for "In Vitro – Mutagenicity Bacterial Reverse Mutation Test (Salmonella typhimurium)," required under REACH Annex VII. However, the Ames test is not applicable to NMs [70]. Generally, the applicability for REACH seems to be most promising for environmental fate models at the moment. Development of many QSPR and QSAR models took place or is ongoing in recent EU research projects. However, not all are finalized or publicly available (yet). However, available models can be used for screening purposes and to support grouping and read-across (e.g., properties such as solubility and cytotoxicity could be used to group NMs).

- ➢ Awareness needs to be raised in the research community that model development should be targeted either to fill specific knowledge gaps or with concrete regulatory applications in mind. Calls for new EU projects should take these needs into account.
- ➢ An organizational mechanism needs to be established for the targeted development of methods and their application to fill data gaps of regulatory relevance.

2.6.4 Model Accessibility and Visibility

Apart from the current limited availability and relevance for regulatory applications, another barrier to the uptake of the models in decision-making is their limited public visibility and accessibility. Many models are "hidden" and have only been described in project reports so far. Even when models are eventually published in the scientific literature, their use is limited without a user-friendly software implementation for their application, such as the planned Nanosolutions project software tool for predicting NM health or environmental hazards. The step from model and tool development within individual research projects to their public availability and sustained maintenance after the end of the projects is not often taken.

At the very least, appropriate and complete documentation of the models should be made available in a public repository so they could be reproduced. Ideally, models should be implemented in the form of a software tool, to increase practical usability.

For the review of the current state of existing NM models (QSPR/QSAR; PBK, physiologically based dynamic, and dosimetry models; environmental fate models [19, 20]) the Nanocomput project created/updated templates to record the models, adapted for NMs: a modified QSAR Model Reporting Format (QMRF) template for reporting nano-related QSAR and a model reporting template for toxicokinetic, toxicodynamic, and environmental fate models (available from Ref. [9]). The model inventories were made available via the JRC Data Catalogue [71].

- ➢ Public dissemination of research and development outputs such as QSPR/QSAR models should be a contractual obligation for EU research projects and facilitated via publicly accessible platforms. Public platforms for dissemination include, for example, the JRC QSAR Model Database[22], the Estonian QSAR Data Bank[23][72], eNanomapper[24] [40], and the OECD QSAR Toolbox[25] [73].
- ➢ Models should be documented in bespoke model description templates and made available in a public repository (see below).

[22]https://qsardb.jrc.ec.europa.eu/qmrf
[23]http://qsardb.org
[24]https://www.enanomapper.net
[25]https://www.qsartoolbox.org

- Models should ideally be implemented in a user-friendly form, for example, as online tools or KNIME[26] workflows, to facilitate their uptake and application.

2.6.5 Practicality of Performing Read-Across for Nanomaterials

In addition to the lack of data and models for regulatory endpoints and knowledge gaps linking the NM properties to their potential toxicity, there is also a disconnection between legal information requirements and the minimal data set recommended to support grouping and read-across. Thus, at the moment no incentives exist for filling the data gaps or for data sharing in view of regulatory applications.

The ECHA guidance for applying read-across for NMs [5] is more a general framework and may be difficult to apply in the light of the above. A concrete and detailed operational guidance is needed to facilitate the uptake of the read-across approach for NMs in the regulatory context. The ECHA RAAF was found to be applicable to NMs and should be used to investigate and document uncertainties and confidence in the read-across argumentation, with some nano-specific adaptions and practical guidance [9, 29–31].

The read-across case studies carried out in the Nanocomput project [9, 29–31] have also shown the applicability of cheminformatics tools to support the grouping hypothesis: There was no agreement in the literature about the genotoxicity mechanism of nano-TiO_2 and MWCNTs, for example, chemical reactivity, reactive oxygen species (ROS) generation, agglomeration, and sedimentation. Furthermore, the variability of the measurement data of the NM characterization hampered linking the characteristics to the observed effects. Therefore, chemoinformatic tools have been used to analyze the NMs' physicochemical properties and find correlations for similar analogues and indications on how they may drive genotoxicity. The read-across hypothesis was based on or verified by this type of analysis.

Cheminformatics tools for modeling of NMs have been made publicly available, for example, from eNanomapper (e.g., Nano-

[26]https://www.knime.com/

Lazar extended for NMs) and NanoBRIDGES (e.g., tools for data pretreatment, variable selection methods, applicability domain, and endpoint-dependent analogue identification software), on the respective project websites. These tools are useful to complement read-across approaches but should be used carefully and adapted to the specific case investigated.

- Detailed practical guidance needs to be provided supplementing the ECHA guidance for NM read-across, with more concrete guidance on how to deal with available and often imperfect data and how to develop the read-across argumentation and address nano-specific uncertainties.
- Cheminformatics tools (software, web applications) need to be collected and made publicly available, and awareness needs to be raised (e.g., through a one-stop hub for NM modeling) and guidance provided for their application in supporting regulatory applications.
- A requirement or recommendation on the use of relevant and reliable in vitro and computational models to fill data gaps in regulatory assessments (e.g., in a weight-of-evidence approach) would be an incentive to generate adequate data. A focused data generation effort could be established to generate such data.

2.6.6 Need for Infrastructure

Centralized infrastructure is needed for information, data, and models to make them publicly available and accessible and allow knowledge exchange. Even when information, data, and models are publicly available, it might be difficult to find them or know that they exist in the first place. Therefore, a pointer is needed to the respective repositories and individual websites to build and assemble the community knowledge and reduce fragmentation of knowledge. This will also prevent duplication of work, which still exists to a certain extent in recent nanosafety projects.

Thus, to ensure visibility and overcome the fragmentation of nanotoxicity-related information (SOPs, data, models, tools, etc.), a one-stop hub is essential. It does not necessarily need to store all information directly, but it needs to assemble the information in a structured manner and link to the individual websites containing the information and/or tools. It should be maintained to accommodate all upcoming information and knowledge in the future and could be filled also retrospectively with existing information, for example, by a systematic review of the literature and past EU project reports. This needs to be resourced and centrally coordinated. In view of the limited quality of data obtained with nonstandardized experimental methods so far, a strategy should also be devised for the selection and use of suboptimal data.

The European Union Observatory for Nanomaterials has been designated to play the role of a central hub, making use of and pointing to existing technologies and compilations. For example, it refers to the technical framework of the eNanomapper project, allowing upload and storage of data from research projects, with some flexibility of data formats [40].

eNanomapper is also a resource for the chemoinformatic tools that could be used to support read-across. Well-established model repositories, such as the JRC QSAR Model Database, the Estonian QSAR Data Bank, and the OECD QSAR Toolbox, should be used additionally to advance the uptake of computational methods in nanosafety assessment. It is important, though, to link them all to the central hub to avoid fragmentation of knowledge. The NSC organization and structure should be further exploited to support scientific coordination across research projects.

- An EU level one-stop hub with links to information sources, databases, modeling tools, and their websites should be used.
- The utility of data produced in EU projects needs to be evaluated and useful data sets uploaded to a centralized repository.
- A knowledge sharing strategy needs to be developed to raise awareness in the research and regulatory communities of relevant initiatives and facilitate access to available methods, tools, databases, and guidance documents.

In conclusion, although considerable progress has been made and is ongoing toward addressing the identified challenges for building and using computational approaches in nanosafety assessment, there is still need for further research and development, as well as for guidance and coordination of the efforts in view of regulatory application.

A number of recommendations have been made to support an increase in availability and uptake of computational methods for regulatory assessment of NMs. They are related to inherent scientific uncertainties in the understanding of NM behavior; data quality and variability; standardization and harmonization; the availability of data and predictive models; and the accessibility and visibility of models, including the need for infrastructure.

Acknowledgment

This work was part of the Nanocomput project, carried out by the European Commission Joint Research Centre (JRC) for the Directorate-General for Internal Market, Industry, Entrepreneurship and SMEs (DG GROW) under the terms of an administrative arrangement between the JRC and DG GROW.

Disclaimer

The report presents the findings and conclusions of the authors but does not represent an official view of the JRC, DG GROW, or other European Commission services.

References

1. European Parliament and Council (2006). Regulation (EC) No 1907/2006 of the European Parliament and of the Council of 18 December 2006 concerning the Registration, Evaluation, Authorisation and Restriction of Chemicals (REACH), establishing a European Chemicals Agency, amending Directive 1999/45/EC and repealing Council Regulation (EEC) No 793/93 and Commission Regulation (EC) No 1488/94 as well as Council Directive 76/769/EEC and Commission Directives 91/155/EEC, 93/67/EEC, 93/105/EC and 2000/21/EC, OJ L 396, 30.12.2006, pp. 1–849.
2. European Parliament and Council (2008). Regulation (EC) No 1272/2008 of the European Parliament and of the Council of 16 December 2008 on classification, labelling and packaging of substances

and mixtures, amending and repealing Directives 67/548/EEC and 1999/45/EC, and amending Regulation (EC) No 1907/2006, OJ L 353, 31.12.2008, pp. 1–1355.

3. European Chemicals Agency, https://echa.europa.eu/regulations/nanomaterials
4. European Chemicals Agency (2016). Practical guide: how to use alternatives to animal testing to fulfil the information requirements for REACH registration, ECHA-16-B-25-EN.
5. European Chemicals Agency (2017). Guidance on information requirements and chemical safety assessment, Appendix R.6-1 for nanomaterials applicable to the Guidance on QSARs and Grouping of Chemicals, ECHA-17-G-17-EN.
6. European Chemicals Agency (2017). Guidance on information requirements and chemical safety assessment, Appendix R7-1 for nanomaterials applicable to Chapter R7a – Endpoint specific guidance, ECHA-17-G-16-EN.
7. European Chemicals Agency (2017). Guidance on information requirements and chemical safety assessment, Appendix R7-1 for nanomaterials applicable to Chapter R7b – Endpoint specific guidance, ECHA-17-G-15-EN.
8. European Chemicals Agency (2017). Guidance on information requirements and chemical safety assessment, Appendix R7-2 for nanomaterials applicable to Chapter R7c – Endpoint specific guidance, ECHA-17-G-14-EN.
9. Worth, A., Aschberger, K., Asturiol, D., Bessems, J., Gerloff, K., Graepel, R., Joossens, E., Lamon, L., Palosaari, T. and Richarz, A.-N. (2017). Evaluation of the availability and applicability of computational approaches in the safety assessment of nanomaterials, EUR 28617 (Publications Office of the European Union, Luxembourg), http://publications.jrc.ec.europa.eu/repository/bitstream/JRC106386/kjna28617enn.pdf.
10. Lynch, I. (ed.) (2016). NanoSafetyCluster, compendium of projects in the European NanoSafety Cluster, 2016 edition, http://www.nanosafetycluster.eu/uploads/files/pdf/2016_NSC_Compendium.pdf.
11. Lynch, I. (ed.) (2017). NanoSafetyCluster, compendium of projects in the European NanoSafety Cluster, 2017 edition, https://www.nanosafetycluster.eu/uploads/files/pdf/2017_NSC_Compendium.pdf.
12. Savolainen, K., Backman, U., Brouwer, D., Fadeel, B., Fernandes, T., Kuhlbusch, T., Landsiedel, R., Lynch, I. and Pylkkänen, L. (2013).

Nanosafety in Europe 2015–2025: towards safe and sustainable nanomaterials and nanotechnology innovations (Helsinki: Finnish Institute of Occupational Health), https://www.nanosafetycluster.eu/uploads/files/pdf/report159.pdf.

13. Stone, V., et al. (2017). NanoSafetyCluster Research Regulatory Roadmap 2017: Research priorities relevant to development or updating of nano-relevant regulations and guidelines, https://www.nanosafetycluster.eu/uploads/files/pdf/RRR_Final_version_090317.pdf.

14. Stone, V., Pozzi-Mucelli, S., Tran, L., Aschberger, K., Sabella, S., Vogel, U., Poland, C., Balharry, D., Fernandes, T., Gottardo, S., Hankin, S., Hartl, M. G. J., Hartmann, N., Hristozov, D., Hund-Rinke, K., Johnston, H., Marcomini, A., Panzer, O., Roncato, D., Saber A. T., Wallin, H. and Scott-Fordsmand, J. J. (2014). ITS-NANO: prioritising nanosafety research to develop a stakeholder driven intelligent testing strategy, *Part. Fibre Toxicol.*, **11**, 9.

15. Cassee, F., Lynch, I., Valsami-Jones, E. and Falk, A. (eds.) (2017). NSC action plan 2017–2021, https://www.nanosafetycluster.eu/uploads/NSC2017-2021ACTIONPLAN_for_comments.pdf.

16. Mustad, A. P., Smeets, B., Jeliazkova N., Jeliazkov V. and Willighagen, E. L. (2014). Summary of the Spring 2014 NSC database survey, https://figshare.com/articles/Summary_of_the_Spring_2014_NSC_Database_Survey/1195888.

17. Marchese Robinson, R. L., Lynch, I., Peijnenburg, W., Rumble, J., Klaessig, F., Marquardt, C., Rauscher, H., Puzyn, T., Purian, R., Åberg, C., Karcher, S., Vriens, H., Hoet, P., Hoover, M. D., Hendren, C. O. and Harper, S. L. (2016). How should the completeness and quality of curated nanomaterial data be evaluated? *Nanoscale*, **8**, 9919–9943.

18. Hendren, C. O., Powers, C. M., Hoover, M. D. and Harper, S. L. (2015). The nanomaterial data curation initiative: a collaborative approach to assessing, evaluating, and advancing the state of the field, *Beilstein J. Nanotechnol.*, **6**(1), 1752–1762.

19. Lamon, L., Asturiol, D., Vilchez, A., Cabellos, J., Damásio, J., Janer, G., Richarz, A. and Worth, A. (2019). Physiologically based mathematical models of nanomaterials for regulatory toxicology: a review, *Comput. Toxicol.*, **9**, 133–142.

20. Lamon, L., Asturiol, D., Vilchez, A., Ruperez-Illescas, R., Cabellos, J., Richarz, A. and Worth, A. (2019). Computational models for the assessment of manufactured nanomaterials: development of model

reporting standards and mapping of the model landscape, *Comput. Toxicol.*, **9**, 143–151.

21. Gajewicz, A., Rasulev, B., Dinadayalane, T. C., Urbaszek, P., Puzyn, T., Leszczynska, D. and Leszczynski, J. (2012). Advancing risk assessment of engineered nanomaterials: application of computational approaches, *Adv. Drug Delivery Rev.*, **64**(15), 1663–1693.
22. Oksel, C., Ma, C. Y., Liu, J. J., Wilkins, T., Wang and X. Z. (2015). (Q)SAR modelling of nanomaterial toxicity: a critical review, *Particuology*, **21**, 1–19.
23. Richarz, A.-N., Madden, J. C., Marchese Robinson, R. L., Lubiński, Ł., Mokshina, E., Urbaszek, P., Kuz'min, V. E., Puzyn, T. and Cronin, M. T. D. (2014). Development of computational models for the prediction of the toxicity of nanomaterials, *Perspect. Sci.*, **3**, 27–29.
24. Richarz, A.-N., Avramopoulos, A., Benfenati, E., Gajewicz, A., Golbamaki Bakhtyari, N., Leonis, G., Marchese Robinson, R. L., Papadopoulos, M. G., Cronin, M. T. and Puzyn, T. (2017). Compilation of Data and Modelling of Nanoparticle Interactions and Toxicity in the NanoPUZZLES Project, in *Modelling the Toxicity of Nanoparticles*, Tran, L., Bañares, M. A. and Rallo, R., eds. (Springer International Publishing, Cham), pp. 303–324.
25. Valsami-Jones, E, Lynch, I. and Charitidis, C. A. (2016). Nanomaterial ontologies for nanosafety: a rose by any other name..., *J. Nanomed. Res.*, **3**(5), 4–6.
26. Klaine, S. J., Alvarez, P. J. J., Batley, G. E., Fernandes, T. F., Handy, R. D., Lyon, D. Y., Mahendra, S., McLaughlin, M. J. and Lead, J. R. (2008). Nanomaterials in the environment: behavior, fate, bioavailability, and effects, *Environ. Toxicol. Chem.*, **27**(9), 1825–1851.
27. Kroll, A., Pillukat, M. H., Hahn, D. and Schnekenburger, J. (2012). Interference of engineered nanoparticles with in vitro toxicity assays, *Arch. Toxicol.*, **86**(7), 1123–1136.
28. European Chemicals Agency (2017). Read-across assessment framework (RAAF), ECHA-17-R-01-EN.
29. Lamon, L., Asturiol, D., Richarz, A., Joossens, E., Graepel, R., Aschberger K. and Worth, A. (2018). Grouping of nanomaterials to read-across hazard endpoints: from data collection to assessment of the grouping hypothesis by application of chemoinformatic techniques, *Part. Fibre Toxicol.*, **15**, 37.
30. OECD (2018). Case study on grouping and read-across for nanomaterials – genotoxicity of nano-TiO_2, Series on Testing and Assessment No. 292, ENV/JM/MONO(2018)2, (OECD Environment, Health and Safety Publications, Paris).

31. Aschberger, K., Asturiol, D., Lamon. L., Richarz, A., Gerloff, K. B. and Worth, A. (2018). Grouping of multi-walled carbon nanotubes to read across genotoxicity: a case study to evaluate the applicability of regulatory guidance, *Comput. Toxicol.*, **9**, 22–35.

32. OECD (2016). Grouping and read-across for the hazard assessment of manufactured nanomaterials, Report from the expert meeting, Series on the Safety of Manufactured Nanomaterials No. 76, ENV/JM/MONO(2016)59, (OECD Environment, Health and Safety Publications, Paris).

33. OECD (2017). Alternative testing strategies in risk assessment of manufactured nanomaterials: current state of knowledge and research needs to advance their use. Series on the Safety of Manufactured Nanomaterials No. 80, ENV/JM/MONO(2016)63, (OECD Environment, Health and Safety Publications, Paris).

34. Hansen, S. F., Hjorth, R., Skjolding, L. M., Bowman, D. M., Maynard, A. and Baun, A. (2017). A critical analysis of the environmental dossiers from the OECD sponsorship programme for the testing of manufactured nanomaterials, *Environ. Sci.: Nano*, **74**, 147–160.

35. Thomas, D. G., Gaheen, S., Harper, S. L., Fritts, M., Klaessig, F., Hahn-Dantona, E., Paik, D., Pan, S., Stafford, G. A., Freund, E. T., Klemm, J. D. and Baker, N. A. (2013). ISA-TAB-Nano: a specification for sharing nanomaterial research data in spreadsheet-based format, *BMC Biotechnol.*, **13**, 2.

36. Marchese Robinson, R. L., Cronin, M. T. D., Richarz, A.-N. and Rallo, R. (2015). An ISA-TAB-Nano based data collection framework to support data-driven modelling of nanotoxicology, *Beilstein J. Nanotechnol.*, **6**(1), 1978–1999.

37. Marchese Robinson, R. L. and Cassano, A. (2015). All NanoPUZZLES ISA-TAB-Nano datasets [Data set], Zenodo, http://doi.org/10.5281/zenodo.35493.

38. Vriens, H., Mertens, D., Regret, R., Lin, P., Locquet, J. P. and Hoet, P. (2017). Case Study III: The Construction of a Nanotoxicity Database – The MOD-ENP-TOX Experience, in *Modelling the Toxicity of Nanoparticles*, Tran, L., Bañares, M. A. and Rallo, R., eds. (Springer International Publishing, Cham), pp. 325–344.

39. https://characterizationmatters.wordpress.com/parameters/

40. Jeliazkova, N., Chomenidis, C., Doganis, P., Fadeel, B., Grafström, R., Hardy, B., Hastings, J., Hegi, M., Jeliazkov, V., Kochev, N., Kohonen, P., Munteanu, C. R., Sarimveis, H., Smeets, B., Sopasakis, P., Tsiliki, G., Vorgrimmler, D. and Willighagen, E. (2015). The eNanoMapper

database for nanomaterial safety information, *Beilstein J. Nanotechnol.*, **6**(1), 1609–1634.

41. Oomen, A., Bleeker, E. A. J., Bos, P. M. J., van Broekhuizen, F., Gottardo, S., Groenewold, M., Hristozov, D., Hund-Rinke, K., Irfan, M.-A., Marcomini, A., Peijnenburg, W. J. G. M., Rasmussen, K., Sánchez Jiménez, A., Scott-Fordsmand, J. J., van Tongeren, M., Wiench, K., Wohlleben, W. and Landsiedel, R. (2015). Grouping and read-across approaches for risk assessment of nanomaterials, *Int. J. Environ. Res. Public Health*, **12**(10), 13415–13434.
42. Simkó, M., Tischler, S. and Mattsson, M. O. (2015). Pooling and analysis of published in vitro data: a proof of concept study for the grouping of nanoparticles, *Int. J. Mol. Sci.*, **16**(11), 26211–26236.
43. Totaro, S., Crutzen, H. and Riego Sintes, J. (2017). Data logging templates for the environmental, health and safety assessment of nanomaterials, EUR 28137 EN, doi:10.2787/505397. (Publications Office of the European Union.)
44. http://publications.jrc.ec.europa.eu/repository/bitstream/JRC103178/nanoreg_templates.zip
45. Marchese Robinson, R. L. (2016). Zenodo copy of NanoPUZZLES ISA-TAB-Nano templates (version 4), Zenodo, http://doi.org/10.5281/zenodo.45283
46. Gottardo, S., Quiros Pesudo, L., Totaro, S., Riego Sintes, J. and Crutzen, H. (2016). NANoREG harmonised terminology for environmental health and safety assessment of nanomaterials, EUR 27808 EN, doi:10.2788/71213. (Publications Office of the European Union.)
47. Hund-Rinke, K., Baun, A., Cupi, D., Fernandes, T. F., Handy, R., Kinross, J. H., Navas, J. M., Peijnenburg, W., Schlich, K., Shaw, B. J. and Scott-Fordsmand, J. J. (2016). Regulatory ecotoxicity testing of nanomaterials – proposed modifications of OECD test guidelines based on laboratory experience with silver and titanium dioxide nanoparticles, *Nanotoxicology*, **10**(10), 1442–1447.
48. Jantunen, A. P. K., Gottardo, S., Rasmussen, L. and Crutzen, H. P. (2018). An inventory of ready-to-use and publicly available tools for the safety assessment of nanomaterials, *NanoImpact*, **12**, 18–28.
49. Jantunen, A. P. K., Gottardo, S., Rasmussen, L. and Crutzen, H. P. (2017). NANoREG Toolbox for the Safety Assessment of Nanomaterials, European Commission, Joint Research Centre (JRC) [Dataset] PID: http://data.europa.eu/89h/jrc-nano-ehs-ring-nanoreg-tb.
50. OECD (2007). Guidance Document on the Validation of (Quantitative) Structure-Activity Relationship [(Q)SAR] Models, Series on Testing

and Assessment No. 69, ENV/JM/MONO(2007)2, (OECD Environment, Health and Safety Publications, Paris).

51. Puzyn, T., Jeliazkova, N., Sarimveis, H., Marchese Robinson, R. L., Lobaskin, V., Rallo, R., Richarz, A.-N., Gajewicz, A., Papadopulos, M. G., Hastings, J., Cronin, M. T. D., Benfenati, E. and Fernández, A. (2018). Perspectives from the NanoSafety Modelling Cluster on the validation criteria for (Q)SAR models used in nanotechnology, *Food Chem. Toxicol.*, **112**, 478–494.

52. ProSafe Project (2017). The ProSafe White Paper: towards a more effective and efficient governance and regulation of nanomaterials, http://www.rivm.nl/sites/default/files/2018-11/ProSafe%20White%20Paper%20updated%20version%2020170922.pdf

53. Van Teunenbroek, T., Baker, J. and Dijkzeul, A. (2017). Towards a more effective and efficient governance and regulation of nanomaterials, *Part. Fibre Toxicol.*, **14**, 54.

54. Gottardo, S., et al. (2017). NANoREG framework for the safety assessment of nanomaterials, EUR 28550 EN, doi:10.2760/245972. (Publications Office of the European Union.)

55. Burello, E. (2017). Review of (Q)SAR models for regulatory assessment of nanomaterials risks, *NanoImpact*, **8**, 48–58.

56. Baun, A., Sayre, P., Steinhäuserc, K. G. and Rose, J. (2017). Regulatory relevant and reliable methods and data for determining the environmental fate of manufactured nanomaterials, *NanoImpact*, **8**, 1–10.

57. Nowack, B. (2017). Evaluation of environmental exposure models for engineered nanomaterials in a regulatory context, *NanoImpact*, **8**, 38–47.

58. Falk, A., Schimpel, C., Haase, A., Hazebrouck, B., López, C. F., Prina-Mello, A., Savolainen, K., Sips, A., Lopez de Ipiña, J. M., Lynch, I., Charitidis, C. and Germ, V. (2016). Research roadmap for nanosafety. Part III: Closer to the market (CTTM), https://www.nanosafetycluster.eu/uploads/files/pdf/CTTM_NSC_Roadmap_final_for_NSC.PDF.

59. Haase, A. and Klaessig, F. (eds.) (2018). EU US roadmap nanoinformatics 2030, doi:10.5281/zenodo.1486012

60. de la Iglesia, D., Harper, S., Hoover, M. D., Klaessig, F., Lippell, P., Maddux, B., Morse, J., Nel, A., Rajan, K., Reznik-Zellen, R. and Tuominen, M. T. (2011). Nanoinformatics 2020 roadmap, http://eprints.internano.org/607.

61. Winkler, D. A., Mombelli, E., Pietroiusti, A., Tran, L., Worth, A., Fadeel, B. and McCall, M. J. (2013). Applying quantitative structure–activity relationship approaches to nanotoxicology: Current status and future potential, *Toxicology*, **313**(1), 15–23.

62. Winkler, D. A. (2016). Recent advances, and unresolved issues, in the application of computational modelling to the prediction of the biological effects of nanomaterials, *Toxicol. Appl. Pharmacol.*, **299**(Suppl C), 96–100.

63. Grassian, V. H., Haes, A. J., Mudunkotuwa, I. A., Demokritou, P., Kane, A. B., Murphy, C. J., Hutchison, J. E., Isaacs, J. A., Jun, Y.-S., Karn, B., Khondaker, S. I., Larsen, S. C., Lau, B. L. T., Pettibone, J. M., Sadik, O. A., Salehm, N. B. and Teaguen, C. (2016). NanoEHS - defining fundamental science needs: no easy feat when the simple itself is complex, *Environ. Sci.: Nano*, **3**(1), 15–27.

64. Bañares, M. A., Haase, A., Tran, L., Lobaskin, V., Oberdörster, G., Rallo, R., Leszczynski, J., Hoet, P., Korenstein, R., Hardy, B. and Puzyn, T. (2017). CompNanoTox2015: novel perspectives from a European conference on computational nanotoxicology on predictive nanotoxicology, *Nanotoxicology*, **11**(7), 839–845.

65. OECD (2017). Test no. 318: dispersion stability of nanomaterials in simulated environmental media, OECD guidelines for the testing of chemicals, Section 3, (OECD Publishing, Paris), https://doi.org/10.1787/9789264284142-en.

66. OECD (2018). Test no. 412: subacute inhalation toxicity: 28-day study, OECD guidelines for the testing of chemicals, Section 4, (OECD Publishing, Paris,) https://doi.org/10.1787/9789264070783-en.

67. OECD (2018). Test no. 413: subchronic inhalation toxicity: 90-day study, OECD Guidelines for the testing of chemicals, Section 4, (OECD Publishing, Paris), https://doi.org/10.1787/9789264070806-en.

68. Gerloff, K., Landesmann, B., Worth, A., Munn, S., Palosaari, T. and Whelan, M. (2017). The adverse outcome pathway approach in nanotoxicology, *Comput. Toxicol.*, **1**, 3–11.

69. Burello, E. and Worth, A. P. (2015). A rule for designing safer nanomaterials: do not interfere with the cellular redox equilibrium, *Nanotoxicology*, **9**(S1), 116–117.

70. OECD (2014). Genotoxicity of manufactured nanomaterials: report of the OECD expert meeting. Series on the safety of manufactured nanomaterials no. 43. ENV/JM/MONO(2014)34, (OECD Environment, Health and Safety Publications, Paris).

71. Worth, A., Aschberger, K., Asturiol, D., Bessems, J., Gerloff, K., Graepel, R., Joossens, E., Lamon, L., Palosaari, T. and Richarz, A. (2017). Computational models for the safety assessment of nanomaterials. European Commission, Joint Research Centre (JRC) [Dataset] PID: http://data.europa.eu/89h/jrc-eurl-ecvam-nanocomput.

72. Ruusmann, V., Sild, S., Maran U. (2015). QSAR DataBank repository: open and linked qualitative and quantitative structure–activity relationship models. *J. Cheminf.*, **7**, 32.

73. Dimitrov, S. D., Diderich, R., Sobanski, T., Pavlov, T. S., Chankov, G. V., Chapkanov, A. S., Karakolev, Y. H., Temelkov, S. G., Vasilev, R. A., Gerova, K. D., Kuseva, C. D., Todorova, N. D., Mehmed, A. M., Rasenberg, M. and Mekenyan, O. G. (2016). QSAR Toolbox – workflow and major functionalities, *SAR QSAR Environ. Res.*, **27**(3), 203–219.

Chapter 3

Physicochemical Properties of Nanomaterials from in silico Simulations: An Introduction to Density Functional Theory and Beyond

Laura Escorihuela,[a] Alberto Fernández,[a] Robert Rallo,[b] and Benjamí Martorell[a]

[a]*Departament d'Enginyeria Química, Universitat Rovira i Virgili, Av. Països Catalans, 26, 43007, Tarragona, Catalonia, Spain*

[b]*Advanced Computing, Mathematics, and Data Division, Pacific Northwest National Laboratory, Richland, WA 99352, United States*

laura.escorihuela@urv.cat

This chapter covers the fundamentals of density functional theory (DFT), used in electronic structure simulation methods for the in silico estimation of physicochemical properties of NMs. These nanodescriptors are used, for instance, in the development of statistical models for toxicological risk assessment of NMs. Depending on the complexity of the approximation for the exchange-correlation (XC) term in the electron density calculation,

Computational Nanotoxicology: Challenges and Perspectives
Edited by Agnieszka Gajewicz and Tomasz Puzyn

ISBN 978-981-4800-64-8 (Hardcover), 978-0-429-34137-3 (eBook)
www.jennystanford.com

several levels of accuracy exist for DFT, forming the so-called Jacob's ladder. Furthermore, the limitations of DFT are discussed and possible solutions for these limitations are examined, showing how challenging in silico methods can be at the nanoscale, even in solvated environments. To conclude, methods beyond classic DFT are also described here to show how to overcome some of the limitations observed in classic DFT methods.

3.1 Introduction

Nanomaterials (NMs), due to their intrinsic properties, are the cornerstone of a wide range of technologically advanced applications, where metal oxide (MeO) nanoparticles (NPs) are the most used in areas such as electronics, optics, optoelectronics, pharmacy, medicine, cosmetics, and textiles [33]. There is increasing presence of NPs in many consumer products; however, there is still an important knowledge gap on how size influences their physicochemical properties and, in turn, their toxicity [14, 42]. For instance, at the nanoscale, several MeOs show significant toxicity whereas they do not show any significant toxicity at the microscale [10]. Nonetheless, toxicity assessment of NMs is a daunting task that involves multiple testing conditions and endpoints and, in particular, for a variety of NP configurations (i.e., different combinations of core, shell, and functionalization layers, etc.).

Developing nanodescriptors to link physicochemical features with toxicological outcomes requires the detailed computational modeling of the NM at multiple scales (electronic, atomistic, mesoscopic, etc.). Figure 3.1 illustrates how an NM descriptor, the electron bandgap, can be obtained from experiments or derived from computational methods. Electronic structure calculation models (based on quantum mechanical calculations), mostly based on the application of density functional theory (DFT), are commonly used in areas such as solid-state physics and materials science for obtaining those nanodescriptors. In silico testing, specifically the establishment of quantitative (nano)structure–activity relationships (QNARs) and quantitative (nano)structure–property relationships (QNPRs), provides a cost-effective approach to fill the existing gaps in nanosafety data. The establishment of QNPRs and QNARs requires a combination of experimental and computational information, including (i) the detailed physicochemical and biological characterization of the NMs and (ii) the development of computational nanodescriptors suitable to represent the electronic, atomic, and molecular structures of NMs.

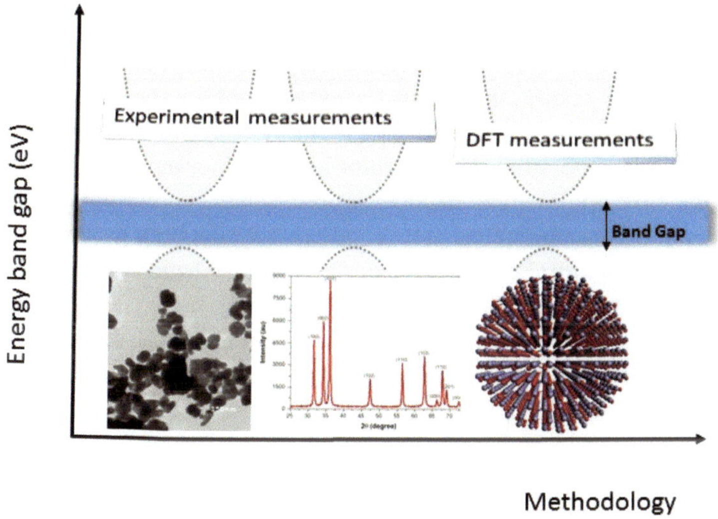

Figure 3.1 Relationship representation of a physical property (electronic bandgap) with respect to possible experimental techniques and computational methods to obtain it.

Size-dependent properties that directly influence a material's environmental impacts can be calculated from the electronic structure and subsequently used to develop statistical models. Properties where the correlation between both methodologies has been used to develop polyparameter statistical models [45, 57] include solubility, Henry's law, partitioning constants, and oxidation potential [43, 55].

The development and validation of standard protocols for the experimental and theoretical characterization of NPs is fundamental to generating the high-quality data required to develop reliable QNPRs and QNARs. Several reference descriptions of experimental and theoretical research protocols have been developed by the Organisation for Economic Co-operation and Development (OECD) [34]. In Europe, the NanoSafety Cluster [44], promoted by the European Commission, monitors and harmonizes actions related to the risk assessment of NMs.

The experimental characterization of the toxicological profiles of NMs is an expensive and time-consuming task. An alternative is to use in silico methods, based on electronic structure calculations, to predict the toxicity using QSAR and QNPR correlations. Advances in electronic structure theory, the improvement in computational chemistry codes, and the increased computing power become the perfect combination to facilitate accurate materials simulations. The ab initio simulations have become a key tool for material developers to help in the prediction and interpretation of experimental data and help in the safe-by-design task. DFT calculations can predict specific physicochemical properties given the chemical and crystal structure of the material [51].

As a result of these advances, given a crystal structure, it is possible to predict thermodynamic and mechanical properties, acoustic and elastic behaviors, and band energies. Even more, spectroscopic and optical intrinsic properties of materials, such as magnetic nuclear resonance, infrared, Raman scattering, and inelastic neutron scattering, can also be modeled with reasonable accuracy.

The aim of this chapter is to give an overview of electronic structure simulations and their applications in computational material science to obtain physicochemical properties that can be used as nanodescriptors for the toxicological assessment of NMs. The chapter focuses on DFT methods since they are the most used for the prediction of material properties and chemical reactions. Other approaches, such as methods based on post-Hartree–Fock (HF) treatment, have been excluded due to their intrinsic computational complexity and their lack of use by the community. Other possible methods, which we don't treat in this chapter, are based on classic mechanics and force fields. They are also extremely useful, though their results depend on the parametrization of the models, and their transferability and use in chemical reactions is more limited than DFT. The remainder of the chapter is structured as follows. Section

3.2 discusses the classic development of DFT and what is called the Jacob's ladder. Section 3.3 introduces alternative methods that go beyond classic DFT to overcome some of their limitations. Finally, Section 3.4 gives some concluding remarks.

3.2 Classic Density Functional Theory: Jacob's Ladder

A formal proof of the notion of DFT came in the 1960s, when Hohenberg and Kohn published their two well-known theorems, where they reported (i) that the total energy and properties of an electronic structure are a unique functional of the electron density (ρ) of a nondegenerate system and (ii) that the electron density that minimizes the total energy is the exact ground-state density [21].

In 1965, Kohn and Sham introduced the atomic orbitals in the DFT theory and proposed the Kohn–Sham equations [23]. This is a set of $N/2$ equations (for closed-shell systems with no spin polarization) describing the wave function of each of the N electrons as a noninteracting particle. This formulation developed the application of DFT in computational fields. In these Kohn–Sham equations, the electron is subjected to a modified effective potential from the nuclei and electrons, which is much easier to solve than a wave function of the whole 3D system, where this noninteracting system has the same ground-state density as the real system [12, 53]. In fact, since the electron-electron interactions are always the same, it is the external potential due to ions that uniquely determines the Hamiltonian. Therefore, the total energy of the noninteracting system is given in Eq. 3.1:

$$E[\rho]_{\mathrm{DFT}} = T_{\mathrm{S}}[\rho] + V_{\mathrm{ext}}[\rho] + V_{\mathrm{H}}[\rho] + E_{\mathrm{xc}}[\rho] \tag{3.1}$$

where T_S is the kinetic energy of the electrons, V_{ext} is the external potential created by the nuclei, V_H is the coulombic repulsion between the electrons, and E_{XC} is the XC term of the electronic energy. In this noninteracting system, the kinetic energy for the noninteracting particles is expressed in two parts: the first part is the kinetic energy exactly calculated for the noninteracting electrons, and the second part considers the interaction between electrons included in the

$E_{XC}[\rho]$ terms. Therefore, the expression of the total energy can be expanded as:

$$E[\rho]_{DFT} = -\frac{1}{2}\sum_{i=1}^{n}\int \varphi_i^*(r_1)\nabla_i^2\varphi_i(r_1)\mathrm{d}r_1$$
$$-\sum_{x=1}^{N}\int \frac{Z_x}{r_{xi}}\rho(r_1)\mathrm{d}r_1 + \frac{1}{2}\iint \frac{\rho(r_1)\rho(r_2)}{r_{12}}\mathrm{d}r_1\mathrm{d}r_2 + E_{xc}[\rho] \quad (3.2)$$

In Eq. 3.2, φ_i (i = 1, 2 . . .) are the Kohn–Sham orbitals; the first term is for the kinetic energy for noninteracting electrons; the second term is for the nuclear-electron interactions; the third part indicates the coulombic interactions corresponding to the energy between the interacting electrons; and at the end there is the XC term, which contains the correction of the kinetic part for the interacting electrons and the classic correction for the electron-electron repulsion, the sum of the error made in using a noninteracting kinetic energy and the error made in treating the electron-electron interaction classically.

The correspondence of the charge density and energy of the many-body and the noninteracting system is exact only if the exact functional is known. For any particular system we could, in principle, solve the Schrödinger equation exactly and determine the energy functional and its associated potential. This, of course, involves a greater effort than a direct solution for the energy. Nevertheless, the ability to determine exact properties of the universal functional in a number of systems allows excellent approximations to the functional to be developed and used in unbiased and thus predictive studies of a wide range of materials—a property usually associated with an ab initio theory. For this reason the approximations to DFT discussed below are often referred to as ab initio or first principles methods.

Depending on the treatment of the XC term, the accuracy of the DFT calculations varies. Classically, these approximations have been classified depending on their complexity. This gave rise to the concept of DFT Jacob's ladder, where each level of approximation adds more complexity in the treatment of the E_{XC} term [41]. The approximations are, in terms of increasing complexity, the local density approximation (LDA), the generalized gradient approximation (GGA), the meta-GGA, the hybrid functionals, and finally the exact electron density functional. These five steps of Jacob's ladder will be explained in more detail in the following sections. The exact expression of the XC functional is not treated here due to the present limitations in its use to only a few systems with a small number of atoms and its high computational cost.

3.2.1 Local Density Approximation

LDA was the first approach (and the simplest one used nowadays) to evaluate the XC energy. In this approximation, the XC energy is evaluated as a function of a uniform gas of electrons [40, 53]. The exchange part for the model of a homogeneous, constant-density gas is:

$$E_{\mathrm{LDA}}^{\mathrm{X}}[\rho]=-C_{\mathrm{x}}\int\rho^{\frac{4}{3}}(r)\mathrm{d}r;\, C_{\mathrm{x}}=-\frac{3}{4}\left(\frac{3}{\pi}\right)^{1/3} \tag{3.3}$$

In LDA, the XC energy density at position r depends only on the particle density at that point, $\rho(r)$. An important property of any density functional is the XC hole, the region around any particle in which the probability of finding another identical particle is reduced. In other words, an electron never interacts with itself, and thus the exact XC energy provides a self-interaction correction to the Hartree electrostatic energy. This hole arises from Pauli's exclusion principle, and the total reduction of the density of the other particles should be equal to −1. LDA satisfies this exclusion principle exactly, meaning the shape of the LDA XC hole is incorrect, although it has the correct average shape [56].

As one can expect, LDA presents some limitations due to its simplicity, especially for the ionization potentials and binding energies for molecules. However, it works well in electronic homogeneous systems such as general solids; it improves the quality of the results with the size of the system, and it is a relatively computationally cheap methodology [16].

3.2.2 GGA and Meta-GGA

Although LDA gave some good results for systems with homogeneous electronic density structures, it failed to describe many materials and properties. This is because one cannot consider most materials as homogeneous gases. The energy is not only dependent on the local value of the electronic density but also dependent on the gradient of the density, $\Delta\rho(r)$. This leads to the formulation of the GGA. The typical expression of the energy for GGA is:

$$E_{\text{GGA}}^{\text{XC}}[\rho] \approx \int \rho(r)\varepsilon_{\text{xc}}(\rho, \nabla\rho)\text{d}r \tag{3.4}$$

The GGA functional ensures the normalization condition of the spherical hole and a negative factor in the XC, leading to an energy value dependent on weak perturbations on the homogeneous electron gas density. In addition, it gives an analytical improvement in the atomization energies and corrects the overbonding of H-bonding solids [16]. Several versions for the GGA functional have been reported depending on the form of the functional: PW91 [37], BLYP [4], PBE [36], etc.

Compared to LDA functionals, GGAs are more accurate for calculating total energies and atomization energies because the method is based more on the principles of the quantum mechanics scaling the densities than on achieving correction in the upper and lower parts of the scale of densities. Several approximations on the gradient have been implemented, and all of them give better results than LDA for geometries, vibrational frequencies, and charge densities, though the computational cost is higher [28, 54]. However, it is still not accurate enough for some chemical properties of the molecules, such as the van de Waals interactions, ionization potentials, and electron affinities.

Beyond the complexity of GGA functionals, the next step in Jacob's ladder are the meta-GGA functionals [39, 40, 52], where semilocal information about the electronic density is introduced in the Laplacian in terms of spin density or kinetic energy density, which includes some derivation of the occupied Kohn–Sham orbitals. The typical form for this functional is:

$$E_{\text{m-GGA}}^{\text{XC}}[\rho] \approx \int \rho(r)\varepsilon_{\text{xc}}(\rho, \nabla\rho, \nabla^2\rho, \tau)\text{d}r, \tag{3.5}$$

where τ is the kinetic energy. With this approximation, some properties, such as atomization energies, are slightly improved at a much larger computational cost with respect to GGA. Among those functionals one can find derivations such as Tao–Perdew–Staroverov–Scuseria (TPSS) [38] or the recent strongly constrained and approximately normed (SCAN) [52].

3.2.3 Hybrid Functionals

Even for GGA and meta-GGA, some chemical features are still not well described. For example, the gap between the bonding band and the conductive band in semiconductors is too small for GGA calculations [16]. In the case of HF calculations, this gap tends to be too large. These two results lead to the idea of mixing these two methods to improve the quality of the results. The exchange part of the XC potential is in these models the exact exchange of the HF method. The correlation part comes from the DFT treatment. This makes these methods a hybrid between DFT and HF. Therefore, in this hybrid methodology one combines the XC of GGA methods with part of the HF exchange, which is fitted empirically. However, in this method the HF exchange potential does not take into account the dynamical correlation, so a new term for dynamical correlation must be added into the functional, and this is done through the adiabatic connection [53].

Among other functionals, one of the most used for atoms and molecules is the B3LYP functional [4, 29], with coefficients fitting the observed atomization energies, ionization potentials, proton affinities, and total atomic energies for a number of small molecules. The resultant energy functional is:

$$E^{\mathrm{XC}}_{\mathrm{B3LYP}}[\rho]=E^{\mathrm{X}}_{\mathrm{LDA}}[\rho]+0.2\left(E^{\mathrm{X}}_{\mathrm{HF}}[\rho]+E^{\mathrm{X}}_{\mathrm{LDA}}[\rho]\right)$$
$$+0.72\left(E^{\mathrm{X}}_{\mathrm{PW91}}[\rho]+E^{\mathrm{X}}_{\mathrm{PW91}}[\rho]\right)+E^{\mathrm{C}}_{\mathrm{LDA}}[\rho]+0.81\left(E^{\mathrm{C}}_{\mathrm{GGA}}[\rho]-E^{\mathrm{C}}_{\mathrm{LDA}}[\rho]\right) \tag{3.6}$$

In the case of solid state, some numerical problems appear due to the periodical conditions and the use of periodic wave basis sets, making these types of simulations computationally very expensive. This is one of the handicaps that does not allow hybrid calculations in very large solid systems. The most common hybrid functionals used in solids are PBE0 [6] and HSE06 [18]. The resultant energy functionals are:

$$E^{\mathrm{XC}}_{\mathrm{PBE0}}[\rho]=\frac{1}{4}E^{\mathrm{X}}_{\mathrm{HF}}[\rho]+\frac{3}{4}E^{\mathrm{X}}_{\mathrm{PBE}}[\rho]+E^{\mathrm{C}}_{\mathrm{PBE}}[\rho] \tag{3.7}$$

and

$$E_{\text{HSE06}}^{\text{XC}}[\rho]=\frac{1}{4}E_{\text{HF-SR}}^{\text{X}}[\rho]+\frac{3}{4}E_{\text{PBE-SR}}^{\text{X}}[\rho]+E_{\text{PBE-LR}}^{\text{X}}[\rho]+E_{\text{PBE}}^{\text{C}}[\rho], \tag{3.8}$$

where SR and LR refer to short- and long-range interactions (cutoff for HSE06 is 0.2). In general, one can affirm that computed binding energies, geometries, and frequencies are systematically more reliable than the best GGA functionals.

3.2.4 The Limits of Classic DFT

DFT methods have very favorable characteristics, and numerous scientists from different areas employ them to compute and predict properties in numerous systems (in materials science, nanotechnology and nanoengineering, chemistry, physics, earth and environmental sciences, etc.). In Table 3.1 we present a summary of the most relevant advantages and disadvantages of each method in the Jacob's ladder. Nevertheless, the first limitation in DFT methods is computational: at present, systems with more than 1000 atoms are difficult to compute, even using large supercomputers. This is due to the cubic increase in the computational requirements with respect to the number of atoms.

There are many fields where the current functionals are known to perform poorly. Current DFT methods describe poorly weak interactions due to dispersed forces such as van der Waals interactions, and even stronger interactions, such as hydrogen bonds, present many difficulties for DFT functionals due to the electrostatic LR interactions [45].

Besides, XC functionals, which do not include self-interaction corrections, lead to loosely bound electrons such as anions arising from systems with relatively low electron affinities. Another issue is the dissociation of charge odd-electron systems, since it takes place with dissociation energy displaying an artificial barrier. Transition state structures are similarly predicted to be stable by functionals that do not include an exact exchange. Though HF overestimates activation barriers, hybrid methods often give reasonable barriers.

Table 3.1 Summary of Jacob's ladder and the different levels of theory on classic DFT

Approximation	Advantages	Disadvantages
LDA [40, 53]	–Results improve with the size of the system. –It is excellent for systems with constant electronic densities. –It is accurate enough for calculations on solids and solid surfaces.	–The results are modest due to approximations on kinetic energy. –It is not accurate enough for atoms and molecules. –It tends to underestimate atomic and ionization energies. –It tends to overestimate binding energies.
GGA [28, 40]	–It gives an analytical improvement in the atomization energies. –It corrects the overbonding of H bonds. –There is more accuracy in total energies and atomization energies than LDA for molecules and solids.	–It is not accurate enough for the van de Waals interactions, ionization potentials, or electron affinities. –In solid state, GGA is not much better than LDA for electron affinities or ionization potentials.
Meta–GGA [39, 40, 52]	–Atomization energies are more accurate than GGA values.	–The increase in accuracy is not substantial with respect to GGA. –There is an increase in computational time with respect to GGA.
Hybrid–DFT [29, 49]	–There is great improvement with respect to GGA for binding energies, geometries, and frequencies. –The functionals are more accurate due to exact exchange energy on the exchange hole.	–In solid–state physics (plane wave basis) it is very difficult to compute the exact XC terms. –It does not properly scale to the exact exchange under uniform density scaling.
Exact XC [30]	–It gives a good description of van der Waals forces –It gives the exact XC energy.	–It requires a large number of unoccupied bands. –It is way more expensive than LDA and GGA. –It has been only tested on small molecules.

Another problem of DFT methods is the prediction of excited states, due to the ground-state description of the Hohenberg–Kohn theorem, as well as the charge transfer description in molecules, since the XC functionals are inherently local, only depending on the local electron density.

Finally, DFT describes the different spin multiplicity by XC functional, and therefore, the relative energy of states is often poorly described. Hybrid methods with a suitable parametrized amount of exact exchange perform better because HF theory describes the exchange integral on occupied orbitals [10, 15].

With all this in mind, one can affirm that new methods developed beyond the DFT are necessary to take into account some of the weaknesses of DFT. This is the main topic of the next section.

3.3 Beyond Classic DFT

Although DFT is a very powerful tool, as explained in the previous section, it still has some limitations in its use. Therefore, new developments based on DFT (or based on wave-function methods, but then joined with the DFT methodology) have been developed to compute and predict better a wider range of properties and materials.

In the next sections we give an overview of computational methods beyond classic DFT, which deal with some of the limitations found in DFT. We also describe these methods and how they are developed.

3.3.1 DFT+*U*

In the HF method, the coulombic interactions are correctly described with the presence of the exact exchange term, cancelling the self-interaction of the electron. However, due to the fittings of the functionals, in DFT that exchange is only approximated, and as a consequence, the self-interaction term is not correctly cancelled. This is known as the self-interaction error. This error causes the poor description of electrons' interaction in strongly correlated systems. The coulomb interaction between electrons is important and dominant for some transition metals, rare earths, or insulating

MeOs, and the idea of a simple correlated electron gas is not enough to properly describe their properties. Because the local coulomb repulsion of the d orbitals is underestimated by the classic LDA or GGA, an extra repulsion term needs to be added to correct this underestimation (the U repulsion term, the Hubbard term). This correction is known as DFT+U (LDA+U or GGA+U, depending on the initial functional used) methodology [9]. Simply said, DFT+U describes the "strongly correlated" electronic states of a system (typically, localized d or f orbitals) using the Hubbard model, whereas the rest of electrons are treated at the level of classic DFT functionals. This is a methodology that is similar to hybrid functionals, in the sense that it tries to correct the exchange term in E_{XC} but in this case without mixing the exact correlation of the HF method.

In the DFT+U method, two new parameters are needed, U and J. The parameter U represents the strengths of the on-site coulomb interaction, and the parameter J adjusts the strengths of the exchange interaction. These two parameters are combined into a new single parameter called effective U or $U_{eff} = U - J$. With this correction, the electron potential is locally reduced for the specified orbitals of the respective atoms or elements, as a consequence reducing the hybridization with the orbitals of other elements and their interaction. The limit of $U_{eff} = 0$ represents the classic DFT [31]. The value of U_{eff} is fitted against experimental data for materials representative of the elements to be corrected with this methodology. Within DFT+U, the total energy is expressed as follows:

$$E_{\mathrm{DFT}+U}[\rho] = E_{\mathrm{DFT}}[\rho] + E_{\mathrm{Hub}}(\sigma_{nn}) - E_{\mathrm{dc}}(\sigma_{nn}) = E_{\mathrm{DFT}}[\rho] + \sum_{n} \frac{U_{\mathrm{eff}}}{2} \mathrm{Tr}(\sigma_n - \sigma_n \sigma_n) \tag{3.9}$$

where ρ is the electronic density, E_{Hub} is the Hubbard correction energy, E_{dc} is the energy that accounts for doubly counted electrons in the E_{DFT} and E_{Hub} terms, and σ is the electron density matrix. E_{DFT} represents the approximate DFT total energy functional being corrected, and E_{Hub} is the term that contains the Hubbard Hamiltonian to model correlated states [19].

The DFT+U method has helped to improve the results on the bandgaps of some insulators and semiconductors that were underestimated with classic DFT. Nevertheless, they depend on an empirical correction that can overlocalize the d or f electrons of the metals. Besides, different parameters U depend on the chemical environment of the element, not being transferable, which is a problem in elements with different oxidation states. Besides, due to the overlocalization, unpaired electrons' interactions and magnetic structures can be wrongly predicted [3].

3.3.2 *GW*

In the previous section, the DFT+U formulation has been explained and it has been shown that it is a good solution for systems that contain localized states in order to describe in a better way bandgaps and geometries of the system. However, DFT+U is just a correction forcing the localization of the d and f electrons. Nevertheless, methods previously described are not efficient when one wishes to compute the quasiparticle energies due to the self-interaction error in the XC functionals [15, 47], and properties such as electronic excitations, ionization potentials, and electron affinities must be computed in a different scheme.

An alternative to classic DFT or DFT+U methods to compute those quasiparticle energies (as well as to obtain adequate bandgaps) is the use of the *GW* approximation. *GW* can be explained with the HF approximation but with a dynamically screened coulomb interaction [2, 59], and then the self-interaction energy in *GW* is a nonlocal property that depends on the energy of the particle. In the GW approach one expands Green's function (G) in terms of the screened coulomb interaction (W) from the HF method, which leads to Hedin's equation [17]:

$$iG(x,x') = \left\langle N \middle| T\left[\varphi(x)\varphi^{T}(x')\right] \middle| N \right\rangle \tag{3.10}$$

where N is the exact ground state of the electronic system, $\varphi(x)$ is an annihilation operator in the Heisenberg representation, and T is the time-ordering operator that arises naturally from the time development operator defined later in the equation.

The physical interpretation of Green's function is that for $t' > t$ it is the probability amplitude that a hole created at x will propagate to x' and for $t > t'$, it is the probability amplitude that an electron added at x' will propagate to x. Thus, Green's function describes the photoemission and inverse photoemission processes. Another key property of Green's function is that its poles define the excitation energies of the quasi particles and therefore it is possible to find those energies in an adequate manner [59].

Although good results have been obtained both in solids and single molecules [59], calculations are computationally expensive and therefore it is desirable to find simplifications that reduce the numerical effort but still maintain the accuracy of the full calculations. However, there is no simple solution that satisfies all systems. Most simplified theories are for semiconductors and insulators, with a good description of the bandgap, but in any case, it is difficult to construct a good approximation for the self-energy that includes both nonlocality and energy dependence, one that is simple enough to be applicable to complex systems without significant loss of accuracy.

3.3.3 Density Functional Tight Binding

As mentioned in the previous sections, both DFT and linear-scaling DFT (LS-DFT) have some limitations. On one hand DFT cannot deal with large numbers of atoms (or electrons, to be fair), has difficulties when performing long molecular dynamics simulations, and is expensive. On the other hand, although LS-DFT can handle up to tens of thousands of atoms, it struggles in metallic systems and requires large high-performance computing systems to take advantage of its development.

A powerful alternative developed in recent years is the application of tight binding (TB) theory in DFT implementation. In classic DFT, the energy of a system of M electrons in the field of N nuclei at positions $\vec{R}$ is based on a functional of the electrons density $n(\vec{r}\,)$, following the Kohn–Sham equation. In density functional tight binding theory (DFTB) [11, 27, 50], the energy is expressed as the second-order expansion of a small charge density fluctuation $\delta n(\vec{r'}\,)$ith respect to the reference density $n_0(\vec{r'}\,)$the zeroth order expansion corresponds to the TB method). The total DFTB energy can be expressed after this expansion as:

$$E_{\text{DFTB}} = \sum_{i=1}^{\text{occ states}} \langle \Psi_i | H_0 | \Psi_i \rangle + E_{ii} - \frac{1}{2} \iint' \frac{n_0(\vec{r}')n_0(\vec{r})}{|\vec{r} - \vec{r}'|} \mathrm{d}\vec{r}'\mathrm{d}\vec{r} + E_{\text{XC}}[n_0(\vec{r})] - \int E_{\text{XC}}[n_0(\vec{r})] n_0(\vec{r}) \mathrm{d}\vec{r} + \frac{1}{2} \iint' \left(\frac{1}{|\vec{r} - \vec{r}'|} + \frac{\delta^2 E_{\text{XC}}}{\delta n(\vec{r}) \delta n(\vec{r}')} \bigg|_{n_0(\vec{r})} \right) \delta n(\vec{r}) \delta n(\vec{r}') \mathrm{d}\vec{r}'\mathrm{d}\vec{r} \tag{3.11}$$

where the first term is the Kohn–Sham energy of the occupied orbitals, the second one is the interatomic repulsion, the third one is the coulombic repulsion, the fourth and fifth terms are the XC energy–derived terms, and the final term is related to the small fluctuation applied to the system $\delta n(\vec{r}')$

The evaluation of E_{DFTB} implies computing several elements depending on the fluctuation of the electronic charge density, which is not simple. The first approximation consists in neglecting the last term in Eq. 3.11, which is the classic form of DFTB. However, in doing this, the charge density on the atomic centers is kept constant in the energy evaluation, and therefore this can only be applied to covalent systems. More commonly, one develops the last term of this expression and truncates the multipole expansion after the monopole term. This approximation is known as the self-consistent charge DFTB, and the energy is expressed as:

$$E_{\text{DFTB}} = \sum_{i=1}^{\text{occ states}} \langle \Psi_i | H_0 | \Psi_i \rangle + E_{\text{rep}} + \frac{1}{2} \sum_{\alpha,\beta}^{N} \Delta q_\alpha \Delta q_\beta \gamma_{\alpha\beta} \tag{3.12}$$

where the first term represents the energy of all the Kohn–Sham orbitals of the reference system (corresponding to the electronic density of neutral atoms), the second term encloses the SR repulsive terms of the nuclei-nuclei interactions plus other XC factors, and the third term accounts for the charge transfer between different atomic constituents. In more detail, E_{rep} is defined as:

$$E_{\text{rep}} = E_{ii} - \frac{1}{2} \iint' \frac{n_0(\vec{r}')n_0(\vec{r})}{|\vec{r} - \vec{r}'|} \mathrm{d}\vec{r}'\mathrm{d}\vec{r} + E_{\text{XC}}[n_0(\vec{r})] - \int E_{\text{XC}}[n_0(\vec{r})] n_0(\vec{r}) \mathrm{d}\vec{r} \tag{3.13}$$

However, the full evaluation of all terms in Eq. 3.13 requires a huge computational effort and, since it is based on the XC functional, the final result will depend on this election. Therefore, the strategy typically used in DFTB is to represent this repulsion energy as a sum of pairwise functions between atoms. These functions are obtained by fitting them to high-level theoretical calculations [27].

The third term of Eq. 3.12 takes into account the charge fluctuations between atoms in the system, and it is derived from the last term in Eq. 3.11. In this last term, for local XC functionals (see Section 3.2), the second-order terms become a Dirac delta function, this last term becoming the electrostatic interaction shown in Eq. 3.12 [27]. For Eq. 3.12, Δq_α represents the Mulliken charge change with respect to the neutral atom α and $\gamma_{\alpha\beta}$ is the extent of that charge interaction between centers α and β. Finally, the $\gamma_{\alpha\beta}$ function is approximated using the Hubbard parameters U_α and U_β of the atoms (directly related to the chemical hardness of the element and used also in the DFT+U methodology) and the distance between those two centers [11]. The Hubbard parameters are obtained from ab initio simulations in DFT calculations, and therefore we do not introduce at this point any empirical data for the fitting of the computational methodology.

All the values parameterized for the V_{rep} of the different atoms required in the simulations, together with the symmetry transformations for the evaluations of the orbital energies, are saved in the Slater–Koster files (SK) [35]. The SK files are tabulated for a large quantity of pairs of atoms (the full periodic table is still not available, but the database is increasing) and available to the general public on the DFTB project webpage, www.dftb.org. Alternatively, one can create his/her own files from ab initio simulations, though at a huge cost in terms of time for computational and results analysis. At present, several codes and quantum chemistry packages include the DFTB methodology and are compatible with the SK files published for general use: DFTB+ [1, 11], deMon, AMBER, ADF, and Gaussian.

With DFTB one can compute systems formed by thousands of atoms in relatively small computational clusters, this being a reliable method. Nevertheless, the SK files available are still relatively limited to organic molecules, some inorganic solids, and transition metals. Besides, the results that one obtains are also determined by the functionals used to implement the SK files. As a consequence, for instance, when liquid water systems are being computed, the hydrogen bonds are not properly described, as in the case of the GGA functionals, and DFTB trends to overestimate the intermolecular forces in water, forming voids in the water bulk [8].

3.3.4 LS-DFT

One of the limitations in the use of DFT methods is the size of the systems to compute. Currently, simulations of 1000 atoms can only be performed on large supercomputers, since the computational costs scale in a factor range of 3–4 with respect to the number of atoms in the simulation (N^{3-4}). This is, of course, an important handicap in performing simulations in biological systems; solids with defects at parts-per-million scale; or NMs in solution, where tens or hundreds of thousands of water molecules must be taken into account for a good description of the liquids.

LS-DFT aims to overcome the cubic order of the computational requirements with the size of the system and tries to obtain ratios close to linearity. In this case, LS-DFT is not a new methodology but a new type of algebraic speed-up to compute the electron density in the Kohn–Sham equations using the properties of sparse algebra (sparse matrices are those in which most elements are zero). In real-space grids one can express the electron density matrix as [20, 22]:

$$\rho(r,r') = \sum_{\alpha\beta}\sum_{R}\varphi_{\alpha R}(r)K^{\alpha\beta}\varphi_{\beta R}(r'), \tag{3.14}$$

where $K^{\alpha\beta}$ is the kernel density. The total energy is then a function of this kernel matrix and the coefficients that it contains. In LS-DFT one cuts the long-range interactions (this is a free parameter to be set in the calculations), and therefore, part of the elements of the kernel matrix becomes zero, speeding up the simulation times and increasing the number of atoms that one can deal with. Only the local interactions determined by the cutoff radii are evaluated.

Therefore, the main advantage of LS-DFT is that one can perform simulations of hundreds of thousands of atoms at a relatively small computational cost. Codes that include this methodology are ONTEP, BigDFT, and CP2K. However, due to the locality of sparse algebra, there is a clear limitation in the use of LS-DFT where metals and some semiconductors are concerned [48]. Results obtained with this method also depend on the cutoff radii, which must be given converged results, and on the DFT functional chosen to compute the energies of the system, containing the same limitations as the corresponding functional.

3.3.5 Time-Dependent DFT

When performing DFT simulations, one of the things one has to take into account is that the Hohenberg theorem states that one can get the exact energy of a system if the exact functional is known, only for the ground state. This has as a consequence: computing the energies of excited states, and therefore the calculations of transitions, spectra, and excitations, is a difficult task in the framework of classic DFT.

The most appropriate methods to describe excitations are the linear response methods because the poles on the linear response electron density correspond to the excitation energies of the system. In particular, in the framework of DFT, one can extend the formalism to a time-dependent domain using an external time-dependent potential to deal with excitations. This is the time-dependent DFT (TD-DFT) formalism [13, 58].

TD-DFT is based on the Runge–Gross theorem [46], who proved that if two N-electron systems start from the same initial state but are subject to two different time-dependent potentials, their respective time-dependent densities will be different [58]. In the case of TD-DFT, the exact time-dependent density, $\rho(r,t)$, is computed from a noninteracting system with N single-particle orbitals:

$$\rho(r,t)=\sum_{i}^{N}\left[\varphi_i(r,t)\right]^2, \tag{3.15}$$

where the orbitals φ_i satisfy the time-dependent Kohn–Sham equations:

$$i\frac{\mathrm{d}}{\mathrm{d}t}\varphi_i(r,t)=\left[-\frac{1}{2}\nabla_i^2+V_s(r,t)\right]\varphi_i(r,t) \tag{3.16}$$

where $V_S(r,t)$ is the time-dependent external potential felt by the electrons and includes all the classic DFT terms plus the time-dependent potential. Nevertheless, to properly describe the time evolution in the theorem, one has to include the action integral in the Eq. 3.16, where the coulomb explicit term defines the XC and the XC kernel, which includes all the quantum effects of the electron time-dependent interaction and is similar to the XC term of the classic Kohn–Sham formalism. Then, the final equation looks like:

$$i\frac{d}{dt}\varphi_i(r,t)=\left[-\frac{1}{2}\nabla_i^2+V_{\text{ext}}(r,t)+\int\frac{\rho(r',t)}{|r-r'|}d^3r'+\frac{dA_{\text{XC}[\rho]}}{d\rho(r,t)}\right]\varphi_i(r,t), \tag{3.17}$$

where A_{XC} represents the XC part in the time-dependent formalism, which is not known and approximations are needed. Commonly, the adiabatic local density approximation is used, where the XC kernel is replaced by the time-independent term of the classic DFT [32, 57].

The TD-DFT approach has been used successfully in the field of molecular excited states, being an accurate method to predict excitation energies and intensities and being a much cheaper method than, for instance, configuration interaction systems methods. TD-DFT is size consistent, and results depend on the XC functional choice, the hybrid functional B3LYP being the most used for molecular species and the GGA PBE for solids. However, due to the linear response character of the TD-DFT development and the use of classic XC functionals, TD-DFT has problems in computing transitions to valence states of molecules with extended π-systems, doubly excited states, long-range charge transfer excitations, and excitations of Rydberg character [13].

3.3.6 Implicit Solvation Models

Up to this point, we have described theoretical methods to compute the energy of a material, describing explicitly all the interactions one can find in the modeled system. This adds an extra difficulty when describing the environment where the material is, especially for liquid solutions. As we have pointed out in the DFTB and LS-DFT sections, one can describe the solvent explicitly, adding in the model the molecules of the liquid and applying molecular dynamics simulations at the desired temperature. Nevertheless, those simulations are very expensive in computational terms and one has to obtain properties on average values on time simulations.

There exists an alternative to the explicit inclusions of solvent molecules in the KS equations, and this is implicitly incorporating the effect of solvents in the DFT energy calculation. A simple method to include this effect in the materials is using the polarizable continuous model (PCM) [5]. In this type of model, the material or particle is embedded in a continuous environment of a certain permittivity, ε_r, which interacts with the electronic structure of the particle. The solute material is inside a cavity formed in the continuous environment or solvent, and that environment affects the particle. There are several approximations for different types of nonpolar (they are treated as dielectrics) or polar solvents (treated as conductor-like systems), the most used one being the COnductor-like Screening MOdel (COSMO) [26].

The COSMO computes the solvent as a dielectric continuum around the solute system, which is in a cavity and derives the polarization charges in the continuum solvent caused by the solute as an ideal conductor. Once the electronic density of the particle is known, then one can compute the charges in the surface segments and the solvent as:

$$q(r) = \mathrm{q}^*(r)\frac{\varepsilon_r - 1}{\varepsilon_r + x}, \tag{3.18}$$

where $\mathrm{q}^*(r)$ is the electronic from the pure DFT simulation and x is a correction factor between $0 < x < 0.5$, depending on the parametrization performed for each solvent. The advantage in the COSMO is that it includes this charge evaluation in the self-consistent field of the KS equations, avoiding the use of different iterative steps in the calculation of the effect of the solvent on the particles. To improve results obtained with COSMO for different solvents, post-treatments using statistical mechanics to improve the solute-solvent interaction have been performed and the method COSMO-RS was implemented [24].

There exist other types of implicit solvation models [7], the SM*x* family and the solvation model based on density, where *x* stands for a number showing the generation of the model. In this case, the solute is also in a cavity formed in the continuous liquid environment, but the effect of the solvent is accounted on the solute as point charges at atomic sites of the solute. Both SM*x* models and COSMOs present similar results for different solvents, and both have demonstrated to be accurate enough [25, 60].

The implicit solvent models have helped to study chemical reactions and materials in solution. However, some difficulties exist with the methods. In the case of the COSMO, in an ideal conductor the potential in the surfaces of the cavity should cease, but in real solvents with low permittivity, this can create instabilities in the cavity surface, making difficult the convergence of the calculations. Besides, the fact that we include corrections in the electronic structure makes the DFT calculation more expensive in computational terms. And, of course, the final results will depend on the XC functional used for the simulation.

Table 3.2 Summary of methods beyond classic DFT

Approximation	Advantages	Disadvantages
DFT+*U* [9, 19, 31]	–It is a low-cost methodology to accurately correct bandgap underestimation in GGA. –It provides accurate results for strongly correlated systems: rare-earth compounds, transition metals, Mott insulators and impurity systems, and superconductors.	- DFT+*U* solutions can overestimate the degree of localization. –The accuracy in results depends on the functional used in the DFT part.
GW [2, 59]	–It provides excellent evaluation of bandgaps in solids. –It gives an accurate reproduction of photoemission spectra. –It corrects the bandgap offset in semiconductor interfaces. –Mott insulators are properly described.	–It is very expensive in computational terms. –It sums over all empty states. Some approximations must be done to avoid this.

Approximation	Advantages	Disadvantages
DFTB [11, 27, 50]	–It simulates systems of tens of thousands of atoms at the quasi-DFT level with little effort (x100 quicker on average). –It has the same accuracy as the DFT reference model. –It is easy to implement for both molecular and periodic systems, as well as including fields in the simulations.	–The accuracy in results depends on the functional used for the parameterization –Only databases for organic interactions, water, and a few solids exist. –The parameterization of repulsions is not an easy task.
LS–DFT [20, 22]	–It scales linearly with the number of atoms for large systems (>1000 atoms) instead of the cubic order of normal DFT. –Systems of up to hundreds of thousands of atoms can be simulated.	–It is not suitable for metals due to the nonexistence of bandgap. –Accuracy depends on the cutoff parameter of the kernel.
TD–DFT [13, 32, 52]	–It accurately computes the excited states properties: bandgap, dielectric properties, and electronic–optical spectra. –It accurately deals with charged systems and the electronic dynamics in "real time." –For small potentials, a linear response perturbation on the time–independent density can be used.	–Results can depend on the DFT functional chosen (normally the adiabatic LDA). – It is very expensive in computational terms.

(*Continued*)

Table 3.2 (*Continued*)

Approximation	Advantages	Disadvantages
PCM [5,26]	–One can model solvents implicitly; no explicit molecules for the solvent are needed. –Chemical reactions in a solvent environment can be performed. –These are highly accurate and extensively tested methods.	–Results depend on the cavity formation and the radii of the cavities. –Electronic convergence is difficult due to the potential gradients in cavity interfaces for nonpolar solvents.

Another problem is the description of the solvent as a continuum in the interaction region (also called solvation shells) presented between the solute and the solvent, since the behavior of solvent molecules in this region is very different to that of the solvent bulk. This is a particular problem in charged systems, where the solute-solvent interaction is even stronger. In such cases, it is better to include a first or even a second solvation shell explicitly in the model to account for the different behaviors of the solvent molecules next to the solute and in the continuous region of the solvent (in the bulk).

As a final reminder of Section 3.3, we present in Table 3.2 a summary of the methods explained here, showing the advantages, possibilities, and disadvantages of the methods.

3.4 Concluding Remarks

In this chapter we have introduced in silico methods based on the DFT framework that allow the computation (together with the application of other techniques from the classic and statistical mechanics) of the physicochemical properties of NMs, properties that are necessary descriptors for QSAR and QNPR models to correlate said properties with the toxicology evaluation or for other correlation goals.

DFT was developed from first principles and states that one can obtain the energy of an electron system if the exact electron density is known. One only needs the crystal structure and the chemical composition for the input of the model. However, the evaluation of the electron density and, in particular, the XC term of the Kohn–Sham equation must be approximated. Depending on the complexity of the XC functional, several levels of theory exist, forming the so-called Jacob's ladder of DFT. The simplest level is the LDA, which is suitable for some solid properties but fails for most molecular structures due to the more complex bonding structure. The next level in theory, GGA improves the description of bond energies but still fails in terms of bandgaps. The next level in theory, the meta-GGA does not suppose a large advance with respect to GGA in terms of quality of results and there is an increase in computational time with respect to GGA. The next level of Jacob's ladder improves the description of the electronic structure and provides good bandgaps and bond energies but is more expensive computationally and difficult to converge in solids. Finally, the last level is using the exact functional, but it is unknown for most systems and only available for very small molecules. That being said, DFT is a very successful methodology, which gives very accurate results, but with some limitations that one must be aware of.

New methodologies based on DFT have been developed to improve the capacities of DFT methods and overcome some of the limitations present in classic DFT methods. To correct the underestimated values of bandgaps due to the overdelocalization of d and f electrons on transition metals, one can force the localization of those ones using the DFT+*U* method. Equally, *GW* also corrects this problem with the bandgap, but it goes beyond DFT+*U*; it also provides proper transition energies of the quasi particle, and therefore one can compute with high accuracy ionization potentials or electron affinities. However, this is an expensive computational method. Another alternative to compute the electronic transitions and reproduce the spectrum of materials is the use of TD-DFT, where a time-dependent potential is applied to the KS formalism and one can obtain the excitations in the electronic spectra.

To increase the size of the systems to compute, as well as perform long molecular dynamics simulations, two different methods have been presented: DFTB and LS-DFT. DFTB is based on the TB expansion of the electron density obtained from DFT; this is a relatively cheap and accurate method, which can deal with complex reactions (although the repulsion between the atoms must be parametrized and not all of them are available at the moment).

LS-DFT is a method that makes use of the properties of sparse algebra. It is a very powerful method, and one can compute systems with hundreds of thousands of atoms, though this method is not suitable for metallic materials.

Finally, we have given a brief overview of the effect of solvents on the electronic simulations. We have shown that COSMOs and SM*x* models can predict the effect of solvents on solute materials or molecules, including a continuous medium that modifies the electron properties of the molecules. Nevertheless, since at the atomic scale, the continuous model is a rough approximation, it is also appropriate to consider explicitly the solvent molecules of the first and second solvation shells, when possible.

Therefore, in this chapter we have given an overview of the electronic structure calculation methods available for NMs for the in silico prediction of the key physicochemical properties.

Acknowledgments

This work was supported by the project NanoDesk (SOE1/P1/E0215) and *Generalitat de Catalunya* (2014 SGR 1352; 2017 SGR 623). RR acknowledges the support provided by the Laboratory Directed Research and Development Program at the Pacific Northwest National Laboratory, a multiprogram national laboratory operated by Battelle for the US Department of Energy.

References

1. Aradi, B., Hourahine, B. and Frauenheim, T. (2007). DFTB+, a sparse matrix-based implementation of the DFTB method, *J. Phys. Chem. A*, **111**(26), 5678–5684.
2. Aryasetiawan, F. and Gunnarsson, O. (1997). The GW method, *Rep. Prog. Phys.*, **61**, 237.
3. Bajaj, S., Sevik, C., Cagin, T. Garay, A., Turchi, P. E. A. and Arróyave, R. (2012). On the limitations of the DFT+U approach to energetics of actinides, *Comput. Mater. Sci.*, **58**, 48–56.
4. Becke, A. D. (1988). Density-functional exchange-energy approximations with correct asymptotic behaviour, *Phys. Rev. A*, **38**, 3098.

5. Bryantsev, V. S., Diallo, M. S. and Goddard III, W. A. (2008). Calculation of solvation free energies of charged solutes using mixed cluster/continuum models, *J. Phys. Chem. B*, **112**, 9709–9719.
6. Carlo, A. and Barone, V. (1999). Toward reliable density functional methods without adjustable parameters: the PBE0 model, *J. Chem. Phys.*, **98**(7), 5648–5652.
7. Cramer, C. J. and Truhlar, D. G. (2006). SMx continuum models for condensed phases, in *Trends and Perspectives in Modern Computational Science*, Maroulis, G. and Simos, T. E., eds. (Brill/VSP, Leiden), pp. 112–139.
8. Choi, T. H., Liang, R., Maupin, C. M. and Voth, G. A. (2013). Application of the SCC-DFTB method to hydroxide water clusters and aqueous hydroxide solutions, *J. Phys. Chem. B*, **117**, 5165–5179.
9. Coccocioni, M. (2012). The LDA+U approach: a simple hubbard correction for correlated ground states, in *Correlated Electrons: From Models to Materials Modeling and Simulation*, Vol. 2, Pavarini, E., Koch, E., Anders, F. and JarrellLie, M., eds. (Forschungszentrum Jülich GmbH Institute for Advanced Simulation, Germany), Chapter 4.
10. Deeth, R. J. and Fey, N. (2004). The performance of nonhybrid density functionals for calculating the structures and spin states of Fe(II) and Fe(III) complexes, *J. Comput. Chem.*, **25**, 1840–1848.
11. Dreher, K. L. (2004). Comparative pulmonary toxicity assessment of single-wall carbon nanotubes in rats, *Toxicol. Sci.*, **77**, 3–5.
12. Elstner, M., Porezag, D., Jungnickel G., Elsner, J., Haugk, M., Frauenheim, T., Suhai, S. and Seifert, G. (1998). Self-consistent-charge density-functional tight-binding method for simulations of complex materials properties, *Phys. Rev. B*, **58**(11), 7260–7268.
13. Engel, E. and Dreizler, R. M. (2011). *Density Functional Theory: An Advanced Course* (Springer-Verlag Berlin Heidelberg).
14. Fabiano, E. (2013). Computational molecular photophysics, in *Handbook of Molecular Plasmonics*, Della Sala, F. and D'Agostrino, S., eds. (Pan Stransford Publishing, Singapore), Chapter 4, pp. 175–212.
15. Fouquau, A., Casida, M. E., Daku, L. M. L., Hause, A. and Neese, F. (2005). Dispersion interactions in density-functional theory: An adiabatic-connection analysis, *J. Chem. Phys.*, **122**, 044110–044123.
16. Gajewicz, A., Rasulev, B., Dinadayalane, T. C., Urbaszek, P., Puzyn, T., Leszczynska D. and Leszczynski, J. (2012). Advancing risk assessment of engineered nanomaterials: application of computational approaches, *Adv. Drug Delivery Rev.*, **64**, 1663–1693.

17. Guangde, T., Carravetta, V., Vahtras, O. and Agren, H. (2007). Core ionization potentials from self-interaction corrected Kohn-Sham orbital energies, *J. Chem. Phys.*, **127**, 174110.
18. Hasnip, P. J., Refson, K., Probert, M. I. J., Yates, J. R. Clark, S. J. and Pickard, C. J. (2014). Density functional theory in the solid state, *Phyl. Trans. R. Soc. A*, **372**, 20130270.
19. Held, K., Taranto, C., Rohringer, G. and Toschi, A. (2011). Hedin equations, GW, GW+DMFT, and all that, in *The LDA-DMFT Approach to Strongly Correlated Materilas Modeling and Simulations*, Vol. 1, Pavanini, E., Koch, E. Vollhardt, D. and Lichtenstein, A., eds. (Foschungzentrum Jülich, Germany), Chapter 13.
20. Heyd, J., Scuseria, G. E. and Ernzerhof, M. (2003). Hybrid functionals based on a screened Coulumb potential, *J. Chem. Phys.*, **118**(18), 8207.
21. Himmetoglu, B., Floris, A., de Gironcoli, S. and Cococcioni, M. (2014). Hubbard-corrected DFT energy functionals: the LDA+U description of correlated systems, *Int. J. Quant. Chem.*, **114**, 14–49.
22. Hine, N. D. M., Haynes, P. D., Mostofi, A. A., Skylaris, C.-K. and Payne, M. C. (2009). Linear-scaling density-functional theory with tens of thousands of atoms: expanding the scope and scale of calulations with ONETEP, *Comput. Phys. Commun.*, **180**, 1041–1053.
23. Hohenberg, P. and Kohn, W. (1964). Inhomogeneous electron gas, *Phys. Rev. B*, **136**, B864.
24. Khaliullin, R. Z., VandeVondele, J. and Hutter, J. (2013). Efficient linear-scaling density functional theory for molecular systems, *J. Chem. Theory Comput.*, **9**, 4421–4427.
25. Khon, W. and Sham, L. J. (1965). Self-consistent equations including exchange and correlation effects, *Phys. Rev. A*, **140**, A1133.
26. Klamt, A. (2005). *COSMO-RS: From Quantum Chemistry to Fluid Phase Thermodynamics and Drug Design* (Elsevier, Amsterdam, The Netherlands).
27. Klamt, A., Mennucci, B., Tomasi, J., Barone, V., Curutcher, C., Orozco, M. and Luque, F. J. (2009). On the performance of continuum solvation methods. A comment on "Universal Approaches to Solvation Modelling, *Acc. Chem. Res.*, **42**, 489–492.
28. Klamt, A. and Schüürmann (1993). COSMO: a new approach to dielectric screening in solvents with explicit expressions for the screening energy and its gradient, *J. Chem. Soc. Perkin Tans.*, **2**, 799–805.
29. Koskinen, P. and Mäkinen, V. (2009). Density-functional tight-binding for beginners, *Comput. Mater. Sci.*, **47**(1), 237–253.

30. Langreth, D. C. and Mehl, M. J. (1983). Beyond the local-density approximation in calculations of ground-state electronic properties, *Phys. Rev. B*, **28**, 1809–1834.

31. Lee, C. T., Yang, W. T. and Parr, R. G. (1988). Development of the Colle-Salvetti correlation-energy formula into a functional of the electron density, *Phys. Rev. B*, **37**, 785–789.

32. Liechtenstein, A. I., Anisimov, V. I. and Zaanen, J. (1995). Denisty-functional theory and strong interactions: orbital ordering in Mott-Hubbard insulators, *Phys. Rev. B*, **52**, R5467–R5470.

33. Loschen, C., Carrasco, J., Neyman, K. M. and Illas, F. (2007). Fist-principles LDA+U and GGA+U study of cerium oxides: dependence on the effective U parameter. *Phys. Rev. B*, **75**, 035115.

34. Marques, M. A. L., Ullrich, C., Nogueira, F., Rubio, A., Burke, K. and Gross, E. K. U. (2011) *Time-Dependent Density Functional Theory*, 2nd ed. (Springer-Verlag Berlin, Germany).

35. Nature Editorial (2009). 'Plenty of room' revisited, *Nat. Nanotechnol.*, **4**, 781.

36. Organisation for the Economic Cooperation and Development (2016). Physical-chemical parameters: measurements and methods relevant for the regulation of nanomaterials, OECD Workshop Report, 63, Env/JM/MONO(2016)2.

37. Papaconstantopoulos, D. A. and Mehl, M. J. (2003). The Slater-Koster tight-binding method: a computationally efficient and accurate approach, *J. Phys.: Condens. Matter*, **15**, R413–R440.

38. Perdew, J. P., Burke, K. and Ernzerhof, M. (1996). Generalized gradient approximation made simple, *Phys. Rev. Lett.*, **77**, 3865.

39. Perdew, J. P., Chevary, J. A., Vosko, S. H., Jackson, K. A., Pederson, M., Singh, D. J. and Fiolhais (1992). Atoms, molecules, solids and surfaces: applications of the generalized gradient approximation for exchange and correlation, *Phys. Rev. B*, **46**, 6671.

40. Perdew, J. P. and Constantin, L. A. (2006). Laplacian-level density functionals for the kinetic energy density and exchange-correlation energy, *Phys. Rev. B*, **75**, 155109.

41. Perdew, J. P., Kurth, S., Zupan, A. and Blaha P. (1999). Accurate density functional with correct formal properties: beyond the generalized gradient approximation, *Phys. Rev. Lett.*, **82**, 2544.

42. Perdew, J. P. and Ruzsinszky, A. (2010). Fourteen easy lessons in density functional theory, *Int. J. Quantum Chem.*, **110**, 2801–2807.

43. Perdew, J. P., Ruzsinszky, A., Tao, J., Staroverov, V. N., Scuseria, E., and Csonka, G. I. (2005). Prescription for the design and selection of

density functional approximations: more constraint satisfaction with fewer fits, *J. Chem. Phys.*, **123**, 062201-9.

44. Puzyn, T., Rasulev, B., Gajewicz, A., Hu, X., Dasari, T. P., Michalkova, A., Hwang, H.-M., Toropov, A., Leszczynska, D. and Leszczynski, J. (2011). Using nano-QSAR to predict the cytotoxicity of metal oxide nanoparticles, *Nat. Nanotechnol.*, **6**, 175–178.

45. Richarz, A.-N., Madden, J. C., Marchese Robinson, R. L., Lubinski, L., Mokshina, E., Urbaszek, P., Kuzmin, V. E., Puzyn, T. and Cronin, M. T. D. (2015). Development of computational models for the prediction of the toxicity of materials, *Perspect. Sci.*, **3**, 27–29.

46. Riediker, M. and Katalagarianakis, G. (2012). *Compendium of Projects in the European Nanosafety Cluster* (Institute for Work and Health, Lausanne, Switzerland).

47. Roldan, A. (2010). Modelling Au catalyst from bulk to nanoscale, PhD Thesis (Universitat Rovira i Virgili, Tarragona).

48. Runge, E. and Gross, E. K. U. (1984). Density-functional theory for time-depnedent systems, *Phys. Rev. Lett.*, **52**, 997–1000.

49. Schmidt, T. and Kümmel, S. (2016). The influence of one-electron self-interaction on d-electrons, *Computation*, **4**(3), 33.

50. Shimojo, F., Kalia, R., Nakano, A. and Vashishta, P. (2008). Divide-and-conquer density functional theory on hierarchical real-space grids: parallel implementation and applications, *Phys. Rev. B*, **77**, 085103.

51. Shultz, N. E., Zhao, Y. and Truhlar, D. G. (2005). Density functionals for inorganometallic and organometallic chemistry, *J. Phys. Chem. A*, **109**, 11127–11143.

52. Siefert, G. (2007). Tight-binding density functional theory: an approximate Kohn-Sham DFT scheme, *J. Phys. Chem. A*, **111**(26), 5609–5613.

53. Sousa, S. F., Fernandes, P. A. and Ramos, M. J. (2007). General performance of density functional, *J. Phys. Chem. A*, **111**, 10439–10452.

54. Sun, J., Remsing, R. C, Zhang, Y., Sun, Z., Ruzsinszky, A., Peng, H., Yang, Z., Paul, A., Waghmare, U., Wu, X., Klein, M. L. and Perdew, J. P. (2016). Accurate first-principles structures and energies of diversly bonded systems from an efficient density functional, *Nat. Chem.*, **8**, 831–836.

55. Tadmor, E. B. and Miller, R. E. (2011). Quantum mechanics of materials, in *Modeling Materials: Continuum, Atomistic and Multiscales Techniques* (Cambridge University Press, Croydon), Chapter 4, pp. 153–235.

56. Tao, J. and Perdew, J. P. (2005). Test of a non-empirical density functional: short-range part of the van-der-waals interaction in rare-gas dimmers, *J. Chem. Phys.*, **122**, 114102.

57. Tratnyek, P. G., Bylaska, E. J. and Weber, E. J. (2017). In silico environmental chemical science: properties and processes from statistical and computational modelling, *Environ. Sci. Processes Impacts*, **19**, 188–202.

58. Tschinke, V. and Ziegler, T. (1989). On the shape of spherically averaged Fermi-hole correlation functions in density functional theory. 1. Atomic systems, *Can. J. Chem.*, **67**(3), 460–472.

59. Ullrich, C. A. (2006). Time-dependent density-functional theory beyond the adiabatic approximation: Insights from a two electron model system, *J. Chem. Phys.*, **125**(23), 234108.

60. Ullrich, C. A. and Yang, Z. (2014). A brief compendium of time-dependent density-functional theory, *Braz. J. Phys.*, **44**, 154–188.

61. Van Setten, M. J., Weigend, F. and Evers, F. (2013). The GW-method for quantum chemistry applications: theory and implementations, *J. Chem. Theory Comput.*, **9**(1), 232–246.

62. Zanith, C. C. and Pliego Jr., J. R. (2015). Performance of the SMD and SM8 models for predicting solvation free energy of neutral solutes in methanol, dimethyl sulfoxide and acetonitrile, *J. Comput.-Aided Mol. Des.*, **29**, 217–224.

Chapter 4

Bionano Interactions: A Key to Mechanistic Understanding of Nanoparticle Toxicity

David Power,[a] Stefano Poggio,[a] Hender Lopez,[b] and Vladimir Lobaskin[a]

[a]*School of Physics, University College Dublin, Belfield, Dublin 4, Ireland*
[b]*School of Physics and Optometric & Clinical Sciences, Technological University Dublin, Kevin Street, Dublin 8, Ireland*
Vladimir.Lobaskin@ucd.ie

The new paradigm in the assessment of toxicity of nanomaterials relies on a mechanistic understanding of the organism's response to an exposure to foreign materials, from the initial, molecular-level interactions to signaling and regulatory cascades. Here, we present a methodology to quantify the essential interactions at the bionano interface, which can be used in combination with the adverse outcome pathway analysis to build mechanism-based predictive schemes for toxicity assessments. We introduce a set of new, advanced descriptors of the nanomaterials, which refer to their ability to bind biomolecules and trigger the pathways via the molecular initiating events.

Computational Nanotoxicology: Challenges and Perspectives
Edited by Agnieszka Gajewicz and Tomasz Puzyn

ISBN 978-981-4800-64-8 (Hardcover), 978-0-429-34137-3 (eBook)
www.jennystanford.com

4.1 Introduction

Predictive toxicology is now experiencing a transition from descriptive histopathological analyses to a data-rich science with a much greater focus on the understanding of biological mechanisms down to the molecular level. The mechanistic approach to toxicity assessment involves analysis of pathways based on particle tracking and on transcriptomics or metabolomics data, reflecting the activated system-level responses. In this paradigm, one assesses the possibility of initiating an adverse outcome pathway (AOP), which covers the evolution of a toxic process from its molecular initiating event (MIE) to downstream cascading key events (KEs), leading eventually to a pathology or adverse outcome. The pathways are recognized via biomarkers in specific bioassays developed to test each of these events.

In this mechanistic picture, AOPs are triggered and steered by molecular-level interactions, including those at the bionano interface, a nanoscale layer where biological fluids meet foreign materials. Quantitative understanding of the interactions and structure of the bionano interface is, therefore, crucial to our ability to predict the probability of an MIE for the specific AOP and to relate it to the nanomaterial's properties. For most nanomaterials studied until now, the standard physicochemical descriptors are not sufficiently informative to predict the outcome of their interactions with biomolecules and the likelihood of the MIEs. The size, charge, and chemical composition cannot immediately predict whether a particle can produce free radicals, bind to certain cell receptor, or penetrate the cell membrane. Therefore, the mechanistic paradigm demands the development of novel characterization techniques and new pathway-oriented descriptors.

4.2 Advanced Descriptors of the Bionano Interface

4.2.1 Protein Corona

The exposure of nanoparticles (NPs) to biological fluids leads to the formation of a protein layer on the surface of the NP, which is known as the protein corona. This is a central concept in the description of the bionano interface as it has been established that the NP-protein complex is what determines the bioactivity of the nanomaterial. NPs tens of nanometers in size can contain hundreds of different biomolecules, and an NP's structure can be quite complex and include two layers, known as hard corona and soft corona [1–5]. Irreversible (or at least strong) binding of proteins on the NP is associated with the concept of a "hard corona," whereas quick, reversible binding of proteins that have faster exchange rates are defined as a "soft corona." The list of proteins present in the corona depends, on the one hand, on the NP chemistry and reflects the bionano interactions. On the other hand, it also depends on the content of the biological fluid the NP is immersed in. The corona formed in blood is extremely sensitive to the patient's health state, that is, to the biological environment surrounding the NP, and cannot be predicted from the intrinsic properties of the NP alone [6, 7]. Nevertheless, while it is clear that the variability of the corona content is immense, one can hope to find regularities if not in the list of molecules as such then at least in the adsorbed proteins' statistics. Certain properties of proteins, such as the presence of charged or hydrophobic patches, aromatic residues, etc., may tend to increase the propensity of molecules to adsorb on specific materials. This observation led to the idea of the construction of nanomaterial fingerprints that would contain the essential relevant information about the interactions and thus discriminate between different materials [8]. These fingerprints appeared to be useful for the prediction of the biological action of a nanomaterial, namely the NP-cell association. Although the initial statistical quantitative structure–activity relationship (QSAR) model developed in Ref. [8] was heavy (64 parameters), with a more thorough analysis, it was possible to demonstrate that only a few of them are actually important [9]. For example, cell association of gold NPs correlated well with the sequence descriptors responsible for protein charge (such as basic, acidic, and aspartic amino acids [AAs] percentage) as well as with molecular weight and the propensity of the protein to aggregate. Despite that it is generally known that the abundance of any particular protein in the corona varies for different nanomaterials (see a review presented in Ref. [10]); the properties responsible for this variance have not been identified. We believe that a significant improvement in the predictive power of the bioinformatics-based models, as compared to the sequence-only descriptor sets, can be achieved with structure descriptors characterizing size, shape, and charge distribution on the protein, that is, the properties relevant for the interactions with NPs.

4.2.2 Nanoparticle Descriptors and QSARs

Although it is clear that the physicochemical properties of NPs determine their interactions with proteins in biological matrices (e.g., blood plasma and alveolar fluid) and with the immune cells [11], there are only a few known structure–activity relationships between the physicochemical properties of NPs and their effects on the immune system that lead to the most common types of immunotoxicity.

QSARs have been considered as a promising step in building the novel NP toxicity assessment strategies [12, 13]. Their predictive power has been demonstrated, for example, for metal oxide NPs, where the conduction band energy levels have shown to be correlated with their toxicological potential at cellular and systemic levels [14]. However, further progress in the construction of predictive QSARs is seen in connection to the mechanism of action of nanomaterials, which depends on a provision of relevant descriptors based on the AOPs. Advanced material characterization that includes both intrinsic and extrinsic properties, such as hydrophobicity, protein adsorption affinity, dissolution rates, and the ability to generate reactive oxygen species, is necessary to elucidate the molecular-level mechanisms of toxicity, to identify the properties of concern and thus provide a basis for material grouping and read-across techniques, which can be used by regulators.

Some descriptors can be derived directly from the physicochemical properties of the NPs, such as bandgap energy (related to the probability of electron transfer and catalytic activity of the material), heat of formation of an NP, energies of the highest occupied and lowest unoccupied molecular orbitals (HOMO and LUMO), total energy, solvent accessible surface area, dipole moment, molecular weight, polarizability, and dielectric constant [15]. But the activity of an NP can be modified by the medium, which can change its surface charge and surface reactivity, and in the case of biological fluids leads to the formation of a protein corona. Therefore, a set of extrinsic parameters should complement the description. So, to achieve the goal of formulating descriptors these should be based on not only properties of the material but also extrinsic properties of the NPs in exposure conditions and properties of relevant biomolecules present in the medium, which together are responsible for the interactions and the structure of the bionano interface.

4.2.3 Biomolecule Descriptors

In contrast to NPs, the development of universal descriptors for biomolecules is relatively straightforward due to their chemical uniformity, for example, the same AAs present in all proteins or nucleic acids in all DNA. For proteins, the simplest descriptors can be constructed using their AA sequence. These can include counts of AAs of different types, net charge, or total mass. Already this characterization is very rich and capable of predicting complex events at the bionano interface [8, 9]. Moreover, descriptors can be obtained from AA sequences by using a wide range of software tools, such as the EMBOSS PepStats tool [16]. More advanced descriptors for proteins can be built by analyzing their structure. In some cases, starting with the AA sequence of the protein, the 3D structure of the molecule can be retrieved from the Protein Data Bank (PDB) and then used to construct the descriptors. When the structure is not available, one can then use structure prediction software. There are multiple automated tools available for this task, such as i-Tasser [17]. Using the measured or predicted 3D structure of the protein, several advanced descriptors can be calculated. We previously developed a one-bead-per-AA (united atom [UA]) model of globular proteins, which is suitable for this purpose. Some examples of advanced descriptors that can be calculated include protein globule dimensions (radius of gyration and hydrodynamic radius), aspect ratio, dipole moment, rotational inertia, dielectric constant, hydrophobicity, and surface charge at different pH levels and salt concentrations. In addition, protein charge at different pH levels can be calculated using the Poisson–Boltzmann cell model with charge regulation as reported by da Silva et al. [18].

4.2.4 Interaction Descriptors

With the known 3D structure of the protein and the nanomaterial, bionano interaction descriptors can be systematically calculated on the basis of how the proteins adsorb onto the surface of the NPs. While a calculation of the precise conformation of adsorbed molecules and a careful evaluation of ensemble averages are definitely challenging tasks, several relevant quantities can be calculated using a simplified approach. We here make two major approximations: assume additivity of the interactions between the building blocks of the biomolecule and the NP and neglect the change of conformation for adsorbed molecules. While these assumptions prevent us from obtaining accurate adsorption energies, they create a possibility of a uniform screening of thousands of molecules and ranking them based on how strongly they will attach to the surface of the NP. This ranking is a statistical measure of the content of the biomolecular corona and constitutes a unique fingerprint of an NP. Using the UA model [19], one can compute preferred adsorbed orientation and evaluate the mean adsorption energy at different conditions. Moreover, using the same bottom-up construction approach, one can engineer an ultra-coarse-grained model (united AA [UAA]) that closely reproduces the total protein-protein pairwise interaction energy profiles obtained in the UA model. In the UAA model, one typically needs between 5 and 30 UAA beads to capture the geometry and reproduce the adsorption characteristics of the original protein. This second coarse-graining can be based on the mass distribution in the complete protein and can be optimized by tuning the protein diffusion coefficients to those obtained using the UA model. The UAA model is then suitable for modeling competitive protein adsorption and formation of protein corona. An example of the all-atom, UA, and UAA coarse-grained (CG) models for the same protein is shown in Fig. 4.1.

4.3 Multiscale Modeling of the Bionano Interface

4.3.1 General Methodology

To model the bionano interface and generate descriptors to predict the toxicity of NPs with molecular simulations one must address length scales of up to hundred nanometers and timescales of at least microseconds. Despite the growth of the computational power in the last decade, these length and time scales are not achievable with atomistic detailed simulations, and for this reason CG models must be used. Coarse-graining of a molecule consists of reducing the number of degrees of freedom (DOFs) by representing the molecule with CG particles (often called "beads") instead of one particle per atom in the molecule. This reduced representation of the structures of the molecules in the simulation allows one to perform simulations with longer length and time scales than with detailed atomistic models. The main challenge in the development of the CG representation of the molecules and the interactions between the segments of the system is how to reduce the DOF, while ensuring that the reduced representation keeps enough detail to capture the relevant physicochemical processes that occur at the bionano interface. For the task of building CG models there are two main approaches: the top-down approach, in which parameters from macroscopic experimental data are used to calibrate the models [20–22] and the bottom-up approach, in which atomic simulations are used as input to generate the CG structures and force fields [9, 23–25]. We refer the reader to the paper by Noid [26] for an overview of these CG techniques.

In the case of bionano interactions, top-down approaches have been used to study the formation of the NP-biomolecule corona [19, 27, 28], but regardless of their success, the parameterization process relies on the availability of experimental data such as adsorption energies or affinity constants for any protein-surface interaction to be simulated. For practical applications, this constraint is certainly the main drawback as, for example, a real NP-protein corona can be composed of hundreds of different proteins. This disadvantage can be overcome by a systematic bottom-up CG strategy in which the basic building blocks of the model chosen are general enough that they can then be used to construct a wide range of molecules. In the case of modeling proteins, the obvious choice is to select the CG beads to be AA. Then, using atomistic simulations of the interaction of AAs with a surface of interest and techniques such as iterative Boltzmann inversion [29, 30] or force matching [21, 31, 32], the AA-surface interactions can be parameterized. In this way, the obtained AA-surface CG force fields can be used to model the interaction of any protein present if the structure or sequence of the molecule is known.

4.3.2 Coarse-Grained Protein Model

As mentioned previously, once an NP is in contact with a biological medium, a protein corona forms on its surface [33] and the nature of the corona is what regulates the interaction between the NP and the other biomolecules. It has been extremely challenging to develop a model that could predict the composition of the protein corona around an inorganic NP, as this depends on a multitude of physicochemical properties of both the protein and the NP, such as NP size, shape, pH, hydrophilicity/hydrophobicity, and electrostatic effects.

Computer simulations of the interactions of NPs with proteins can offer great support to experiments because of their great speed and flexibility [34]. Full-atomistic simulations have already proven to be a valuable tool in elucidating the binding mechanisms of proteins on metallic NPs [35–37]. However, their performance is severely hindered by the inefficiency of simulating systems with large NPs due to the high number of pair interactions that need to be evaluated. To speed up the calculations, a cutoff of the order of a nanometer is often introduced, but this results in an underestimation of the adsorption energies of proteins on NPs, mainly because the cores of the NPs (especially at sizes of over 10 nm) contribute much to their mutual attraction. In this section, we describe a CG model of protein-NP interactions that overcomes most of the challenges in the inclusion of the bulk part of NPs in the interaction. In the final section, we test the model by simulating the adsorption of the most abundant protein in the blood plasma, human serum albumin (HSA), on TiO_2 NPs.

Our UA protein model [19], in which every AA in the protein is substituted by a single bead whose center is placed at the position of the α-carbon atom, is illustrated in Fig. 4.1. The AA beads in this model are connected by rigid harmonic springs if the distance between them is less than 0.7 nm. The springs take care of the molecule shape, so the conformation does not change during the simulation. The model preserves two main structural features that guide the binding mechanisms, that is, the overall shape of the protein, which is treated as a rigid body, and the overall charge. The super-CG UAA model that typically contains 5–20 beads, presenting a further simplification and allowing one to model competitive protein adsorption, is also shown in Fig. 4.1 along with the full-atom and UA presentations.

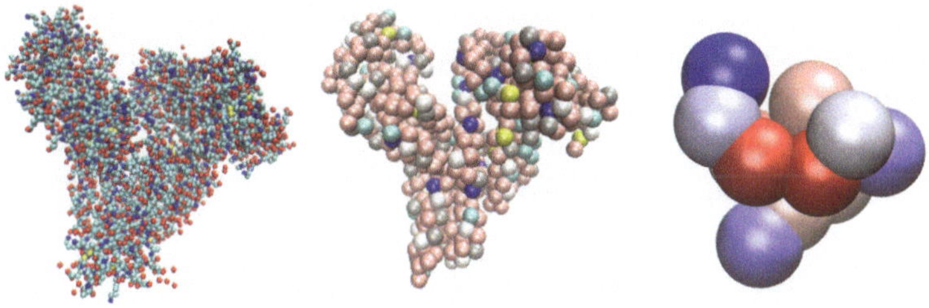

Figure 4.1 All-atomistic (left), UA (center), and UAA models (right) of HSA.

4.3.3 Coarse-Grained Nanoparticles

The protein model described above allows us to reduce the number of components in treating the protein. NP size, however, also plays a fundamental role in the formation of the corona and in the interactions at the bionano interface. The number of atoms needed to represent an NP is again a severe limitation to all atomistic calculations. Simulation of NPs of sizes greater than 10 nm is an unfeasible task even for modern computers, and a CG model for describing NPs is therefore highly needed.

Our model starts by considering the contributions that different atoms in the NP give to the binding interaction and proposes to partition the NP into two segments—a core and a surface segment. The outer layer on the NP surface is directly in contact with the solvent, and the pair interaction with the protein residues must include solvent effects as well as the chemical composition, charge, and hydrophilicity/hydrophobicity of the NP surface. Therefore, the interaction of each residue with the nearest part of the surface must be parameterized to reflect these details, using full-atomistic simulations. The size of the surface segment is thus determined by the cutoff r_c used in the full-atomistic simulation (typically 1 to 2 nm). Geometrically, the surface segment is a lens formed by an intersection of a sphere of radius r_c, centered on the AA bead, and a sphere of the radius of the NP, R, centered on the NP itself (Fig. 4.2). The core comprises the majority of the atoms, but these only interact with the protein via long-range forces, for which we assume that a continuum-level description is sufficient. The core of the NP is then modeled as a single bead of the shape of a sphere of radius R with a cut-out surface lens. The potential between the core and the AA beads in our model is calculated using the Lifshitz theory [38] for interaction between two macroscopic bodies. The overall interaction energy between the NP and the protein is then estimated by a sum over all AAs in the protein:

$$U(D,\theta,\phi)=\sum_{i=1}^{N_{AA}}U_i\left(h_i\left(D,\theta,\phi\right)\right) \tag{4.1}$$

where the energy contributions from individual AAs are evaluated as a sum of the interaction potentials with the core and surface of the NP:

$$U_i(h_i(D,\theta,\phi)) = (U_s)_i + (U_c)_i \tag{4.2}$$

Here, h_i, D, θ, and ϕ are the variables desribing the distance between the AA center of mass (COM) and the surface of the NP, the distance between the protein's COM and the NP's COM, and the orientation of the protein globule, respectively.

In the next two sections, we describe how the potentials are parameterized for the AA–NP interaction.

4.3.4 Generation of Surface Pair Potentials

Here, we assume pairwise additivity of AA-NP interactions. To calculate the adsorption energy of a whole protein, we need to determine the pair interactions for each AA type with the NP. The potentials of mean force (PMFs) for an AA-NP interaction must be first calculated at the atomistic level in order to capture the surface- and protein-specific details. These potentials can be obtained using the metadynamics method [39]. Our starting point is the PMF of an AA with a flat slab of the material of interest. Details of how the atomistic PMFs used in our model were calculated, as well as the parameters used to model the slabs, are described in Ref. [40]. Note that the PMFs obtained in this way include solvation effects.

Generally, the interaction with a convex surface of an NP of a finite radius is less than that for the flat slab due to the lesser number of atoms of the NP within the interaction cutoff distance. To account for this reduction, we correct the PMFs for the flat surface by a distance-dependent multiplicative function $f(h)$ that reflects also the cutoff radius r_c used in the calculations as well as the radius R of the NP.

$$U_s(h, R) = U_s(h, R \to \infty) f(h) \tag{4.3}$$

Here, $U_s(h, R)$ and $U_s(h, R \to \infty)$ are the PMFs for the curved and flat surfaces, respectively, and h is the minimum distance between the AA-bead center and the NP surface. By taking the appropriate limits for R, it is possible to calculate a correction factor for any geometry. A diagram showing how an AA bead interacts with a CG NP is shown in Fig. 4.2.

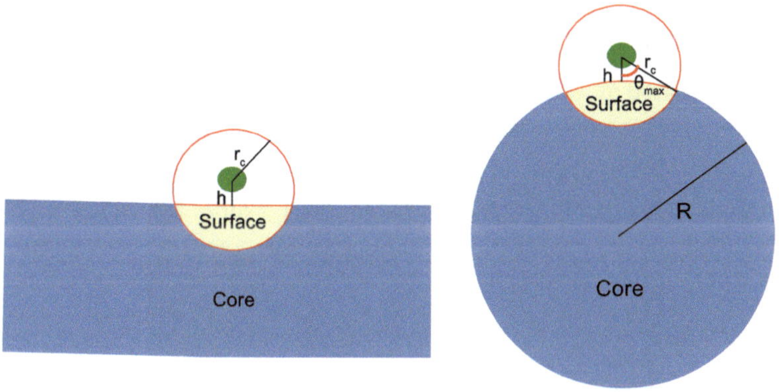

Figure 4.2 Schematic representation of an AA bead (green sphere) interacting with a slab (left) and with an NP of radius R (right) with a cutoff radius r_c. h is the minimum distance between the AA and the NP surface.

We assume that the relevant point-point interaction is dominated by dispersion forces that scale as r^{-6}, where r is the distance between the interaction centers (e.g., atoms). In this case, the attraction energy for a particle to a sphere of radius R, given the finite cutoff r_c, can be calculated via the following equations:

$$U_s(h,R) = \varepsilon V \int_h^{r_c} \int_0^{\theta_{\max}} \int_0^{2\pi} \frac{r^2 \sin\theta}{r^6} d\phi d\theta dr$$

$$\theta_{\max} = \cos^{-1}\left(\frac{r^2 - R^2 + (R+h)^2}{2r(R+h)}\right)_{\max}$$

$$U_s(h,R) = -\frac{\pi\varepsilon V}{h+R}\left(\frac{h-2R}{12h^3} + \frac{-6r_c^2 + 8r_c(h+R) - 3h(h+2R)}{12r_c^4}\right) \quad (4.4)$$

Here, ε is the interaction energy per unit volume. For $R \to \infty$, Eq. 4.4 reduces to that of a flat surface:

$$U_s(h, R \to \infty) = \pi\varepsilon V\left(\frac{1}{6h^3} - \frac{2}{3r_c^3} + \frac{h}{2r_c^4}\right) \quad (4.5)$$

and the correction factor is calculated as

$$f = \frac{U_s(h,R)}{U_s(h,\infty)} = -\frac{r_c^2(h-2R) + 2r_c h(h-2R) - 3h^2(h+2R)}{2(r_c^2 + 2r_c h + 3h^2)(h+R)} \quad (4.6)$$

In the equations above, V is the volume of the AA, which can be represented by a volume of equivalent sphere of radius R_2: $V = \frac{4\pi R_2^3}{3}$ reflecting the size of the respective AA.

Figure 4.3 shows how the volume correction factor changes with the distance from the AA to the surface for a set of NP radii. One can see that, as the radius increases approaching the flat surface limit, $f \to 1$ for all values of h.

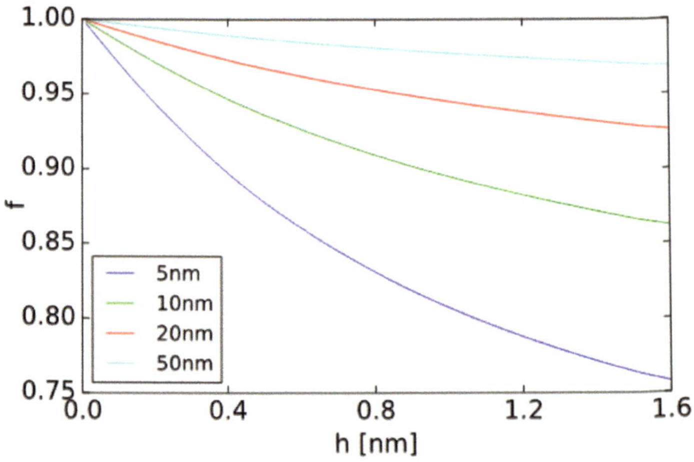

Figure 4.3 Correction factor f versus distance from the surface h for a range of NP radii.

4.3.5 Generation of the Core Potential

The NP core plays a crucial role in the protein adsorption as it contains most of the nanomaterial. A serious limitation of the all-atom models is the difficulty of a correct account for the attraction by the core atoms. This problem is mainly due to the short-range cutoff employed in simulations, leading to a considerable underestimation of the adsorption energies. The latter, however, can be easily calculated in the continuum approximation, which is commonly used in colloid science. The correction we propose in this report is to evaluate the contribution of the core of the NP at distances $r < r_c$ by treating the remote part of the NP as a single sphere less the part within the cutoff distance from the specific AA. The interaction energy between an AA and the core can be computed using the Hamaker method for dispersion forces [38]. For this, we assume that the spheres representing AAs (sphere 2) are small compared to those representing the NP core (sphere 1) and perform an integration over the volume of AAs first. Then, depending on the distance D between the sphere centers, the interaction energy becomes

$$U_C(D) = \begin{cases} -\dfrac{16A_{123}}{9}\dfrac{R_2^3R^3}{(D^2-R^2)^3}, & R+r_c < D \\ -\dfrac{4A_{123}R_2^3}{9}\left(\dfrac{4R^3}{(D^2-R^2)^3} - \dfrac{8Dr_c - 6r_c^2 - 3D^2 + 3R}{4Dr_c^4} - \dfrac{3R-D}{4D(R-D)^3}\right), & R < D < R+r_c \end{cases} \tag{4.7}$$

where the distance to the AA center exceeds the reach of the PMFs, $R + r_c$, the interaction is just the Hamaker potential for two spheres in the approximation $R_2 << R$, while where a part of the NP is covered by the PMF, the van der Waals force on the AA is calculated only for the part of the sphere beyond the cutoff distance.

In these expressions, A_{123} is the Hamaker constant for interaction between material 1 (e.g., the NP) and material 2 (e.g., the protein) through material 3 (e.g., water), the only material-dependent term in the equation. The Hamaker constant for the AA (phase 1)–material (phase 2) interactions through a medium (phase 3) can be obtained as follows:

$$A_{123} = \frac{3}{4} k_B T \frac{(\varepsilon_1 - \varepsilon_3)(\varepsilon_2 - \varepsilon_3)}{(\varepsilon_1 + \varepsilon_3)(\varepsilon_2 + \varepsilon_3)} + \frac{3hv_e}{8\sqrt{2}} \frac{\left(n_1^2 - n_3^2\right)\left(n_2^2 - n_3^2\right)}{\left(n_1^2 + n_3^2\right)^{\frac{1}{2}} \left(n_2^2 + n_3^2\right)^{\frac{1}{2}} \left(\left(n_1^2 + n_3^2\right)^{\frac{1}{2}} + \left(n_2^2 + n_3^2\right)^{\frac{1}{2}}\right)} \tag{4.8}$$

Here n_i are the refractive indices of the materials in the visible region, v_e is the main electronic absorption frequency in the UV (3×10^{15} s^{-1}) range for the materials, and the ε_i values refer to the dielectric constants of the materials, which are equal to n^2 in the visible part of the spectrum. If one of the materials is a conductor, Eq. 4.13 must be modified to take into account the high values of the polarizability and, therefore, of the dielectric constant. The equation for a dielectric-conductor interaction in a medium is given by

$$A_{123} = \frac{3}{8\sqrt{2}} \left(\frac{n_1^2 - n_3^2}{n_1^2 + n_3^2} \right) \frac{h\sqrt{v_1 v_3} \cdot v_2}{\sqrt{v_1 v_3} + \frac{v_2}{\sqrt{n_1^2 - n_3^2}}} \tag{4.9}$$

where v_i are the frequencies of maximum absorption for the material in the UV region; for metals, this frequency corresponds to the plasma frequency.

4.3.6 Calculation of the Adsorption Energy

To evaluate the average adsorption energy for a protein globule on an NP we scan the configurational space (i.e., all possible orientations in which a protein can get adsorbed on the surface of the NP) by systematic rotation of the protein and calculate the Boltzmann-averaged energy. There are three DOFs that must be scanned [19]. The orientation of the protein can be described by a vector from the COM to an arbitrary point of the molecule. It is characterized by two angles, ϕ and θ, and rotating the molecule by an angle $-\phi$ about the z direction and then by an angle $-\theta + 180°$ about the y axis will make the position vector point toward the surface (along the negative z axis). The third DOF is the distance from the COM to the closest point of the surface, d_{COM}. Instead of obtaining the actual adsorption-free energy by calculating the PMF for all orientations, we only calculate the potential energy U, which is the sum of all the pairwise interactions between the surface and the AAs. As the net adsorption energies are expected to be well over $k_B T$ and as the proteins are assumed to be rigid, neglecting thermal fluctuations is justified. For each configuration (ϕ_i, θ_j), the total potential energy is calculated as a function of the distance of the COM, $U(D, \phi_i, \theta_i)$, to the surface for the case of a slab or to the center of the NP for the case of an NP. Following a similar approach as in Kokh et al. [41] with denoting the reaction coordinate $D = z$, the mean interaction energy for any particular orientation for the case of a protein interacting with a spherical NP is given by

$$E\left(\phi_i,\theta_j\right)= -k_B T \ln\left[\frac{3}{\left(R+a\left(\phi_i,\theta_j\right)\right)^3 - R^3}\int_R^{R+a\left(\phi_i,\theta_j\right)} D^2 \exp\left(-\frac{U\left(D,\phi_i,\theta_j\right)}{k_B T}\right) dD\right] \tag{4.10}$$

$$E_{ad} = \frac{\sum_i \sum_j P_{ij} E(\phi_i,\theta_j)}{\sum_i \sum_j P_{ij}} \tag{4.11}$$

$E(\phi_i, \theta_j)$ is the total adsorption energy for a fixed set of angles (ϕ_i, θ_i). P_{ij} in Eq. 4.11 is given by

$$P_{ij} = \sin\theta_j \exp\left(-\frac{E\left(\phi_i,\theta_j\right)}{k_B T}\right) \tag{4.12}$$

The term a (ϕ_i, θ_i) in Eq. 4.10 is the maximum interaction distance from the COM of the protein to the surface for the given orientation.

4.3.7 From United-Atom to United–Amino Acid Description

As mentioned above, to study the formation of a realistic protein corona a more aggressive coarse-graining approach is needed, which we defined as the UAA model. To parameterize the UAA beads, the data output from the UA model described in the previous section is used as a reference. The required output from this UA model consists of a 2D adsorption energy map for the average adsorption energies $E(\phi_i, \theta_i)$.

The generation of the UAA model is done in two steps: (i) a reduced representation of a protein is created on the basis of its all-atom 3D structure, and (ii) the effective bead-bead potentials are derived for the UAA beads and NP beads. In the first part, we used VMD Coarse-Grain Builder (CGB) to create the reduced representation. For this we chose the CGB shape-based method, where a neural network learning algorithm is used to determine the placement of neurons (or CG beads). The CG beads have masses correlated to the clusters of atoms that the beads are representing [41].

At the second stage, a pseudorandom piecewise potential, describing the interaction with the NP, is created for each of these beads. A population of these test potentials is generated, and a genetic algorithm is used to minimize the overall difference between the original CG adsorption map from the UA model and the new CG adsorption map. The genetic algorithm minimizes the function

$$S = \Sigma_{i,j}\left(E_{\text{trial}}\left(\phi_i,\theta_j\right) - E\left(\phi_i,\theta_j\right)\right)^2 \qquad (4.13)$$

Once our CG UAA solution had reached the optimum, we are left with a simplified model of the original protein that has the same (or, at least, similar) characteristic adsorption map as the original UA model.

4.4 Application of the Method

4.4.1 Protein Descriptors

To perform a simulation of adsorption, we use a 3D folded globular structure of the molecule. The 3D structures of the proteins can be obtained from the protein data bank or calculated using an automated predictor system. Here, we demonstrate how advanced descriptors can be calculated. For illustration, we take the most common plasma protein, HSA. The structure of the HSA protein used in this work corresponds to the Protein Data Bank ID 1N5U. We assume physiological conditions, which correspond to a monovalent electrolyte concentration of 100 mM and pH 7. Residue charges in these conditions are +*e* for LYS and ARG, –*e* for ASP and GLU, and +0.5*e* for HIS. The rest of the residues are neutral. Thus, the total charge of the HSA molecule is –6*e*. The resulting protein descriptors are reported in Table 4.1.

Table 4.1 Descriptors of HSA protein

$\frac{I_{zz}}{I_{yy}}$	$\frac{I_{yy}}{I_{zz}}$	$\frac{I_{zz}}{I_{zz}}$	Rotational inertia (g Å^2/mol)	Total charge (*e*)	Dipole moment, d_x (*e* Å)	Dipole moment, d_y (*e* Å)	Dipole moment, d_z (*e* Å)	Solvent accessible surface area (nm^2)
4.955	0.3233	0.6240	2.883×10^7	−6	386	128	98.7	315.9

In the above example, the three descriptors are the principal moments of inertia, obtained by calculating the three eigenvalues of the following inertial matrix:

$$\hat{\text{I}} = \begin{bmatrix} \sum_k m_k\left(y_k^2 + z_k^2\right) & \sum_k m_k x_k y_k & \sum_k m_k x_k z_k \\ \sum_k m_k x_k y_k & \sum_k m_k\left(x_k^2 + z_k^2\right) & \sum_k m_k y_k z_k \\ \sum_k m_k x_k y_z & \sum_k m_k y_k z_k & \sum_k m_k\left(x_k^2 + y_k^2\right) \end{bmatrix} \tag{4.14}$$

Here, m_k represents the masses of the AAs and (x_k, y_k, z_k) are the coordinates of their centers of mass. The eigenvectors now form a new reference system, which is protein specific and independent from the arbitrary axes used within each PDB file. It is along these axes that we calculate the dipole moments. The dipole moment is simply the sum of all the residue charges by the position vector from the origin along each axis:

$$\mathbf{d} = \sum_k q_k \mathbf{r}_k \tag{4.15}$$

The solvent accessible surface area is calculated using the Shrake–Rupley algorithm.

4.4.2 Bionano Interface Descriptors

The model has been validated for a set of common proteins in blood plasma on TiO_2 NPs of radii ranging from 2.5 nm to 640 nm; see Ref. [19] for details on the validation procedure. We sample protein orientations ϕ from 0° to 355° in steps of 5° and θ from 0° to 175° in steps of 5°. Table 4.2 summarizes the Hamaker constants for AA-TiO_2 interaction in water calculated from Eq. 4.8, the UV absorption frequencies for AA are taken from Ref. [43] and the refractive indices from the procedure described in Ref. [44], the refractive index of TiO_2 was taken to be 2.5.

Table 4.2 Hamaker constants for AA-TiO_2 interaction in water

ARG	HIS	LYS	ASP	GLU	SER	THR	ASN	GLN
5.36	5.87	4.65	5.87	5.22	5.53	4.69	5.74	5.44
CYS	GLY	PRO	ALA	VAL	ILE	LEU	MET	PHE
5.66	5.66	4.37	4.52	4.00	3.96	3.91	5.10	5.61
TYR	TRP							
5.06	6.62							

The potentials generated were then used to calculate the binding energies of a test protein, HSA (PDB code: 1N5U) on TiO_2 NPs of radii ranging from 2.5 nm to 640 nm, without surface charges. The binding energies were calculated using the ESPResSo package [45], as described in Section 4.3.6. To evaluate the energy for each binding orientation, the protein was systematically rotated along two angles (ϕ_i, θ_i) around the position vector that joined the protein's COM to the center of the NP. Energies were thus evaluated for 2592 different orientations. A more detailed account of the rotation procedure is given in Refs. [19] and [46].

In Table 4.3, we present a set of descriptors based on the AA–TiO_2 PMFs from Ref. [40].

Table 4.3 Mean adsorption energies for AA–TiO_2 interaction (uncharged flat surface) through water, calculated using Eq. 4.10. All values are reported in units of k_BT.

ALA	ARG	ASN	ASP	CYS	GLN	GLU
0.327	−0.864	−0.493	−0.197	0.145	−0.696	−0.193
GLY	HIS	ILE	LEU	LYS	MET	PHE
−0.024	−0.156	−0.127	−0.138	0.054	0.000	−0.169
PRO	SER	THR	TRP	TYR	VAL	
−0.040	−0.710	−0.234	−0.871	−1.220	0.107	

Figure 4.4 shows the binding energy for one particular orientation of HSA on 5 nm and 50 nm TiO_2 NPs. The contributions from the core and the surface have been separated.

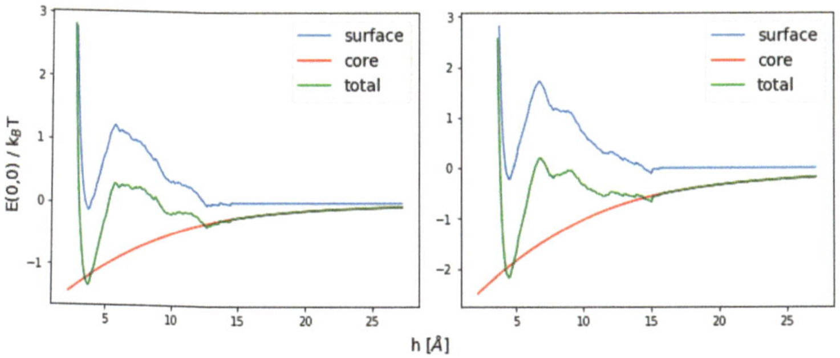

Figure 4.4 HSA E_{ad} profile on a 5 nm TiO_2 NP (left) and a 50 nm (right) TiO_2 NP. The protein orientation corresponds to the PDB configuration.

The contribution from the NP core is evident in both NP sizes. However, it increases as the NP radius is increased and for some orientations becomes the dominant contribution. We also see a much structured short-range part of the potential, which includes a contact minimum at about zero energy compared to the bulk, a maximum of $1k_BT$ – $2k_BT$ that corresponds to a hydration layer between the protein and the NP surface, and a much shallower secondary minimum.

Figure 4.5 shows a binding energy map ($E(\phi_i, \theta_j)$) for HSA on 5 nm and 50 nm TiO_2 particles. We see that the binding energy changes in magnitude with the NP size as the large NP produces deeper minima. Moreover, the map for the 50 nm particle shows more features: in addition to the minima at (230°, 40°) and (340°, 80°) for the 5 nm NP, we find a preferred orientation at (200°, 140°) and another minimum at (70°, 130°).

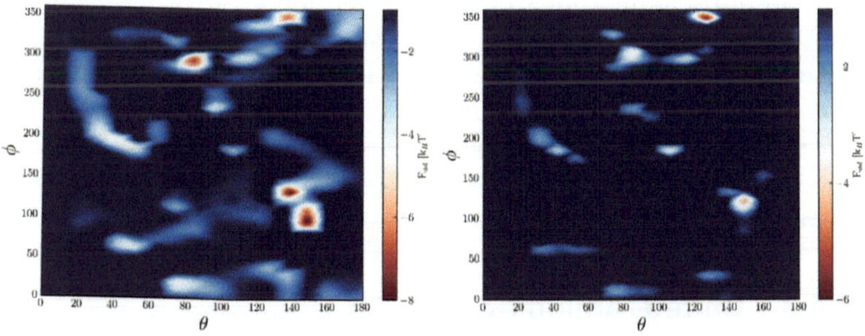

Figure 4.5 $E(\phi_i, \theta_j)$ map of HSA on the surface of a 5 nm TiO_2 NP (left) and a 50 nm TiO_2 NP (right).

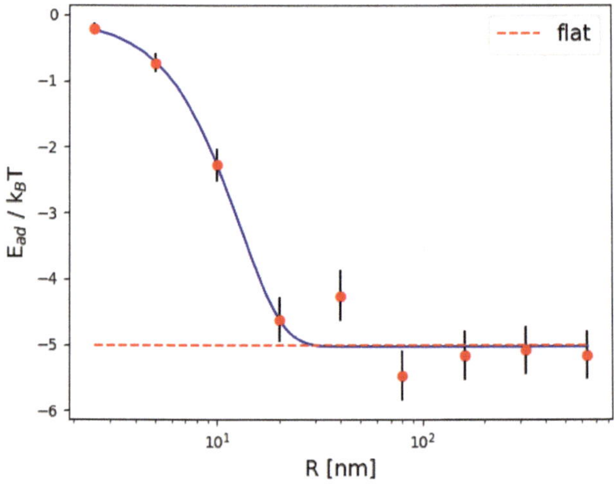

Figure 4.6 Average E_{ad} for HSA (left) on (100) rutile TiO_2 with an increasing NP radius. The dots represent a multiscale calculation according to Eq 4.11, and the red dashed line is for the flat surface ($R = \infty$). The blue solid line is a fit with $E_{ad} = -5.0\left(1.0 - e^{aR^2}\right)$.

The dependence of the adsorption energy of HSA on the NP size is shown in Fig. 4.6. The maximum binding energy is about $-5k_BT$, which corresponds to relatively weak adsorption. This means that the HSA molecules, if adsorbed, will be easily displaced by larger ones, according to the Vroman effect. As we can see from the data in Table 4.3, the contributions to the adsorption energy from most AAs are positive or slightly negative. The largest negative figures come from the charged ARG and large SER, TYR, and TRP. These observations are in agreement with results of direct atomistic and ab initio simulations of AA and peptide adsorption on titania surfaces, which confirm that the adsorption is driven by charged residues [47–49]. We also note two other important features of the size dependence of the adsorption energy: the saturation at large NP radii, such that limiting value corresponds to a flat surface, and the stronger binding at large radii due to the greater van der Waals attraction.

4.4.3 United–Amino Acid Model

The UAA model of the HSA protein, as obtained from our UA model using the procedure from Section 4.3.7, is described in Table 4.4.

Table 4.4 Ultra-coarse-grained (UAA) model for HSA protein as obtained from VMD CG builder

Bead	*X* position (nm)	*Y* position (nm)	*Z* position (nm)	Radius (nm)
Bead 1	5.8252	0.9764	2.1574	1.2905
Bead 2	4.0639	0.9955	1.2708	1.1097
Bead 3	0.2843	0.7511	4.7677	1.1667
Bead 4	1.1033	−0.4196	2.1635	1.1767
Bead 5	1.4744	0.407	−1.2432	1.1042
Bead 6	4.8305	−0.64	2.6579	1.1598
Bead 7	0.9128	1.3064	1.5903	1.1918
Bead 8	3.2678	1.2892	3.5817	1.1686
Bead 9	2.5456	0.0073	0.6177	1.1631
Bead 10	−0.2788	−0.2642	3.5709	1.1036
Bead 11	2.8617	1.859	0.0009	1.1923

The presentation we obtained contains 11 beads of a radius of about 1 nm each. This model is visualized in Fig. 4.1. The corresponding interaction potentials for the UAA blocks with a TiO_2 NP are shown in Fig. 4.7. All the obtained potentials are smooth, and most of them have a single minimum.

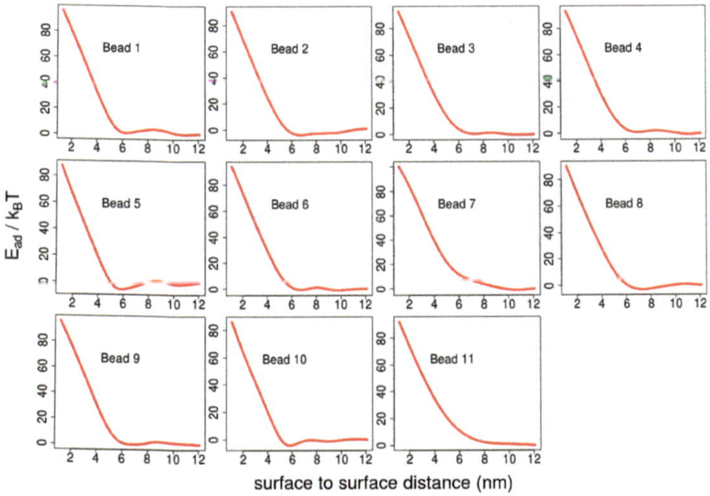

Figure 4.7 PMFs for beads 1–11 of the UAA presentation of HSA protein (see Table 4.3) with a TiO_2 nanoparticle of size 50 nm.

The UAA presentation preserves the shape of the protein and is sufficient to reproduce the detailed structure of the adsorption energy map. The energy maps obtained within UA and UAA presentations are shown in Figs. 4.8 and 4.9. A comparison of these shows that the UAA models capture all the main energy minima (corresponding to the preferred orientations of the protein on the NP surface): (230°, 40°) and (340°, 80°) for 5 nm NPs, although they become more diffuse (Fig. 4.8). For 50 nm NPs (Fig. 4.9), the structure of the energy surface in the UAA model is also more diffuse: the minima at (230°, 40°) and (150°, 50°) are now represented by a single extensive minimum spanning an area of 100° × 20°. Although the minima seen in the UAA presentation are more diffuse, they match the magnitude of those in the UA model and are ranked in the same order.

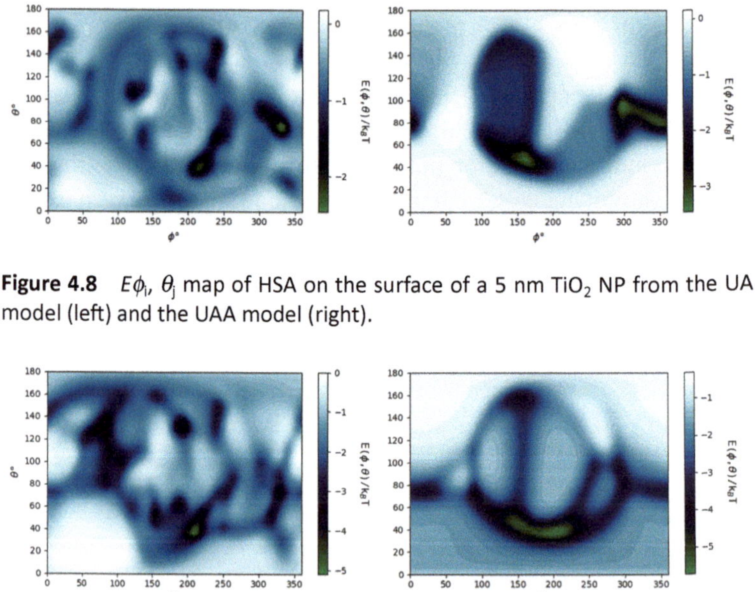

Figure 4.8 $E\phi_i$, θ_j map of HSA on the surface of a 5 nm TiO_2 NP from the UA model (left) and the UAA model (right).

Figure 4.9 $E\phi_i$, θ_j map of HSA on the surface of a 50 nm TiO_2 NP from the UA model (left) and the UAA model (right).

4.5 Conclusions

We presented a universal methodology for the evaluation of advanced descriptors of the bionano interface, which can be used to characterize the interaction of a specific NP with essential biomolecules constituting the protein corona of the NP. Although we made a number of strong approximations on the way, the main benefit we achieved is that the calculation can be applied in a uniform fashion to a large number of biomolecules in a short time and to add more descriptors to elucidate further the mechanism of corona formation and its dependence on the protein's and NP's physicochemical properties. The new descriptors include those for proteins (principal moments of inertia, charge, dipole moment, and the solvent accessible area) and those for interaction with NPs (Hamaker constants for residues, their mean adsorption energies, and the overall adsorption energy for the protein globule). In addition to this, our method delivers preferred orientations for proteins on a specific NP, thus systematically taking into account the NP shape and size, and allows one to rank arbitrary proteins by the adsorption affinity. Thus, dozens of essential descriptors can be routinely calculated. Another achievement of our methodology is that the core of the NP is included in the calculation, which makes the size dependence of the interaction more credible. These descriptors can be used to produce interaction fingerprints for arbitrary nanomaterials with respect to specified interactions and, therefore, provide key information for gauging their biological activity, for example, their ability to produce molecular initiating events (MIEs) or disturb the key events (KEs) of the adverse outcome pathways (AOPs).

Acknowledgments

This research was funded by EU FP7 Grant No. 310465 (MembraneNanoPart project) and EU H2020 Grant No. 686098 (SmartNanoTox project).

References

1. Cedervall, T., Lynch, I., Foy, M., Berggad, T., Donnelly, S., Cagney, G., Linse, S. and Dawson, K. (2007). Detailed identification of plasma proteins adsorbed on copolymer nanoparticles, *Angew. Chem. Int. Ed.*, **46**, 5754–5756.

2. Landsiedel, R., Ma-Hock, L., Kroll, A., Hahn, D., Schnekenburger, J., Wiench, K. and Wohlleben, W. (2010). Testing metal-oxide nanomaterials for human safety, *Adv. Mater.*, **22**, 2601–2627.

3. Lundqvist, M., Sethson, I. and Jonsson, B. H. (2004). Protein adsorption onto silica nanoparticles: conformational changes depend on the particles curvature and the protein stability, *Langmuir*, **20**, 10639–10647.

4. Lundqvist, M., Sethson, I. and Jonsson, B. H. (2005). Transient interaction with nanoparticles "freezes" a protein in an ensemble of metastable near-native conformations, *Biochemistry (Mosc)*, **44**, 10093–10099.

5. Karajanagi, S. S., Vertegel, A. A., Kane, R. S. and Dordick, J. S. (2004). Structure and function of enzymes adsorbed onto single-walled carbon nanotubes, *Langmuir*, **20**, 11594–11599.

6. Dobrovolskaia, M. A., Patri, A. K., Zheng, J., Clogston, J. D., Ayub, N., Aggarwal, P., Neun, B. W., Hall, J. B. and McNeil, S. E. (2009). Interaction of colloidal gold nanoparticles with human blood: effects on particle size and analysis of plasma protein binding profiles, *Nanomedicine*, **5**, 106–117.

7. Dobrovolskaia, M. A., Neun, B. W., Man, S., Ye, X., Hansen, M., Patri, A. K. Crist, R. M. (2014). Protein corona composition does not accurately predict hematocompatibility of colloidal gold nanoparticles, *Nanomed.: Nanotechnol. Biol. Med.*, **10**, 1453–1463.

8. Walkey, C. D., Olsen, J. B., Song, F., Liu, R., Guo, H., Olsen, D. W. H., Cohen, Y., Emili, A. and Chan, W. C. W. (2014). Protein corona fingerprinting predicts the cellular interaction of gold and silver nanoparticles, *ACS Nano*, **8**, 2439–2455.

9. Kamath, P., Fernandez, A., Giralt, F. and Rallo, R. (2015). Predicting cell association of surface-modified nanoparticles using protein corona structure - activity relationships (PCSAR), *Curr. Top. Med. Chem.*, **15**, 1930.

10. Saptarshi, S. R., Duschl, A. and Lopata, A. L. (2013). Interaction of nanoparticles with proteins: relation to bio-reactivity of the nanoparticle, *J. Nanobiotechnol.*, **11**, 26.

11. Dobrovolskaia, M. A., Shurin, M. and Shvedova, A. A. (2016). Current understanding of interactions between nanoparticles and the immune system, *Toxicol. App. Pharmacol.*, **299**, 78–89.

12. Puzyn, T., Rasulev, B., Gajewicz, A., Hu, X., Dasari, T. P., Michalkova, A., Hwang, H. M., Toropov, A., Leszczynsk, D. and Leszczynski, J.

(2011). Using nano-QSAR to predict the cytotoxicity of metal oxide nanoparticles, *Nat. Nanotechnol.*, **6**, 175–178.

13. Gajewicz, A., Schaeublin, N., Rasulev, B., Hussain, S., Leszczynska, D., Puzyn, T. and Leszczynski, J. (2015). Towards understanding mechanisms governing cytotoxicity of metal oxides nanoparticles: hints from nano-QSAR studies, *Nanotoxicology*, **9**, 313–325.
14. Zhang, H., Ji, Z., Xia, T., Meng, H., Low-Kam, C., Liu, R., Pokhrel, S., Lin, S., Wang, X., Liao, Y. P., Wang, M., Li, L., Rallo, R., Damoiseaux, R., Telesca, D., Mädler, L., Cohen, Y., Zink, J. I. and Nel, A. E. (2012). Use of metal oxide nanoparticle band gap to develop a predictive paradigm for oxidative stress and acute pulmonary inflammation, *ACS Nano*, **6**, 4349–4368.
15. EU US Roadmap Nanoinformatics 2030, doi: 10.5281/zenodo.1486011.
16. Rice, P., Longden, I. and Bleasby, A. (2000). EMBOSS: the European molecular biology open software suite, *Trends Genet.*, **16**, 276–277.
17. Roy, A., Kucukural, A. and Zhang, Y. (2010). I-TASSER: a unified platform for automated protein structure and function prediction, *Nat. Protoc.*, 725–738.
18. da Silva, F. L. B., Boström, M. and Persson, C. (2014). The effect of charge regulation and ion-dipole interactions on the selectivity of protein-nanoparticle binding, *Langmuir*, **30**, 4078–4083.
19. Power, D., Rouse, I., Poggio, S., Brandt, E., Lopez, H., Lyubartsev, A. and Lobaskin, V. (2019). A multiscale model of protein adsorption on a nanoparticle surface, *Modelling Simul. Mater. Sci. Eng.*, **27**, 084003.
20. Monopoli, M., Aberg, C., Salvati, A. and Dawson, K. A. (2012). Biomolecular coronas provide the biological identity of nanosized materials, *Nat. Nanotechnol.*, **7**, 779.
21. Lynch, I., Dawson, K. A. and Linse, S. (2006). Detecting cryptic epitopes created by nanoparticles, *Sci. Signaling*, **2006**, pe14.
22. Cedervall, T., Lynch, I., Lindman, S., Berggard, T., Thulin, E., Nilsson, H., Dawson, K. A. and Linse, S. (2007). Understanding the nanoparticle-protein corona using methods to quantify exchange rates and affinities of proteins for nanoparticles, *Proc. Natl. Acad. Sci. U. S. A.*, **104**, 2050.
23. Lindman, S., Lynch, I., Thulin, E., Nilsson, H., Dawson, K. A. and Linse, S. (2007). Systematic investigation of the thermodynamics of HSA adsorption to N-iso-propylacrylamide/N-tert-butylacrylamide copolymer nanoparticles. Effects of particle size and hydrophobicity, *Nano Lett.*, **7**, 914.
24. Allen, L., Tosetto, M., Miller, I., O'Connor, D., Penney, S., Lynch, I., Keenan, A., Pennington, S., Dawson, K. and Gallagher, W. (2006). Surface-

induced changes in protein adsorption and implications for cellular phenotypic responses to surface interaction, *Biomaterials*, **27**, 3096.

25. Salata, O. V. (2004). Applications of nanoparticles in biology and medicine, *J. Nanobiotechnol.*, **2**, 3.
26. Noid, W. G. (2013). Coarse-grained models for biomolecular systems, *J. Chem. Phys.*, **139**, 090901.
27. Vilaseca, P., Dawson, K. A. and Franzese, G. (2013). Understanding and modulating the competitive surface-adsorption of proteins through coarse-grained molecular dynamics simulations, *Soft Matter*, **9**, 6978–6985.
28. Vilanova, O., Mittag, J. J., Kelly, P. M., Milani, S., Dawson, K. A., Rädler, J. O. and Franzese, G. (2016). Understanding the kinetics of protein-nanoparticle corona formation, *ACS Nano*, **10**, 10842–10850.
29. Lyubartsev, A. P. and Laaksonen, A. (1995). Calculation of effective interaction potentials from radial distribution functions: a reverse Monte Carlo approach, *Phys. Rev. E*, **52**, 3730.
30. Reith, D., Pütz, M. and Müller-Plathe, F. (2003). Deriving effective mesoscale potentials from atomistic simulations, *J. Comput. Chem.*, **24**, 1624 -1636.
31. Izvekov, S. and Voth, G. A. (2005). A multiscale coarse-graining method for biomolecular systems, *J. Phys. Chem. B*, **109**, 2469–2473.
32. Noid, W. G., Chu, J. W., Ayton, G. S., Krishna, V., Izvekov, S., Voth, G. A., Das, A. and Andersen, H. C. (2008). The multiscale coarse-graining method. I. A rigorous bridge between atomistic and coarse-grained models, *J. Chem. Phys.*, **128**, 244114.
33. Rahman, M., Laurent, S., Tawil, N., Yahia, L. and Mahmoudi, M. (2013). *Protein-Nanoparticle Interactions*, Vol. 15, Springer Series in Biophysics (Springer-Verlag).
34. An, D., Su, J., Li, C. and Li, J. (2015). Computational studies on the interactions of nanomaterials with proteins and their impact, *Chin. Phys. B*, **24**, 120504.
35. Brancolini, G., Kokh, D. B., Calzolai, L., Wade, R. and Corni, S. (2012). Docking of ubiquitin to gold nanoparticles, *ACS Nano*, **6**, 9863–9878.
36. Ding, F., Radic, S., Chen, R., Chen, P., Geitner, N., Brown, J. and Ke, P. (2013). Direct observation of a single nanoparticle–ubiquitin corona formation, *Nanoscale*, **5**, 9162–9169.
37. Khan, S., Gupta, A. and Nandi, C. (2013). Controlling the fate of protein corona by tuning surface properties of nanoparticles, *J. Phys. Chem. Lett.*, **4**, 3747–3752.

38. Israelachvili, J. N. (2011). *Intermolecular and Surface Forces*, 3rd ed. (Academic Press), p. 704.

39. Duan, Y., Wu, C., Chowdhury, S., Lee, M. C., Xiong, G., Zhang, W., Yang, R., Cieplak, P., Luo, R., Lee, T., Caldwell, J., Wang, J. and Kollman, P. (2003). A point-charge force field for molecular mechanics simulations of proteins based on condensed-phase quantum mechanical calculations, *J. Comput. Chem.*, **24**, 1999–2012.

40. Brandt, E. and Lyubartsev, A. P. (2015). Molecular dynamics simulations of adsorption of amino acid side chain analogues and a titanium binding peptide on the TiO_2 (100) surface, *J. Phys. Chem. C*, **119**, 18126–18139.

41. Kokh, D., Corni, S., Winn, P., Hoefling, M., Gottschalk, K. and Wade, R. (2010). ProMetCS: an atomistic force field for modelling protein-metal surface interactions in a continuum aqueous solvent, *J. Chem. Theory Comput.*, **6**, 1753–1768.

42. Shih, A. Y., Freddolino, P. L., Arkhipov, A. and Schulten, K. (2007). Assembly of lipoprotein particles revealed by coarse-grained molecular dynamics simulations, *J. Struct. Biol.*, **157**, 579–592.

43. Glasel, J. A. (1995). Validity of nucleic acid purities monitored by a260/a280 absorbance ratios, *Biotechniques*, **18**, 62–63.

44. Zhao, H., Brown, P. H. and Schuck, P. (2011). On the distribution of protein ractive index increments, *Biophys. J.*, **100**, 2309–2317.

45. Limbach, H., Arnold, A., Mann, B. and Holm, C. (2006). ESPResSo - an extensible simulation package for research on soft matter systems, *Comput. Phys. Commun.*, **174**, 704.

46. Sun, Y., Welsh, W. and Latour, R. (2005). Prediction of the orientations of adsorbed protein using an empirical energy function with implicit solvation, *Langmuir*, **21**, 5616–5626.

47. Raffaini, G. and Ganazzoli, F. (2012). Molecular modelling of protein adsorption on the surface of titanium dioxide polymorphs, *Phil. Trans. R. Soc. A*, **370**, 1444–1462.

48. Polimeni, M., Petridis, L., Smith, J. C. and Arcangeli, C. (2017). Dynamics at a peptide–TiO_2 anatase (101) interface, *J. Phys. Chem. B*, **121**, 8869–8877.

49. Liu, S., Meng, X.-Y., Perez-Aguilar, J. M. and Zhoua, R. (2016). An in silico study of TiO_2 nanoparticles interaction with twenty standard amino acids in aqueous solution, *Sci Rep.*, **6**, 37761.

Chapter 5

From Modeling Nanoparticle–Membrane Interactions toward Nanotoxicology

Karandeep Singh,[a] Qingfen Yu,[a] Sabyasachi Dasgupta,[a,b] Gerhard Gompper,[a] and Thorsten Auth[a]
[a]*Theoretical Soft Matter and Biophysics, Institute of Complex Systems and Institute for Advanced Simulation, Forschungszentrum Jülich, 52425 Jülich, Germany*
[b]*Department of Physics, University of Toronto, 60 St. George St., Toronto, Ontario, M5S 1A7, Canada*
t.auth@fz-juelich.de

Biological cells are the basic building blocks of life. The cells' plasma membranes serve as communication interfaces with their environment, and they are also the first barriers against the interaction of the cells with nanomaterials. For nanoparticles larger than the thickness of the lipid bilayers, studies of wrapping by membranes are therefore an important step toward achieving a systematic understanding of nanotoxicology. We first introduce a continuum membrane model and focus on the roles of specific nanodescriptors for wrapping, such as particle size, shape, softness, and surface functionalization. We then review the effects

Computational Nanotoxicology: Challenges and Perspectives
Edited by Agnieszka Gajewicz and Tomasz Puzyn

ISBN 978-981-4800-64-8 (Hardcover), 978-0-429-34137-3 (eBook)
www.jennystanford.com

of cooperative interactions between particles, multicomponent biological membranes, membrane curvature prior to wrapping, and spontaneous membrane curvature on wrapping. Finally, we provide examples for experimental validation and discuss the implications of particle-wrapping predictions for nanotoxicology.

5.1 Particles at Membranes

The choice of the appropriate modeling approaches for particle and membranes depend mainly on the nanoparticle size and the desired chemical specificity of the results. For large nanoparticles, stable states of membrane-particle systems can be used as endpoints for quantitative nanostructure–activity relationship (QNAR) models.

5.1.1 Penetration vs. Wrapping

In general, two fundamentally different modes of nanoparticle-membrane interactions can be distinguished. For nanoparticles with sizes of the order of the thickness of the membrane, the interaction between a nanoparticle and a membrane takes place at the atomistic and molecular level [12]. Small nanoparticles can penetrate the lipid bilayer and directly translocate into the cytosol [16] or integrate themselves into the membrane [18]. In the latter case, the particles distort the lipid organization and cause enhanced penetrability or lysis, which has also been reported for peptides [19]. For nanoparticles larger than the thickness of the lipid bilayer, the membrane wraps around the particles [1, 20]. The required adhesion energy between membranes and particles can be due to van der Waals, electrostatic, or molecular interactions, such as receptor-ligand bonds. The particles that interact with the cell remain partially wrapped and bound to the membrane or get endocytosed. Here, the interaction with nanostructures can generate membrane tension and induce lysis [21] or stimulate reorganization of the cytoskeleton [22].

Figure 5.1 shows examples for both wrapping and penetration of ellipsoidal nanoparticles. In the example for penetration, the nanoparticle is pulled across the membrane by an external force. In the example for wrapping, the nanoparticle is enveloped because of an adhesion-energy gain between particle and membrane.

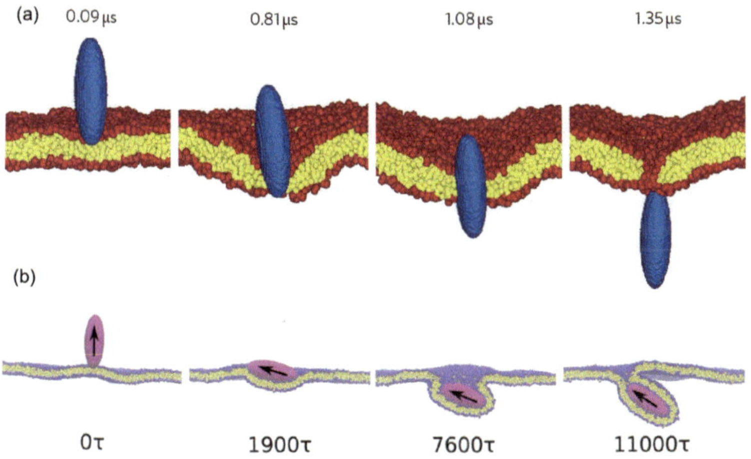

Figure 5.1 Molecular simulations for the interaction of prolate nanoparticles with membranes. (a) A nanoparticle penetrating a lipid-bilayer membrane aided by an external force. Reprinted by permission from Springer Nature Customer Service Centre GmbH: Springer Nature, *Nature Nanotechnology*, Ref. [5], Copyright (2010). (b) A nanoparticle getting wrapped by a lipid bilayer membrane aided by an attractive nanoparticle–membrane interaction. The arrow marks the direction of the long axis of the particle. Republished with permission of The Royal Society of Chemistry, from Ref. [13]; permission conveyed through Copyright Clearance Center, Inc.

5.1.2 Chemically Specific vs. Generic Models

Models for nanoparticle–membrane interactions can be chemically specific for a particular system or generic for systematic studies of specific descriptors, such as nanoparticle shape and size. Computational studies using a chemically specific model often involve atomistic or coarse-grained molecular dynamics simulations. In particular for systems with large nanoparticles, such calculations require a significant amount of computational resources. Therefore, systematic studies for large nanoparticles are usually not feasible using chemically specific approaches. Instead, generic approaches are required.

For penetration, generic models for lipid-bilayer membranes to study nanoparticle penetration and insertion can be based on self-consistent field theory for membranes [16, 24] or on a director field of the lipid molecules [19, 25]. For membrane wrapping, computer simulations that discretize a continuum membrane model are most appropriate because they provide the possibility to study wrapping for a wide variety of nanoparticle shapes [8, 26]. In meshless membrane models the scale of the discretization is comparable with the membrane thickness [23], whereas in triangulated membrane models single vertices can represent thousands of lipids [4, 27, 28]. Table 5.1 summarizes the main characteristics of chemically specific and continuum models, as well as the parameter regimes where these models are applicable.

Table 5.1 Computer simulation approaches

	Computational chemistry	Mesoscale simulations
Type of model	Chemically specific	Generic
Interactions	Atomistic or coarse grained	Coarse-grained or contact energy
Basic parametrization	Force field	Force field or elastic constants
Typical length scales	$\lesssim 15$ nm	15 nm to μm
Typical time scales	μs	ms to s

Note: For an all-atomistic membrane simulation, see Ref. [12]. For a coarse-grained membrane simulation, see Refs. [5, 13]. For simulations using continuum membrane models, see Refs. [4, 8, 23].

5.1.3 Nanoparticle-Wrapping Endpoints

For QNAR models, stable, minimal-energy states can be used as endpoints for nanoparticle–membrane systems: nonwrapped, partial-wrapped, and complete-wrapped particles. The systems can be most conveniently characterized using wrapping diagrams. Figure 5.2 shows the wrapping states for a rod-like nanoparticle whose shape is described by

$$\left(\frac{x^2+y^2}{a^2}\right)^{n/2}+\left(\frac{z}{b}\right)^n=1, \tag{5.1}$$

with $b/a = 1.5$, along with typical system snapshots. In the shallow-wrapped state, the particle is in submarine orientation, with its long axis parallel to the membrane, while in the deep-wrapped state, it is in rocket orientation, with its long axis perpendicular to the membrane. The wrapping transitions between the shallow-wrapped and deep-wrapped states and between the deep-wrapped and complete-wrapped states are discontinuous and separated by energy barriers that arise because the highly curved edges are wrapped. If the adhesion strength is sufficiently high, the energy barriers vanish and the particles are wrapped spontaneously; the thresholds for spontaneous wrapping are indicated by spinodals in the wrapping diagram.[1]

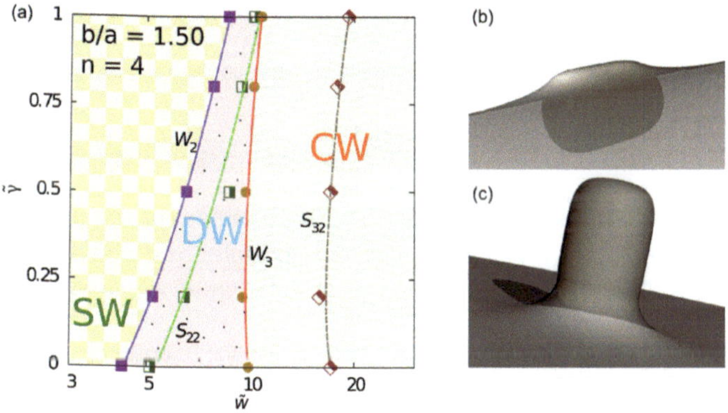

Figure 5.2 Wrapping diagram and typical states for rod-like particles with an aspect ratio $b/a = 1.5$ calculated using triangulated membranes. The particles are represented by Eq. 5.1 with $n = 4$. (a) The wrapping diagram shows stable shallow-wrapped (SW), deep-wrapped (DW), and complete-wrapped (CW) states for various reduced adhesion strengths $\tilde{w} = wA/(2\pi\kappa)$ and reduced membrane tensions $\tilde{\gamma} = \gamma a^2/\kappa$ where A is the surface area of the nanoparticles and a the length of their small axis. W_1 and W_2 indicate wrapping transitions, and S_{22} and S_{32} indicate spinodals where the corresponding energy barrier vanishes. (b, c) System snapshots for submarine and rocket states. Adapted with permission from Ref. [8]. Copyright (2014) American Chemical Society. (https://pubs.acs.org/doi/abs/10.1021/nl403949h). Further permissions related to the material excerpted should be directed to the ACS.

[1]Unwrapping spinodals are not shown in Fig. 5.2.

In vivo, detachment of complete-wrapped nanoparticles from the plasma membrane followed by endosomal release is required to deliver the particles to the cytosol. Here, they can directly interact with organelles, which may cause cytotoxic effects. For cellular uptake, endosomal release thus seems an appropriate choice for an endpoint. However, the steps beyond complete wrapping involve complex biological and physicochemical processes [29]. For example, lysosomal damage and release of the particles into the cytosol can be achieved using engineered nanoparticles with attached polymers that swell at typical endosomal pH values [30, 31]. An alternative mechanism that can lead to lysosomal damage is the interaction of pristine nanoparticles with the membranes of the lysosomes after degradation of their organic corona or their organic functionalization [32]. Because such multistep processes are very challenging for modeling and simulation studies, nanoparticle-wrapping states may thus be nevertheless the most appropriate endpoints for in silico studies.

5.2 The Helfrich Model

For length scales much larger than the thickness of the lipid-bilayer membrane, the membrane can be modeled as a mathematical surface. The deformation energy is calculated using the Helfrich Hamiltonian [33],

$$\mathcal{E}_{\text{def}} = \int_S \mathrm{d}S\left[\gamma + 2\kappa\left(H - c_0\right)^2 + \bar{\kappa}K\right], \tag{5.2}$$

where the integral is performed over the entire membrane area. Here, the mechanical properties of the membrane are characterized by the curvature-elastic parameters: membrane tension γ, bending rigidity κ, spontaneous curvature c_0, and Gaussian saddle-splay modulus $\bar{\kappa}$ The membrane conformation is described by the mean curvature $H = (c_1 + c_2)/2$ and the Gaussian curvature $K = c_1c_2$, where c_1 and c_2 are the principal curvatures.[2]

Many calculations for nanoparticle–membrane interactions focus on the bending-energy term only. The area of the lipid membrane

[2]The principal curvatures are the smallest and the largest curvatures at a point of the membrane.

is assumed to be fixed by the number of the lipid molecules. The spontaneous curvature vanishes for a perfectly symmetric lipid-bilayer membrane. Finally, the integral over the Gaussian curvature is constant if the topology of the system does not change, that is, until the final pinch-off of a wrapped nanoparticle; for homogeneous membranes this term does not contribute to the nanoparticle-membrane interaction.[3]

In addition to the deformation-energy cost, an adhesion-energy gain upon contact between nanoparticle and membrane determines nanoparticle wrapping [8, 20]. The adhesion energy is calculated as the integral over the membrane area S_{ad} that is in contact with the nanoparticle,

$$\mathcal{E}_{\mathrm{adh}} = -\int_{S_{\mathrm{ad}}} \mathrm{d}S\, w, \tag{5.3}$$

using a constant adhesion strength $w > 0$. The formation of protein coronas around nanoparticles can be taken into account by an effective adhesion strength [34–37].

5.3 Predicting Wrapping

Theoretical calculations based on generic models predict the wrapping behavior for various nanoparticle–membrane model systems. The characteristics of the particles that are studied in these calculations—and that can be used as nanoparticle descriptors in risk assessments—are size and surface functionalization, aspect ratio, local surface curvature, and deformability. In addition, we discuss cooperative interactions between nanoparticles, multicomponent biological membranes, and membrane curvature. Table 5.2 provides an overview of descriptors for nanoparticle–membrane systems: particle descriptors, membrane descriptors, and further system descriptors, such as particle dosage.

[3]The value of $\bar{\kappa}$ required to predict the pinch-off of a complete-wrapped nanoparticle from the membrane is often not well known. Furthermore, the also-not-well-known heights of the energy barriers between wrapping states may lead to very high stability of metastable partial-wrapped states. In biological systems, assisted processes, for example, involving dynein, are an outlook for extending computational studies for membrane model systems toward more biological systems.

Table 5.2 Descriptors for nanoparticle–membrane systems and their effect on wrapping for planar membranes

	Complete wrapping is facilitated by	**Membrane attachment is facilitated by**
Particle descriptors		
Particle size	Larger size[i]	Larger size[i]
Surface functionalization	Higher adhesion strength[ii]	Higher adhesion strength[iii]
Aspect ratio	Smaller aspect ratio	[Depends on local curvature only]
Local curvature	More homogeneous local curvature	Locally flat particle surface
Deformability	Less deformable particles	More deformable particles
Membrane descriptors		
Deformability	Floppier membranes	Floppier membranes
Composition	Higher adhesion strength[ii]	Higher adhesion strength[iii]
Spontaneous curvature	Spontaneous curvature toward the particle	Spontaneous curvature away from the particle
Cortical cytoskeleton	Wider cytoskeletal mesh size	Wider cytoskeletal mesh size
Other descriptors		
Particle dosage	Unknown	Unknown
Osmolalities	Small osmotic pressure difference between the interior and exterior of a cell/a vesicle	[Independent of osmotic pressure]
Assisted wrapping	Recruitment of favorably curved components, for example, clathrin	Recruitment of favorably curved components, for example, clathrin

[i]Preferred wrapping for specific nanoparticle curvatures has been observed experimentally [38] and is predicted to occur for receptor-ligand-bond-mediated adhesion [39] or for finite-membrane spontaneous curvature [40].

[ii]In addition to a homogeneous adhesion strength mediated by van der Waals interaction, receptor-ligand bonds or opposite electrical charges on particle and membrane can be modeled by an effective adhesion strength. However, for discrete binding sites, in addition to the strength of the attraction, the density and homogeneity of the surface coverage have to be taken into account [41].

[iii] For attachment, a sparse ligand density on a particle or an inhomogeneous (e.g., Janus particle-like) surface coating is sufficient for attachment [41, 42].

5.3.1 Spherical Nanoparticles

For a spherical nanoparticle at an infinitely large tensionless planar membrane, Eqs. 5.2 and 5.3 can be simplified. The deformation energy

$$\mathcal{E}_{\text{def}} = 8\pi\kappa f_{\text{w}} \tag{5.4}$$

and the adhesion energy

$$\mathcal{E}_{\text{adh}} = -4\pi R^2 w f_{\text{w}} \tag{5.5}$$

are both proportional to the wrapping fraction f_{w} of the nanoparticle. Equating the deformation-energy costs with the adhesion-energy gain for a complete-wrapped nanoparticle therefore allows one to estimate the adhesion strength for the transition between a nonwrapped and a complete-wrapped nanoparticle, $w = 2\kappa/R^2$. For higher adhesion strengths, the nanoparticles are completely wrapped, whereas for smaller adhesion strengths, they remain unwrapped. Similarly, for fixed adhesion strength, nanoparticles that are larger than $R = \sqrt{w/(2\kappa)}$ are completely wrapped whereas smaller nanoparticles remain unwrapped.

Figure 5.3 shows a partial-wrapped nanoparticle with a catenoidal deformation of the surrounding free and tensionless membrane. The wrapping phase diagram shows that the particle remains unwrapped below a threshold adhesion strength. For a tensionless membrane and higher adhesion strengths, it transitions directly to a complete-wrapped state. For a tensed membrane, the binding transition between the nonwrapped and partial-wrapped states is continuous, while the envelopment transition between the partial-wrapped and complete-wrapped states is discontinuous. At sufficiently high adhesion strength the energy barrier vanishes and the particle is wrapped spontaneously.

> **Particle radius:** For a homogeneous adhesion strength, a larger particle size facilitates nanoparticle wrapping.
>
> **Particle surface properties:** Surface functionalization of nanoparticles determines the nanoparticle–membrane adhesion strength. Higher adhesion strength facilitates nanoparticle wrapping.

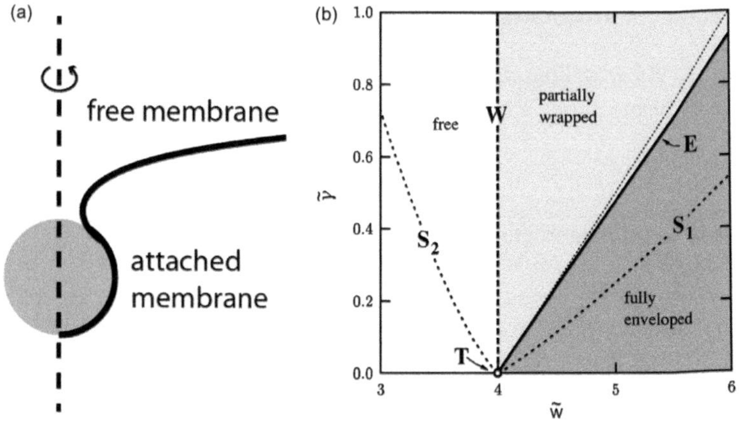

Figure 5.3 Wrapping of spherical nanoparticles. (a) Catenoidal deformation of a membrane around a spherical nanoparticle that is partially wrapped at an infinitely large planar membrane [1]. OA CC BY 3.0 (https://creativecommons.org/licenses/by/3.0/). (b) Wrapping phase diagram in the plane of reduced adhesion constant $\tilde{w}$ and reduced lateral tension $\tilde{\gamma}$ in the low-tension regime $\tilde{\gamma}<1$ at the triple point **T** ($\tilde{w}=4$, $\tilde{\gamma}=0$), the nonwrapped, partial-wrapped, and complete-wrapped phases meet; states predicted by numerical calculations. The dashed line **W** marks the continuous transition at which partial wrapping sets in, the bold solid line **E** indicates the discontinuous transition between partially wrapped and fully enveloped, and the short-dashed lines $\mathbf{S_1}$ and $\mathbf{S_2}$ are the spinodals belonging to **E**. The fine dotted line $\tilde{w}=4+2\tilde{w}$ close to **E** indicates where the complete-wrapped state has the same energy as the nonwrapped state. Reprinted figure with permission from Ref. [9]. Copyright (2004) by the American Physical Society.

5.3.2 Nonspherical Nanoparticles

For nonspherical particles, not only particle size and aspect ratio, but also the local particle surface curvature is important for wrapping [8]. In Fig. 5.4, uptake processes for elongated nanoparticles are sketched. For example, ellipsoidal nanoparticles attach with the least-curved part of their surface to a membrane. This orientation with the long axis parallel to the membrane provides the largest contact area and the smallest membrane deformation. For fast wrapping, such that reorientation is suppressed, the nanoparticle remains in this so-called submarine orientation for the entire wrapping process. The minimal-energy orientation in which rod-like nanoparticles with flat ends attach is the orientation with their longest axis perpendicular to the membrane, the so-called rocket orientation. Whereas for a short aspect ratio and highly curved edges the particle remains in rocket orientation until complete wrapping, for higher aspect ratios and rounded edges, the particle temporarily reorients to submarine orientation in the shallow-wrapped state (see Fig. 5.2).

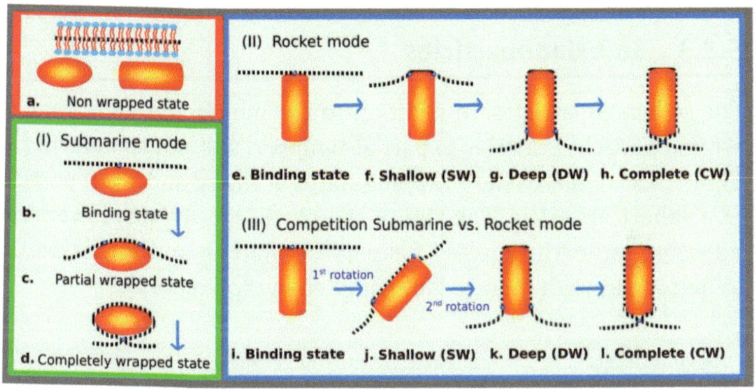

Figure 5.4 Modes of entry for nanoparticle uptake by membrane wrapping: (I) submarine mode with the long axis of the particles oriented parallel to the membrane, (II) rocket mode with the long axis oriented perpendicular to the membrane, and (III) competition between submarine and rocket mode, as observed for rod-like particles with high aspect ratios. The complete-wrapped particle is connected by an infinitely small catenoidal neck to the membrane; the particle orientation in this state is irrelevant. Adapted with permission from Ref. [8]. Copyright (2014) American Chemical Society. (https://pubs.acs.org/doi/abs/10.1021/nl403949h). Further permissions related to the material excerpted should be directed to the ACS.

The wrapping pathways sketched in Fig. 5.4 do not only show sequences of states that can be attained temporarily on the way to complete wrapping. Depending on the value of the membrane–particle adhesion strength, the indicated partial-wrapped states can also be stable, minimal-energy states. For example, the potential existence of two stable partial-wrapped states is characteristic for rod-like nanoparticles due to the inhomogeneous curvature distribution of their surface. Even at infinitely large tensionless planar membranes, an inhomogeneous curvature of the nanoparticle surface can thus stabilize several partial-wrapped, membrane-bound states [8, 26].

Particle aspect ratio: A higher particle aspect ratio stabilizes partial-wrapped states.
Particle shape: An inhomogeneous particle surface curvature stabilizes partial-wrapped states.

5.3.3 Soft Nanoparticles

For soft nanoparticles, in addition to the membrane deforming to wrap a particle, particles in partial-wrapped states also deform [2]. Figure 5.5 compares the wrapping states of a hard and a soft particle at a spherical vesicle. The ability of a particle to deform lowers the adhesion strength required for particle binding and increases the adhesion strength required for complete wrapping [43].

Figure 5.5 Budding of a rigid and a deformable nanoparticle out of a spherical vesicle; particle and vesicle shapes obtained by numerical calculations. Whereas for the rigid nanoparticle only the vesicle deforms, for soft particles both particle and vesicle deform during wrapping. In the nonwrapped and complete-wrapped states both rigid and soft particles are spherical. Reprinted with permission from Ref. [2]. Copyright (2017) American Chemical Society.

Particle deformability: A deformable particle binds at lower adhesion strengths and is completely wrapped at higher adhesion strengths compared with a hard particle with equal size and surface properties. Partial-wrapped states are stabilized for soft particles compared with hard particles.

5.3.4 Dosage Effects: Cooperative Wrapping

The interaction of single nanoparticles with lipid-bilayer membranes, which has been discussed so far, is not the only important case for understanding nanoparticle–membrane interactions. Computer simulations predict that membrane-mediated interactions between nanoparticles can alter the wrapping degree of the particles and that partial wrapping leads to membrane-mediated nanoparticle interactions (see Fig. 5.6). For example, linear aggregation of nanoparticles has been reported for biologically relevant regimes of membrane bending rigidity and nanoparticle–membrane adhesion strength, both for spherical and

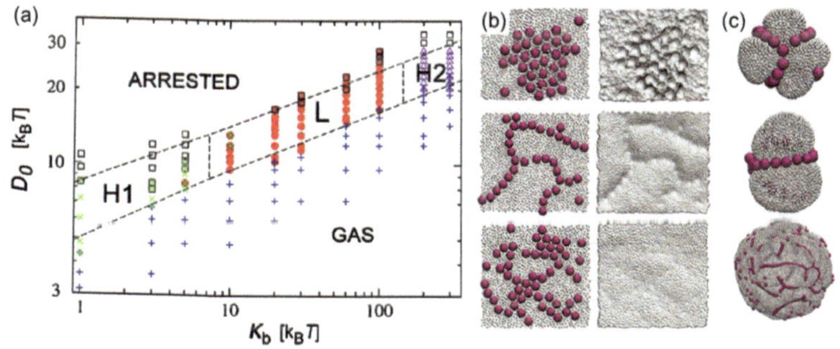

Figure 5.6 Membrane-deformation induced nanoparticle attraction as predicted by computer simulations. (a) Phase diagram for particle self-assembly on a fluid membrane in terms of the membrane bending rigidity $K_b = 2\sqrt{3}\kappa$ and particle binding energy D_0. (b) Snapshots with typical aggregates in the *H*1, *L*, and *H*2 phases in a top-to-bottom order and the deformation patterns that they leave on the membrane. The membrane is modeled using beads of size σ; the membrane area $A = (40 \times 40)\sigma^2$, the nanoparticle size $\sigma_{np} = 4\sigma$, and the nanoparticle surface fraction $\rho = 0.27$. (c) Snapshots of the linear aggregates on spherical vesicles. The upper two snapshots show the system of $R = 15\ \sigma$, $\sigma_{np} = 4\ \sigma$, and $\rho = 0.11$, and the bottom snapshot depicts $R = 45\ \sigma$, $\sigma_{np} = 3\ \sigma$, and $\rho = 0.16$. Reprinted figure with permission from Ref. [11]. Copyright (2012) by the American Physical Society.

nonspherical nanoparticles at membranes that are planar prior to wrapping [11, 44]. At higher adhesion strengths, the aggregates "reorient" to form particle-filled membrane tubes [6, 14]—analogously to the reorientation of elongated nanoparticles in the deep-wrapped state [8] (see Fig. 5.7). Finally, at very high adhesion strengths single-particle buds are observed.

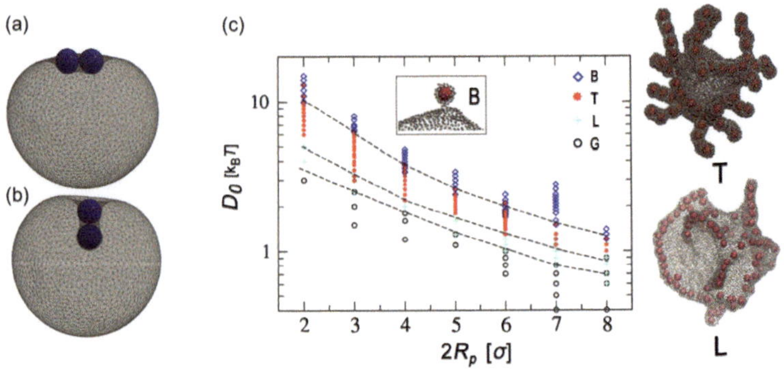

Figure 5.7 Nanoparticle-induced tube formation as predicted by computer simulations. (a) Bound minimal-energy state of two particles for a reduced volume v = 0.96 of the vesicle and rescaled adhesion energy $wR_p^2/\kappa = 2$ of the particle. (b) Bound minimal-energy state of two particles for v = 0.92 and $wR_p^2/\kappa = 2.33$. Reprinted figures with permission from Ref. [6]. Copyright (2012) by the American Physical Society. (c) $D_0 - R_p$ phase diagram for the particle aggregates on vesicles and protrusions induced by spherical particles and snapshots of the linear (**L**) and tubular (**T**) phases. The inset shows a typical single-particle bud conformation (**B**) that occurs at large D_0. The bottom region of the phase diagram is the gaseous phase (**G**). The radius of the vesicles is R = 30 σ, where the membrane is modeled by particles of size σ. The particle surface fraction is kept constant at 0.15. Reprinted figure with permission from Ref. [14]. Copyright (2012) by the American Physical Society.

An important contribution for calculating interactions between partial-wrapped nanoparticles at membranes is the deformation energy of the free membrane, similarly to the interaction between conical inclusions within or curved proteins attached to lipid membranes [45–48]. For particles, the particle-membrane adhesion energy also has to be taken into account. The interaction between two partial-wrapped particles on a planar membrane has been reported to be attractive due to an increasing contact area between membrane and particle with decreasing distance between the particles: the resulting adhesion-energy gain overcomes the energy costs for deforming the free membrane [11, 49].

> **Cooperative wrapping:** Membrane-mediated attraction induces aggregation and tube formation of partial-wrapped nanoparticles.

5.3.5 Multicomponent Biological Membranes

Although biological membranes—unlike model membranes—usually consist of many components [50]. From a mechanical point of view, they can often be well described using effective, homogeneous curvature-elastic parameters. However, nanoparticle adhesion is qualitatively different from a homogeneous van der Waals attraction. In multicomponent membranes, more adhesive lipids, proteins, or rafts enrich near the particles. Enrichment of favorable components has also been reported for membranes with charged proteins and lipids [10], as well as for membranes with conical proteins that locally decrease the bending energy for wrapping [40, 51].

The classical example of multicomponent membranes is systems with specific binding of receptors to ligands [52, 53]. For models that take into account the diffusion of receptors to nanoparticles a major finding is a preferred particle radius for wrapping [39], contrary to models with homogeneous adhesion strength where wrapping is facilitated for increasing nanoparticle size [8, 20]. Furthermore, in models based on receptor-ligand bonds, multivalent binding can lead to super-selective nanoparticle adhesion [54].

In biological systems with high salt concentrations and therefore strong screening of electrostatic interactions, adhesion mediated by oppositely charged components can be modeled analogously to receptor-ligand bonds. Figure 5.8 shows computer simulation results for a charged nanoparticle getting wrapped by a vesicle with charged lipids [10]. Depending on the number of charged lipids, the wrapping fraction for the stable partial-wrapped state increases. Above a threshold value, for wrapping fractions above 60%, the particle jumps to a complete-wrapped state because the bending energy for further wrapping is only weakly dependent on the wrapping fraction.

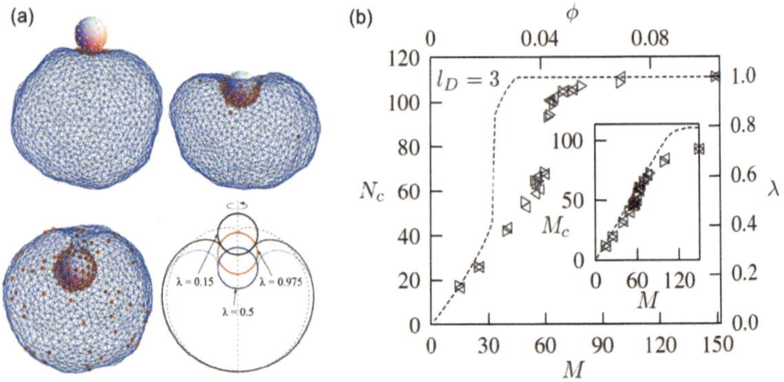

Figure 5.8 Wrapping of a spherical nanoparticle with radius R_p by a spherical vesicle with radius of R_v/R_p = 4.4 as predicted by computer simulations. (a) Simulation snapshots for M = 15, M = 60, and M = 150 charged vertices on the membrane. The charges on the vesicle are mobile, and the charges on the particle are fixed. The cross sections show vesicle shapes for wrapping fractions f_w = 0.15, f_w = 0.5, and f_w = 0.975. (b) The number of vertices in the wrapping region N_c and the wrapping fraction f_w as functions of the total number of charged vertices on the vesicle M. The fraction of charged vertices is denoted by ϕ. The different symbols indicate whether the initial state of the Monte Carlo simulation was nonwrapped or complete-wrapped. At $\phi \approx 0.04$ a discontinuous wrapping transition occurs; the dashed line is a prediction based on a phenomenological model. The inset shows the number of charged vertices in the wrapping region, M_c. All results correspond to a Debye screening length $l_D = R_p$ and a membrane bending rigidity = 10 k_BT. Reprinted from Ref. [10], with the permission of AIP Publishing.

> **Lipid domains:** Lipid rafts or larger domains may constitute adhesive patches on a multicomponent lipid-bilayer membrane.
> **Specific adhesion:** Receptor-ligand bonds mediate specific adhesion. The discrete nature of the interaction can stabilize partial-wrapped states.
> **Charged membranes:** For the high salt concentrations in biological systems, electrostatic attraction can be modeled analogously to receptor-ligand bonds.

5.3.6 Actual and Spontaneous Membrane Curvature

Finite membrane curvature prior to wrapping and spontaneous membrane curvature can facilitate or hinder nanoparticle binding and envelopment. For example, for a spherical nanoparticle at the outside of a spherical vesicle with a comparable size, numerical calculations predict an energy barrier for the combined binding-envelopment transition [3, 55, 56]. On the contrary, for a nanoparticle inside a vesicle, binding is facilitated but complete wrapping is hindered. Similarly, for a membrane with spontaneous curvature, numerical calculations predict an energy barrier for the combined binding-envelopment transition if the spontaneous curvature agrees with the curvature of the particle surface, whereas binding is facilitated but complete wrapping is hindered if the sign of the spontaneous curvature is opposite [3, 40].

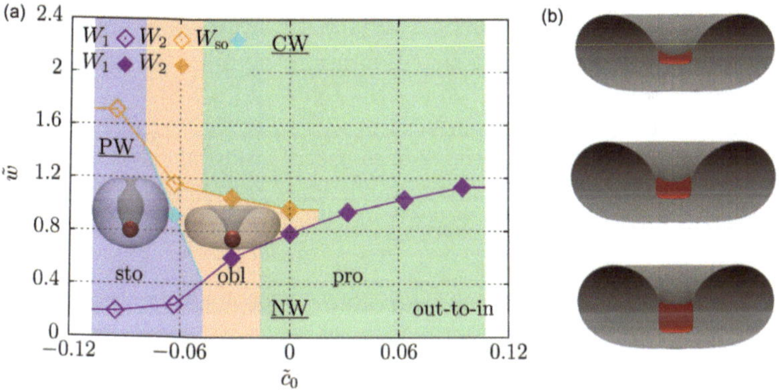

Figure 5.9 Wrapping of a nanoparticle at a vesicle predicted using triangulated membranes. (a) Wrapping diagram for a spherical nanoparticle with a radius R_p that attaches to the outside of a spherical vesicle with a radius R_v; for R_p/R_v = 0.2, reduced vesicle volume v = 0.7. The wrapping transitions are indicated for various membrane spontaneous curvatures $c_0 = c_0R_p$ and reduced adhesion strengths $\tilde{w} = wR_p^2/(2\kappa)$. Filled symbols indicate a discontinuous transition with an energy barrier, open symbols a continuous transition. Regimes for nonwrapped (NW), partial-wrapped (PW), and complete-wrapped (CW) particles are delimited by the wrapping transitions W_1 and W_2, stomatocyte, oblate, and prolate vesicle shapes are indicated by the background color. The transition W_{so} indicates a shape transition between two partial-wrapped states. (b) Wrapping of a cube-like nanoparticle at an oblate vesicle. Republished with permission of The Royal Society of Chemistry, from Ref. [3]; permission conveyed through Copyright Clearance Center, Inc.

Figure 5.9 shows the complex interplay between actual and spontaneous membrane curvature for nanoparticle wrapping at nonspherical vesicles. For high positive spontaneous curvature, opposite to the particle surface curvature, a direct, discontinuous transition from the nonwrapped to the complete-wrapped state is predicted at prolate vesicles [3]. For high negative spontaneous curvature, which agrees with the particle surface curvature, a parameter regime for stable partial-wrapped states between a continuous binding transition and a continuous envelopment transition is predicted at the initially stomatocytic vesicles. Here, partial wrapping of the nanoparticles can induce vesicle shape changes. However, Fig. 5.9 also illustrates that for wrapping of nonspherical particles at small vesicles, the curvature of the particle surface is often higher than the curvature of the vesicle membrane. Therefore, the effect of particle surface curvature on wrapping often dominates the effects of the membrane actual and spontaneous curvatures [3].

Curved membranes: Membrane curvature toward the particle prior to wrapping facilitates binding and hinders complete wrapping. For opposite curvature, an energy barrier separates nonwrapped and complete-wrapped states.

Spontaneous membrane curvature: Spontaneous membrane curvature toward the particle leads to an energy barrier between nonwrapped and complete-wrapped states. Opposite spontaneous curvature facilitates binding and hinders complete wrapping.

5.4 Experimental Validation

Experimental validation is crucial for developing computer models for nanoparticle–membrane interactions. In particular, membrane model systems allow for detailed comparisons between experiments and simulations (see Fig. 5.10). With electron microscopy the bilayer that envelopes endocytosed nanoparticles can be resolved in micrographs. For larger particle sizes, fluorescence microscopy can be applied to separately visualize particle aggregates and lipid structures that mediate adhesion. The attraction of micrometer-sized particles on lipid-bilayer membranes can even be captured dynamically. Figure 5.10e–g shows spherical particles that initially are 4 μm apart and get into contact after about 1 s.

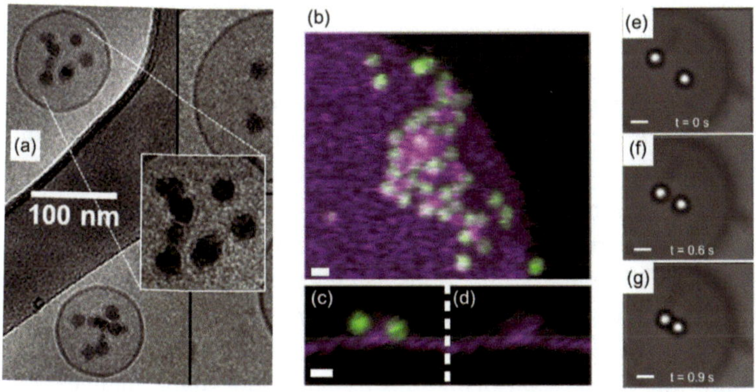

Figure 5.10 Particle-wrapping experiments. (a) Cryo-TEM micrographs taken 3 weeks after sample preparation. The enlargement shows the presence of supported lipid bilayers around internalized Si nanoparticles. Reprinted with permission from Ref. [7]. Copyright © 2014 WILEY-VCH Verlag GmbH & Co. KGaA, Weinheim. (b–d) Confocal images of particles bound to lipid membranes by streptavidin-biotin-PEG-DOPE linkers. (b) Large-particle aggregate. (c) Two- particle interaction at a membrane; the membrane fluorescence from the image in (c) is shown separately in (d). The scale bars correspond to 1 μm [15]. OA CC BY 4.0 (https://creativecommons.org/licenses/by/4.0/). (e–g) Particle binding and interaction on giant unilamellar vesicles (GUVs), observed using bright-field microscopy. Two particles self-assemble when bound to the same GUV. The time sequence of images shows particles at the top of a GUV (18 μm diameter) interacting across a distance of over 4 μm and quickly moving toward each other in a time of about 1 s. The scale bar corresponds to 2 μm. Reprinted figures with permission from Ref. [17]. Copyright (2016) by the American Physical Society.

Whereas the strength of membrane model systems is their well-defined composition, it is also important to test the applicability and transferability of generic computer simulation results for biological systems with higher complexity. Here, well-characterized cells can serve as biological model systems because effects of nanoparticle wrapping can be quantified in detail. Their multicomponent lipid-bilayer membranes that are often supported by a cortical cytoskeleton distinguish them from artificial membrane-only systems. For example, human erythrocytes have been quantitatively studied for decades [57]. Nowadays, equilibrium shapes, membrane flickering, and flow properties are well understood [58–60].

Systematic experiments for nanoparticles at erythrocytes revealed that carboxylated polystyrene nanoparticles get attached to the erythrocyte cell membranes only sparsely (~2% coverage) [61]. Although the numbers of adsorbed particles for various nanoparticle concentrations in solution resemble Langmuir isotherms, a thermodynamic equilibrium between the particle concentrations at the membrane and in solution is not consistent with the irreversible attachment reported in the experiments. Instead, computer simulations using a receptor model suggest that binding kinetics, receptor availability and diffusion, and multivalent binding determine the number of bound nanoparticles [61]. In addition, numerical calculations based on the shape changes of the nanoparticle-decorated erythrocytes observed in the experiments allow determining particle-induced effective spontaneous membrane curvatures.

5.5 Toward Nanotoxicology

The biological fate and toxicity of nanomaterials depend on particle size, shape, aspect ratio, mechanical properties, and surface functionalization and patterning [62]. For example, a high risk of toxicity is reported for graphene-based nanoplatelets in the respiratory system, whereas graphene sheets appear to be less harmful. Similarly, the toxicity may differ for metal oxide nanoparticles with different physicochemical properties due to their corrugated shapes. Furthermore, chemical modifications of carbon nanotubes can reduce toxicity and lead to accumulation in different parts on an organism. In vitro studies also show that toxicity of nanoparticles depends on size, dose, and cell type or organism [63, 64]. Predictive nanotoxicology, therefore, requires a systematic characterization of these descriptors, using machine-learning methods to establish links between them. Generic computer simulation studies that characterize the interaction of nanomaterials with the plasma membranes, which separate cells from their environment, can help to achieve systematic understanding.

Nanoparticle size and pristine surface properties are perhaps the most natural parameters to be included in a QNAR model [65]. However, perspectives for nanosafety suggest that many more aspects besides the chemical similarity, primary size, and size distribution of particles shall be considered [66], for example, shape, porosity, crystal structure, and surface coatings, including the formation of lipid and protein coronas. Model development based on in vitro experiments can help to advance, in particular, two aspects that are important for in silico toxicology in this multiparameter space [67]: (i) improve our understanding of the biological mechanisms and (ii) help to screen compounds. For example, computer simulations have been used to develop the hypothesis of receptor-mediated adhesion on the basis of experimental data for nanoparticle adhesion to erythrocytes [61]. Here, simulations and numerical calculations have also been able to predict multivalent binding, as well as to calculate an effective spontaneous curvature for the nanoparticle-decorated cell membranes.

Studies of nanoparticle–membrane interactions using continuum membrane models, such as those presented in this chapter, enable predictions regarding how nanodescriptors on the particle scale influence the biophysical responses of cells. Mesoscale approaches and techniques, therefore, complement quantum chemical calculations and all-atomistic molecular dynamics simulations for computational risk assessment.

Acknowledgments

Our research on the interaction of particles with biological membranes has been supported by the EU FP7 NMP collaborative project PreNanoTox (309666). We thank Rafi Korenstein and Alexander Barbul (Tel Aviv) and Emilio Benfenati (Milan) for helpful discussions.

References

1. Dasgupta, S., Auth, T. and Gompper, G. (2017). Nano- and microparticles at fluid and biological interfaces, *J. Phys.: Condens. Matter*, **29**, 373003.
2. Yi, X. and Gao, H. J. (2017). Budding of an adhesive elastic particle out of a lipid vesicle, *ACS Biomater. Sci. Eng.*, **3**, 2954–2961.
3. Yu, Q. F., Othman, S., Dasgupta, S., Auth, T. and Gompper, G. (2018). Nanoparticle wrapping at small non-spherical vesicles: curvatures at play, *Nanoscale*, **10**, 6445–6458.

4. Gompper, G. and Kroll, D. M. (2004). Triangulated surface models of fluctuating membranes, in *Statistical Mechanics of Membranes and Surfaces*, Nelson, D. R., Piran, T. and Weinberg, S., eds. (World Scientific, Singapore).
5. Yang, K. and Ma, Y. Q. (2010). Computer simulation of the translocation of nanoparticles with different shapes across a lipid bilayer, *Nat. Nanotechnol.*, **5**, 579–583.
6. Bahrami, A. H., Lipowsky, R. and Weikl, T. R. (2012). Tubulation and aggregation of spherical nanoparticles adsorbed on vesicles, *Phys. Rev. Lett.*, **109**, 188102.
7. Michel, R., Kesselman, E., Plostica, T., Danino, D. and Gradzielski, M. (2014). Internalization of silica nanoparticles into fluid liposomes: formation of interesting hybrid colloids, *Angew. Chem. Int. Ed.*, **53**, 12441–12445.
8. Dasgupta, S., Auth, T. and Gompper, G. (2014). Shape and orientation matter for the cellular uptake of nonspherical particles, *Nano Lett.*, **14**, 687–693.
9. Deserno, M. (2004). Elastic deformation of a fluid membrane upon colloid binding, *Phys. Rev. E*, **69**, 031903.
10. Fosnaric, M., Iglic, A., Kroll, D. M. and May, S. (2009). Monte Carlo simulations of complex formation between a mixed fluid vesicle and a charged colloid, *J. Chem. Phys.*, **131**, 105103.
11. Saric, A. and Cacciuto, A. (2012). Fluid membranes can drive linear aggregation of adsorbed spherical nanoparticles, *Phys. Rev. Lett.*, **108**, 118101.
12. Barnoud, J., Rossi, G. and Monticelli, L. (2014). Lipid membranes as solvents for carbon nanoparticles, *Phys. Rev. Lett.*, **112**, 068102.
13. Chen, L. P., Xiao, S. Y., Zhu, H., Wang, L. and Liang, H. J. (2016). Shape-dependent internalization kinetics of nanoparticles by membranes, *Soft Matter*, **12**, 2632–2641.
14. Saric, A. and Cacciuto, A. (2012). Mechanism of membrane tube formation induced by adhesive nanocomponents, *Phys. Rev. Lett.*, **109**, 188101.
15. van der Wel, C., Vahid, A., Saric, A., Idema, T., Heinrich, D. and Kraft, D. J. (2016). Lipid membrane-mediated attraction between curvature inducing objects, *Sci. Rep.*, **6**, 32825.
16. Guo, Y. C., Terazzi, E., Seemann, R., Fleury, J. B. and Baulin, V. A. (2016). Direct proof of spontaneous translocation of lipid-covered hydrophobic nanoparticles through a phospholipid bilayer, *Sci. Adv.*, **2**, e1600261.

17. Sarfati, R. and Dufresne, E. R. (2016). Long-range attraction of particles adhered to lipid vesicles, *Phys. Rev. E*, **94**, 012604.

18. Pogodin, S., Werner, M., Sommer, J. U. and Baulin, V. A. (2012). Nanoparticle-induced permeability of lipid membranes, *ACS Nano*, **6**, 10555–10561.

19. Zemel, A., Ben-Shaul, A. and May, S. (2005). Perturbation of a lipid membrane by amphipathic peptides and its role in pore formation, *Eur. Biophys. J.*, **34**, 230–242.

20. Deserno, M. and Bickel, T. (2003). Wrapping of a spherical colloid by a fluid membrane, *EPL (Europhys. Lett.)*, **62**, 767–773.

21. Ivanova, E. P., Hasan, J., Webb, H. K., Gervinskas, G., Juodkazis, S., Truong, V. K., Wu, A. H. F., Lamb, R. N., Baulin, V. A., Watson, G. S., Watson, J. A., Mainwaring, D. E. and Crawford, R. J. (2013). Bactericidal activity of black silicon, *Nat. Commun.*, **4**, 2838.

22. Santoro, F., Dasgupta, S., Schnitker, J., Auth, T., Neumann, E., Panaitov, G., Gompper, G. and Offenhausser, A. (2014). Interfacing electrogenic cells with 3D nanoelectrodes: position, shape, and size matter, *ACS Nano*, **8**, 6713–6723.

23. Noguchi, H. and Gompper, G. (2006). Meshless membrane model based on the moving least-squares method, *Phys. Rev. E*, **73**, 021903.

24. Ginzburg, V. V. and Balijepailli, S. (2007). Modeling the thermodynamics of the interaction of nanoparticles with cell membranes, *Nano Lett.*, **7**, 3716–3722.

25. Noguchi, H. and Takasu, M. (2002). Adhesion of nanoparticles to vesicles: a Brownian dynamics simulation, *Biophys. J.*, **83**, 299–308.

26. Dasgupta, S., Auth, T. and Gompper, G. (2013). Wrapping of ellipsoidal nano-particles by fluid membranes, *Soft Matter*, **9**, 5473–5482.

27. Auth, T., Fedosov, D. A. and Gompper, G. (2019). Simulating membranes, vesicles, and cells, in *The Giant Vesicle Book*, Dimova, R. and Marques, C., eds. (CRC Press, Boca Raton).

28. Kroll, D. M. and Gompper, G. (1992). The conformation of fluid membranes - Monte-Carlo simulations, *Science*, **255**, 968–971.

29. Canton, I. and Battaglia, G. (2012). Endocytosis at the nanoscale, *Chem. Soc. Rev.*, **41**, 2718–2739.

30. Chen, S. Y., Wang, S. Q., Kopytynski, M., Bachelet, M. and Chen, R. J. (2017). Membrane-anchoring, comb-like pseudopeptides for efficient, pH mediated membrane destabilization and intracellular delivery, *ACS Appl. Mater. Interfaces*, **9**, 8021–8029.

31. Motornov, M., Roiter, Y., Tokarev, I. and Minko, S. (2010). Stimuli-responsive nanoparticles, nanogels and capsules for integrated multifunctional intelligent systems, *Prog. Polym. Sci.*, **35**, 174–211.

32. Wang, F. J., Yu, L., Monopoli, M. P., Sandin, P., Mahon, E., Salvati, A. and Dawson, K. A. (2013). The biomolecular corona is retained during nanoparticle uptake and protects the cells from the damage induced by cationic nanoparticles until degraded in the lysosomes, *Nanomed. Nanotechnol. Biol. Med.*, **9**, 1159–1168.

33. Helfrich, W. (1973). Elastic properties of lipid bilayers: theory and possible experiments, *Z. Naturforsch. C*, **28**, 693–703.

34. Tenzer, S., Docter, D., Kuharev, J., Musyanovych, A., Fetz, V., Hecht, R., Schlenk, F., Fischer, D., Kiouptsi, K., Reinhardt, C., Landfester, K., Schild, H., Maskos, M., Knauer, S. K. and Stauber, R. H. (2013). Rapid formation of plasma protein corona critically affects nanoparticle pathophysiology, *Nat. Nanotechnol.*, **8**, 772–781.

35. Lundqvist, M., Stigler, J., Elia, G., Lynch, I., Cedervall, T. and Dawson, K. A. (2008). Nanoparticle size and surface properties determine the protein corona with possible implications for biological impacts, *Proc. Natl. Acad. Sci. U. S. A.*, **105**, 14265–14270.

36. Salvati, A., Pitek, A. S., Monopoli, M. P., Prapainop, K., Bombelli, F. B., Hristov, D. R., Kelly, P. M., Aberg, C., Mahon, E. and Dawson, K. A. (2013). Transferrin-functionalized nanoparticles lose their targeting capabilities when a biomolecule corona adsorbs on the surface, *Nat. Nanotechnol.*, **8**, 137–143.

37. Churchman, A. H., Wallace, R., Milne, S. J., Brown, A. P., Brydson, R. and Beales, P. A. (2013). Serum albumin enhances the membrane activity of ZnO nanoparticles, *Chem. Commun.*, **49**, 4172–4174.

38. Chithrani, B. D., Ghazani, A. A. and Chan, W. C. W. (2006). Determining the size and shape dependence of gold nanoparticle uptake into mammalian cells, *Nano Lett.*, **6**, 662–668.

39. Gao, H. J., Shi, W. D. and Freund, L. B. (2005). Mechanics of receptor-mediated endocytosis, *Proc. Natl. Acad. Sci. U. S. A.*, **102**, 9469–9474.

40. Agudo-Canalejo, J. and Lipowsky, R. (2015). Critical particle sizes for the engulfment of nanoparticles by membranes and vesicles with bilayer asymmetry, *ACS Nano*, **9**, 3704–3720.

41. Vacha, R., Martinez-Veracoechea, F. J. and Frenkel, D. (2011). Receptor-mediated endocytosis of nanoparticles of various shapes, *Nano Lett.*, **11**, 5391–5395.

42. Gao, Y. and Yu, Y. (2013). How half-coated Janus particles enter cells, *J. Am. Chem. Soc.*, **135**, 19091–19094.

43. Yi, X., Shi, X. H. and Gao, H. J. (2011). Cellular uptake of elastic nanoparticles, *Phys. Rev. Lett.*, **107**, 098101.

44. Xiong, K., Zhao, J. Y., Yang, D. W., Cheng, Q. W., Wang, J. L. and Ji, H. B. (2017). Cooperative wrapping of nanoparticles of various sizes and shapes by lipid membranes, *Soft Matter*, **13**, 4644–4652.

45. Bahrami, A. H. and Weikl, T. R. (2018). Curvature-mediated assembly of Janus nanoparticles on membrane vesicles, *Nano Lett.*, **18**, 1259–1263.

46. Auth, T. and Gompper, G. (2009). Budding and vesiculation induced by conical membrane inclusions, *Phys. Rev. E*, **80**, 031901.

47. Reynwar, B. J., Illya, G., Harmandaris, V. A., Muller, M. M., Kremer, K. and Deserno, M. (2007). Aggregation and vesiculation of membrane proteins by curvature-mediated interactions, *Nature*, **447**, 461–464.

48. Kim, K. S., Neu, J. and Oster, G. (1998). Curvature-mediated interactions between membrane proteins, *Biophys. J.*, **75**, 2274–2291.

49. Weikl, T. R., Kozlov, M. M. and Helfrich, W. (1998). Interaction of conical membrane inclusions: effect of lateral tension, *Phys. Rev. E*, **57**, 6988–6995.

50. Sackmann, E. (2005). Biological membranes - architecture and function, in *Structure and Dynamics of Membranes: From Cells to Vesicles*, Lipowsky, R. and Sackmann, E., eds. (Elsevier, Amsterdam).

51. Curk, T., Wirnsberger, P., Dobnikar, J., Frenkel, D. and Saric, A. (2018). Controlling cargo trafficking in multicomponent membranes, *Nano Lett.*, **18**, 5350–5356.

52. Fenz, S. F., Bihr, T., Schmidt, D., Merkel, R., Seifert, U., Sengupta, K. and Smith, A. S. (2017). Membrane fluctuations mediate lateral interaction between cadherin bonds, *Nat. Phys.*, **13**, 906–913.

53. Hu, J. L., Lipowsky, R. and Weikl, T. R. (2013). Binding constants of membrane-anchored receptors and ligands depend strongly on the nanoscale roughness of membranes, *Proc. Natl. Acad. Sci. U. S. A.*, **110**, 15283–15288.

54. Martinez-Veracoechea, F. J. and Frenkel, D. (2011). Designing super selectivity in multivalent nano-particle binding, *Proc. Natl. Acad. Sci. U. S. A.*, **108**, 10963–10968.

55. Bahrami, A. H., Lipowsky, R. and Weikl, T. R. (2016). The role of membrane curvature for the wrapping of nanoparticles, *Soft Matter*, **12**, 581–587.

56. Agudo-Canalejo, J. and Lipowsky, R. (2015). Adhesive nanoparticles as local probes of membrane curvature, *Nano Lett.*, **15**, 7168–7173.

57. Brochard, F. and Lennon, J. F. (1975). Frequency spectrum of the flicker phenomenon in erythrocytes, *J. Phys. (France)*, **36**, 1035–1047.

58. Lim, H. W. G., Wortis, M. and Mukhopadhyay, R. (2002). Stomatocyte-discocyte-echinocyte sequence of the human red blood cell: evidence for the bilayer-couple hypothesis from membrane mechanics, *Proc. Natl. Acad. Sci. U. S. A.*, **99**, 16766–16769.

59. Fedosov, D. A., Noguchi, H. and Gompper, G. (2014). Multiscale modeling of blood flow: from single cells to blood rheology, *Biomech. Model. Mechanobiol.*, **13**, 239–258.

60. Turlier, H., Fedosov, D. A., Audoly, B., Auth, T., Gov, N. S., Sykes, C., Joanny, J. F., Gompper, G. and Betz, T. (2016). Equilibrium physics breakdown reveals the active nature of red blood cell flickering, *Nat. Phys.*, **12**, 513–519.

61. Barbul, A., Singh, K., Horev-Azaria, L., Dasgupta, S., Auth, T., Korenstein, R. and Gompper, G. (2018). Nanoparticle-decorated erythrocytes reveal that particle size controls the extent of adsorption, cell shape, and cell deformability, *ACS Appl. Nano Mater.*, **1**, 3785–3799.

62. Gajewicz, A., Rasulev, B., Dinadayalane, T. C., Urbaszek, P., Puzyn, T., Leszczynska, D. and Leszczynski, J. (2012). Advancing risk assessment of engineered nanomaterials: Application of computational approaches, *Adv. Drug Delivery Rev.*, **64**, 1663–1693.

63. Murugadoss, S., Lison, D., Godderis, L., Van den Brule, S., Mast, J., Brassinne, F., Sebaihi, N. and Hoet, P. H. (2017). Toxicology of silica nanoparticles: an update, *Arch. Toxicol.*, **91**, 2967–3010.

64. Ivask, A., Kurvet, I., Kasemets, K., Blinova, I., Aruoja, V., Suppi, S., Vija, H., Kakinen, A., Titma, T., Heinlaan, M., Visnapuu, M., Koller, D., Kisand, V. and Kahru, A. (2014). Size-dependent toxicity of silver nanoparticles to bacteria, yeast, algae, crustaceans and mammalian cells in vitro, *Plos One*, **9**, e102108.

65. Manganelli, S., Leone, C., Toropov, A. A., Toropova, A. P. and Benfenati, E. (2016). QSAR model for predicting cell viability of human embryonic kidney cells exposed to SiO_2 nanoparticles, *Chemosphere*, **144**, 995–1001.

66. Puzyn, T., Jeliazkova, N., Sarimveis, H., Robinson, R. L. M., Lobaskin, V., Rallo, R., Richarz, A. N., Gajewicz, A., Papadopulos, M. G., Hastings, J., Cronin, M. T. D., Benfenati, E. and Fernandez, A. (2018). Perspectives from the NanoSafety Modelling Cluster on the validation criteria for (Q)SAR models used in nanotechnology, *Food Chem. Toxicol.*, **112**, 478–494.

67. Gleeson, M. P., Modi, S., Bender, A., Robinson, R. L. M., Kirchmair, J., Promkatkaew, M., Hannongbua, S. and Glen, R. C. (2012). The challenges involved in modeling toxicity data in silico: a review, *Curr. Pharm. Des.*, **18**, 1266–1291.

Chapter 6

Descriptors in Nano-QSAR/QSPR Modeling

Ewelina Wyrzykowska,[a] Karolina Jagiello,[a,b] Bakhtiyor Rasulev,[c] and Tomasz Puzyn[a]

[a]*Laboratory of Environmental Chemometrics, Faculty of Chemistry, University of Gdansk, Wita Stwosza 63, Gdansk, 80-308, Poland*

[b]*QSAR Lab Ltd, Trzy Lipy 3, Gdansk, 80-172, Poland*

[c]*Department of Coatings and Polymeric Materials, North Dakota State University, Fargo, ND 58108-6050, USA*

wyrzykowska@qsar.eu.org

The quantitative structure–activity/property relationships (QSAR/QSPR) models are one of the efficient methods supporting the experimental investigations of chemicals, including nanomaterials and their risk assessment, as well as their application potential. So far, nano-QSAR/QSPR was successfully applied in the prediction of biological activity and physicochemical properties for nanomaterials. However, wider application of QSAR/QSPR for nanomaterials is challenging for several reasons. First, there is a lack of methods to provide the complete characterization of nanoparticles, which is required in a standard procedure of QSAR/

Computational Nanotoxicology: Challenges and Perspectives
Edited by Agnieszka Gajewicz and Tomasz Puzyn

ISBN 978-981-4800-64-8 (Hardcover), 978-0-429-34137-3 (eBook)
www.jennystanford.com

QSPR model development. The second issue is related to the lack of a reliable computational methodology to accurately reflect the specific nature of nanostructures, in other words, a lack of reliable nanodescriptors. This chapter is focused on the computational aspects of characterization of nanostructures in nano-QSAR/QSPR modeling and discusses various applied nanodescriptors.

6.1 Nano-QSAR/QSPR Modeling: Benefits and Challenges

The unique properties of nanomaterials are used in many fields where the development of advanced technology and sophisticated products requires special miniaturization and precision to achieve the specific properties [1–12]. Nevertheless, the potential application of nanoparticles still requires to be preceded by comprehensive investigations to verify their physicochemical features. In addition, due to the ubiquity of nanomaterials, it is important to take a closer look at their influence on the living organisms to avoid any unforeseen negative effects in the longer term. It is necessary to make detailed investigations of nanomaterials' features. There are many approaches in this area [13–20], and one of the solutions involves in silico methods to support laboratory experiments. Computational modeling adheres to the idea of 3Rs—replacement, refinement, and reduction—in animal testing. Reducing animal involvement in laboratory experiments results in lower costs and reduced time of research. Moreover, computational models allow estimation of the physicochemical and biological properties of substances even before their synthesis. Often, computational methods are able to provide a reason for any toxicity and give a basis for a possible mode of action. Among the recommended computational methods, one is quantitative structure–activity/property relationship (QSAR/QSPR) modeling. So far, QSAR/QSPR modeling was successfully applied in the prediction of physicochemical properties of pharmaceuticals [21], persistent organic pollutants (POPs) [22–24], ionic liquids [25–27], etc. Also, research in nanomaterials is already supported by quantitative methods [28–30]. However, nano-QSAR/QSPR modeling is a huge challenge for scientists for several reasons, related to two main areas of this issue.

First, there is a lack of complete structural characteristics of nanoparticles obtained by applying the same procedures. A standard procedure of QSAR/QSPR model development should be based on a homogeneous data set obtained in the same conditions. Obviously, there are opportunities to use information obtained under various conditions to draw conclusions about nanomaterials' features, but these conclusions are definitely more difficult to draw and less reliable. Furthermore, physicochemical properties of nanoparticles, that is, the tendency to aggregate or surface charge, change substantially according to external conditions [31–33]. As a consequence, temperature, pressure, suspended chemical individuals in solutions, and other conditions modify nanomaterial features, including their toxic effect toward living organisms [34, 35]. In this context, experiments obtained in the same conditions could be crucial in the detailed description of the relationship between nanostructures and their features.

The second issue is related to the lack of any efficient methodology to accurately reflect the specific nature of nanostructures. In the case of nanomaterials such as nanocrystals, nanotubes, nanosheets, polymer nanoparticles, nanodendrimers, and nanoliposomes, a change in the physical structure significantly impacts the chemical structure and, consequently, the properties. This means that nanomaterials with the same chemical composition, that is, the same metal oxide but different in physical aspects, such as size, shape, porosity, surface area, and aspect ratio, will exhibit entirely different properties [8, 15, 16, 36–38]. Therefore, there are a lot of questions regarding the procedures of descriptor calculations. For example, what kind of molecular model should be used in the computational procedure to reflect the actual properties of nanostructures? How do you distinguish nanomaterials that have the same chemical composition but differ in size, shape, porosity, etc.? For example, it is possible to build a molecular model of a nanostructure as a cluster of a specific size but how do you decide what an appropriate size of this cluster would be? Also, a molecular model of the real size of a nanomaterial could certainly reflect its properties but for quantum-mechanical (QM) calculations, a nano-sized particle could be an extremely large system for calculation and, as a result, ergonomically inadequate due to excessive consumption of computational resources. In addition, the issue of nanostructure description becomes more complex when we deal with modified/decorated nanomaterials, namely, with different cores, shells, coatings, attached organic ligands, or surface functional groups.

However, it is important to point out that nano-QSAR/QSPR models have already been successfully applied in the prediction of physicochemical properties [29, 38–40] as well as in explaining possible causes of nanomaterial toxicity [41–45]. A substantial number of these models are based on data sets regarding nanomaterials with varying chemical compositions. Thus, the applied descriptors distinguished nanoparticles according to their chemical characters and did not consider any physical aspect. Nevertheless, the development of descriptors suitable for physical properties of nanomaterials is also observed. The aim of this chapter is to review the evolution of descriptors for nanomaterials. Next, we present a number of already developed approaches for describing the specific physicochemical properties of nanostructures.

6.2 Idea of Descriptors

Descriptors are the quantitative parameters that express the properties of chemical. They aim to provide specific information (structural, intrinsic, or extrinsic properties) of chemicals in simplified form. In QSAR/QSPR modeling, descriptors enable one to determine relationships between chemical structures of analyzed substances with their biological or physicochemical features. However, a perfect descriptor that could be applied in a significant QSAR/QSPR model should display appropriate qualities [46], as follow:

- A descriptor should correlate with a modeling response.
- A descriptor should be suitable for a variety of classes of compounds.
- A descriptor should emphasize even the small structural dissimilarity between compounds.
- The calculation of a descriptor should be easy and fast.
- The physical interpretation of a descriptor should be possible and intuitive in relation to the modeling feature.

There are a few types of descriptor classifications, but the main ones describe in detail (i) the chemical structures and (ii) the physicochemical properties. Descriptors in both groups can be obtained by experimental measurements and theoretical calculations. For instance, the chemical structure can be described using absorption spectra data (UV-Vis, nuclear magnetic resonance, infrared, X-ray diffraction, etc.) as well as by 2D-graphical topology expressed as calculated indexes. In the case of properties, some can be expressed as measurements, that is, surface area (BET), and others as the outcome of theoretical calculations, that is, ionization potential.

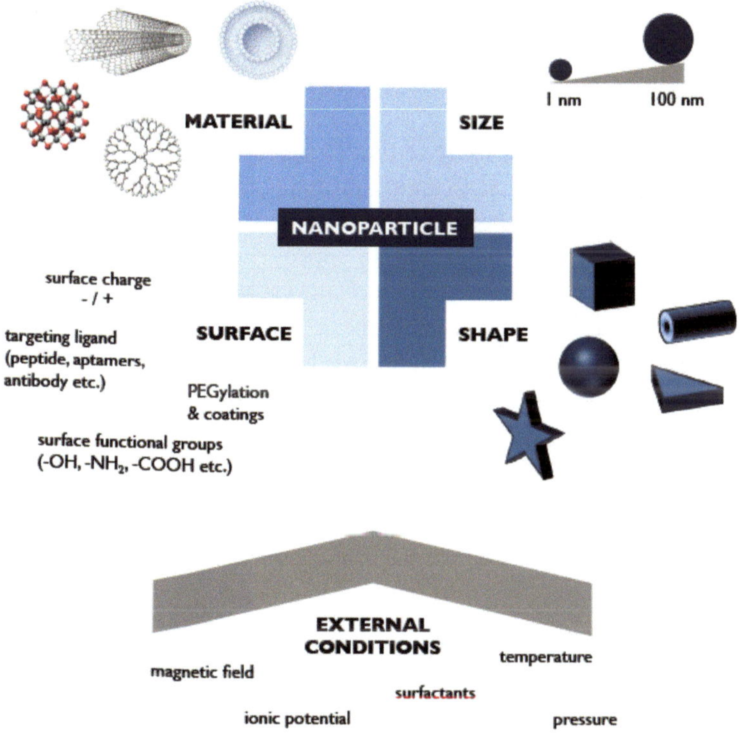

Figure 6.1 Factors determining nanomaterials features.

In the context of descriptors for nanomaterials (nanodescriptors), the idea of a simplified form of structural and physicochemical parameters is not trivial. The reflection of a nanostructure's characteristic requires a different approach than what has been applied so far. In the classical approach, the scope and level of detailed information increase with the considered dimensionality of the chemical structure. In the case of nanostructures, the ways of thinking have to be substantially extended because next to the chemical composition, any physical aspect, that is, size, as well as the surrounding, of chemical individuals plays a significant role in nanomaterials features (Fig. 6.1). Thus, these aspects are discussed in detail in the context of theoretical calculations in the following sections.

6.2.1 Size Aspect

From the theoretical chemistry point of view, higher and more accurate levels of theory are preferred to obtain reliable results in computational modeling. However, their application is limited by the size of the investigated system because when the number of atoms is large, it significantly prolongs the time of calculations or even precludes them. In the case of nanostructures, it is an enormous challenge to choose an appropriate molecular model to reflect properties of nanostructures in computational calculations. Frequently, it is impossible to use the actual size of a nanoscale system (1–100 nm, 10–1000 Å) due to overloading of the computational resources. The only alternative is to perform calculations based on a smaller part of the nanostructure. However, the question about the proper size that represents properly a whole cluster is still open.

There are several theoretical studies [36, 38, 47] that discuss the influence of nanocluster size of pristine metal oxides on their intrinsic properties (Fig. 6.2). The results indicate that some of the properties are linearly correlated with the size of the cluster, that is, such properties as the heat of formation, total energy, electronic energy, and solvent accessible surface area. In turn, for the other properties such as bandgap energy, chemical hardness, chemical softness, and electrophilicity the relationship with the cluster size is nonlinear and increases/decreases until it reaches the saturation point, which usually corresponds to a 5 nm cluster [38]. Thus, depending on the descriptor, it is possible to perform calculations using a small-sized cluster, even equal to a unit cell, and then extrapolate the value to the actual size of the nanostructure. However, these observations are based solely on results obtained for cubic [36, 38] and spherical [47] nanoclusters of metal oxides and what is the appropriate and optimal size of other nanostructures, such as nanotubes and nanosheets, in QM calculations is still unknown.

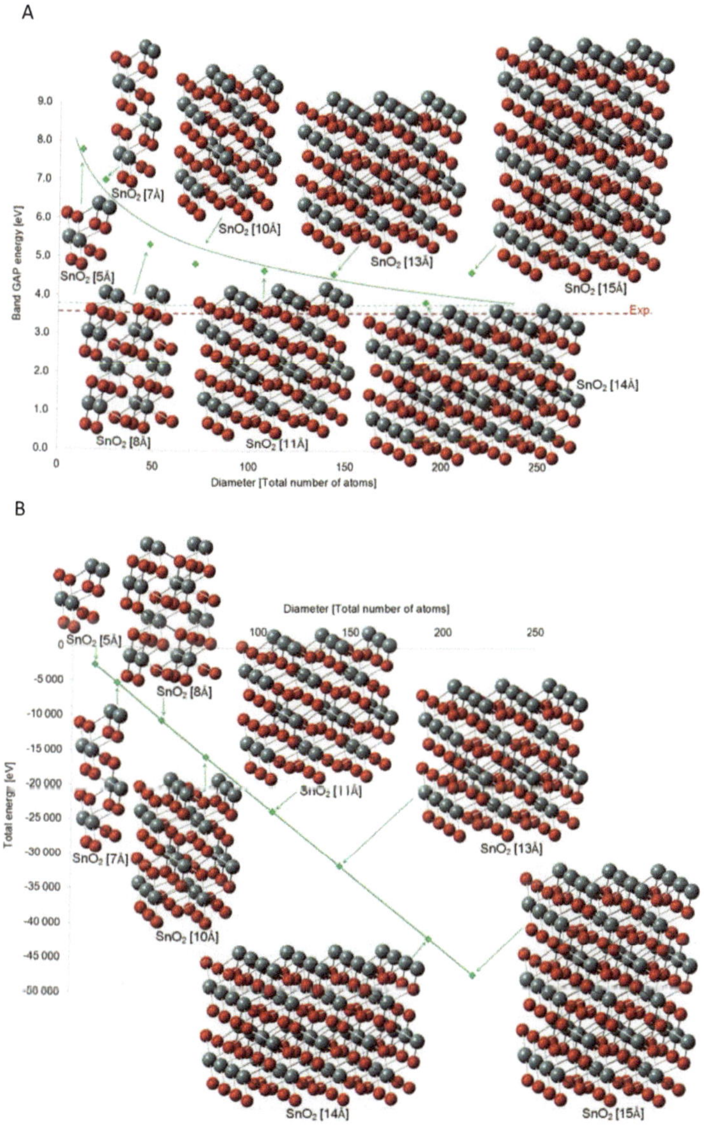

Figure 6.2 Size dependence of bandgap energy (A) and total energy (B) of SnO_2 nanoparticle clusters. Republished with permission of Bentham Science Publishers, Ltd, from Ref. [38]; permission conveyed through Copyright Clearance Center, Inc.

Moreover, it is important to note that the choice of the proper nanocluster size determines not only the level of theory in quantum-chemical calculations and, consequently, reliability of results but also the load of computational resources. Increasing a cluster's size will affect the duration of calculation and increase the total cost. However, it may not always result in better outcomes. Therefore, it is necessary to develop a balanced solution that allows one to take into account the actual nanomaterial properties in optimal time, postulate QM theory of an acceptable level, and simultaneously lower the cost of computing power.

6.2.2 Chemical Composition Aspect

There are several ways to define the chemical structure, for instance, a simplified molecular input line entry system (SMILES) notation, a 2D structure, a 3D structure, or information about a crystalline structure. So far, these approaches were sufficient to characterize the chemical nature of substances such as bioactive substances [48], solvents [26], organic/inorganic pollutants [22, 24], minerals [49], and polymers [50]. However, the description of the chemical composition of a nanomaterial is more complex [51, 52]. First of all, each physical parameter of the nanomaterial, such as size, length, diameter, number of layers, and density, determines the size of the chemical structure. For example, nanotubes of the same length and chemical composition but of different diameters will have quite a different description of the chemical structure. Secondly, some types of nanomaterials are composed of a core, a shell, a coating, or other surface modification, such as functional groups and targeting ligands. As mentioned above, each element of a nanomaterial's structure plays a crucial role in its final features. Thus, a detailed reflection on a nanomaterial's structure in computational modeling is definitely required.

In the case of nano-QSAR/QSPR modeling, various nanostructure descriptions (SMILES, 2D, 3D, and diversified size of crystalline structure) have already been used [30, 45, 47, 53–55]. Mostly, they were applied to distinguish pristine metal and metal oxide nanoparticles; thus, the authors have focused on the importance of chemical composition and have disregarded the effect of the size of the investigated nanoparticle. Nevertheless, it is important to highlight that different manners of descriptions carry different amounts of information about the chemical structure. Consequently, this is reflected in a different level of descriptor sophistication, as well as in obstacles and time for their calculation. For example, descriptors from a SMILES notation or 2D structure can be obtained directly from available data and do not require any additional data processing. On the other hand, in order to calculate descriptors from a 3D or crystal structure, optimization of the structure's geometry is required, which leads to the need for selecting the appropriate method for this process—quantum mechanics (QM), molecular mechanics, molecular dynamics—and next selecting theory of a suitable level. Effectively, these kinds of descriptors involve much more workload, time, and expert knowledge in theoretical chemistry and modeling. However, these descriptors can be especially useful in distinction of nanomaterials differing to a slight extent in chemical composition, as well as in the interpretation of differences in their features. Simultaneously, it is noteworthy that more often data sets of nanoparticles with the same chemical composition are used in modeling, for instance, a data set of titanium dioxide nanoparticles diversified in size, shape, surface area etc. [43, 44]. This kind of approach is possible due to the development of new, specific descriptors encoding the physical aspects of nanomaterials as well as due to the inclusion of experimental data in the modeling process.

Also, a quite different strategy is implemented for functionalized nanomaterials. It assumes that descriptors for each part of a nanostructure, that is, core, shell, coatings, and surface modifiers, could be calculated separately. Then, if nanomaterials have the same core, their diversification is based solely on the structures of modifiers [56, 57]. In such cases, when surface modifiers are common organic or inorganic compounds, conventional tools and typical descriptors can be applied for modeling [46]. The choice of the description of a nanostructure's chemical composition is arbitrary. However, when neglecting the size information of a nanostructure, it is important to choose features that help in identifying differences between diversified nanostructures and explaining their property of interest at the lowest cost and in the shortest possible time.

6.2.3 Surrounding Aspect

Another important aspect that should be taken into account when considering properties of nanomaterials is the surrounding environment/conditions. The properties of nanoparticles closely depend on the presence of chemical individuals such as ions and surfactants, as well as on physical conditions, such as temperature, pressure, and electric/magnetic field [32, 58, 59]. These factors significantly influence the particle-particle interaction through the change of their surface charge, steric barrier, influence on the Brownian dynamic, etc. [32, 58, 59]. Effectively, the aggregation tendency of nanoparticles is changed and consequently, the size and the final features of nanoparticles also get changed. There are experimental reports [31–33, 58, 60–63] that investigate these factors in detail. However, so far, to the best of our knowledge, no quantitative relationship between external conditions and nanoparticle features has been applied in published nano-QSAR/QSPR models.

Nevertheless, the first attempts to include external conditions in modeling for nanomaterials have been already made. For example, one study describes the relationship between the zeta potentials of pristine metal oxide nanoparticles in neutral and ionic mediums [29]. However, this study only emphasizes the importance of the medium's ionic character in the behavior of suspended nanoparticles and indicates the property of nanoparticles that makes them more vulnerable to a change in the zeta potential. The other study [43, 44] describes the approach that includes external conditions in classification mode. In this way, measurement conditions are included in nano-QSAR modeling in a qualitative manner, not quantitative.

The general difficulty in describing and including external conditions in a quantitative way is that it is hard to investigate nanomaterial properties in a complex range of changing conditions. Moreover, performing such kinds of experiments could be very expensive and time consuming, running counter to the idea of efficient, relatively cheap, and quick computational modeling. Moreover, often, the kind of medium, the degree of dilution, or the measurement time is the subject of any research on nanomaterial features, for example, toxicity (not yet published results of European projects). Hence, it is possible to define the optimal conditions for measurements, namely, conditions in which nanomaterials exhibit the toxicity, that is, the highest toxic potential. This knowledge helps avoid experiments that do not reflect the actual situation. When this strategy will be incorporated into nano-QSAR/QSPR modeling is a matter of time.

6.3 The First Nano-QSAR Model and Its Recalculations

Nano-QSAR modeling was initiated as a result of research on the role of chemical composition in the final nanomaterial features. The first nano-QSAR model was developed to investigate the cytotoxicity effect of 17 different metal oxide nanoparticles on the bacteria *Escherichia coli* [41]. The sole descriptor applied in this model was obtained through a QM calculation and allows one to distinguish the analyzed metal oxides by the difference in electronic structures. However, the experimental data set of the cytotoxicity of nanoparticle was reused several times and constituted the basis of the progress in nanodescriptor development [45, 54, 64, 65] (Table 6.1). Consequently, there were many approaches proposed to simplify the method to describe nanoparticles. For example, simple descriptors derived from the periodic table or SMILES notation were applied. Effectively, the diversity in chemical composition was described in a far simpler and faster way than QM calculations. We present here the first applied kind of nanodescriptors as well as other simplified approaches developed on the basis of the same data set.

6.3.1 Quantum-Mechanical Descriptors

QM descriptors are based on theoretical calculations and reflect the electronic structure of compounds [46, 66, 67]. Depending on the level of theory, these methods come closer to the real values of the intrinsic properties of the analyzed compounds. Among the QM descriptors, there are highest occupied molecular orbital–lowest unoccupied molecular orbital (HOMO-LUMO) energy, ionization potential, electron affinity, chemical hardness/softness, dipole moment, etc. These descriptors help observe and precisely determine even very small variances between structurally similar compounds. Obviously, the precision depends on the approximations and simplifications used in the particular method. However, it is noteworthy that the choice of a more advanced method (also, a higher

Table 6.1 Summary of the recalculations of the first nano-QSAR

Descriptors	Chemical description	Statistics						Reference
		R^2	$RMSE_C$	Q^2_{CV}	$RMSE_{CV}$	Q^2_{EXT}	$RMSE_{EXT}$	
Quantum-mechanical descriptors	Crystal structure	0.862	0.20	0.77	0.24	0.83	0.19	Puzyn et al. [41]
SMILES-based optimal descriptors	SMILES notation	0.74–0.83	0.17–0.23	-	-	0.83–0.96	0.14–0.33	Toropov et al. [65]
Improved SMILES-based optimal descriptors	SMILES notation + physicochemical properties	0.89–0.98	-	0.84–0.97	-	-	-	Pan et al. [45]
Periodic table descriptors	Physicochemical properties from the periodic table	MLR models:						Kar et al. [54]
		0.81–0.90	0.16–0.23	0.74–0.85	-	0.72–0.90	0.15–0.27	
		PLS model:						
		0.82	-	0.82	-	0.79	-	
SiRMS descriptors	Simplex encoding with physicochemical properties	0.93	0.13	-	-	0.78	0.32	Sizochenko et al. [64]
LDM descriptors	Drop model in liquid solutions							
Metal–ligand descriptors	Interactions characteristic							
Full-particle descriptors	Crystal structure, which regards shell and surface regions	0.87	-	0.81	-	-	-	Tamm et al. [47]

level of theory) leads to the time prolongation of calculations and higher consumption of computational resources. Simultaneously, these aspects are also closely related to the size of the analyzed structure, to be more precise, the number of atoms. Puzyn et al. [41] performed the first predictive nano-QSAR model, which considered the effect of the chemical composition of 17 different types of nanoparticles on the cytotoxicity of *E. coli* cells. In this study Puzyn et al. used a semiempirical level of theory method PM6 to calculate QM descriptors for metal oxide clusters (12 Å × 12 Å × 12 Å) as well as for single atoms. Finally, a single descriptor, ΔH_{Me+}, was applied in the model, which represents the enthalpy of the gaseous cation formation of a proper metal from the metal oxide nanoparticle. The applied descriptor helped obtain a strong correlation between nanoparticles' intrinsic property and the modeled endpoint. Results were confirmed with very good statistical parameters of the model (R^2 = 0.862, $RMSE_C$ = 0.20, Q^2_{CV} = 0.77, $RMSE_{CV}$ = 0.24, Q^2_{EXT} = 0.83, and $RMSE_{EXT}$ = 0.19). Simultaneously, the applied descriptor was able to estimate a possible mechanism of the nanomaterial's toxicity, which was earlier proposed by the experts attending the 2008 NATO workshop [68]. Also, the results of modeling are consistent with the results obtained by experimental groups [69, 70], which identified that among the most common nanomaterials, ZnO is one of the most toxic metal oxides, whereas TiO_2 is one of the metal oxides exhibiting the lowest toxicity. The model developed by Puzyn et al. helped explain this difference through the different capability of the cation formation.

The considerable advantage of QM descriptors is a great opportunity to link the intrinsic properties of nanostructures with their external features. Nevertheless, the application of QM descriptors requires the users to have a background in theoretical chemistry, which is crucial in selecting a molecular model of an appropriate size, as well as the proper method for calculations. The first nano-QSAR model did not indicate the difficulties involved because it was able to estimate the toxicity of the analyzed metal oxides on the basis of just the information regarding the metal atoms, the smallest unit that can characterize pristine nanostructures. Thus, it can be said that to some extent all methods at various theoretical levels could be applied. At the same time, the specificity of the given and available information can help explain the mode of action only to some extent.

6.3.2 SMILES-Based Optimal Descriptors

The first recalculation of groundbreaking nano-QSAR model was performed by Toropov et al. [65]. They proposed descriptors that were previously developed for organic compounds [71, 72]. The approach is based on the SMILES notations, which are easily available in any software dedicated to building molecular models. The SMILES-based optimal descriptors are calculated with the available online CORAL software [73] as follows:

$$\mathrm{DCW}(T)=\sum_{k=1}^{E}W\left({}^{1}\mathrm{S}_{k}\right)+\sum_{k=1}^{E-1}W\left({}^{2}\mathrm{S}_{k}\right)+\sum_{k=1}^{E-2}W\left({}^{3}\mathrm{S}_{k}\right) \quad (6.1)$$

Here E represents the total number of SMILES elements for an appropriate molecular structure and is equal to the ${}^{1}\mathrm{S}_k$ value. ${}^{1}\mathrm{S}_k$, ${}^{2}\mathrm{S}_k$, and ${}^{3}\mathrm{S}_k$ represent one-, two-, and three-element SMILES attributes. If a SMILES representation takes the form 'ABCDE', then ${}^{1}\mathrm{S}_k$, ${}^{2}\mathrm{S}_k$, and ${}^{3}\mathrm{S}_k$ are represented as follows:

'ABCDE'→'A', 'B', 'C', 'D', 'E' (${}^{1}\mathrm{S}_k$)
'ABCDE'→'AB', 'BC', 'CD', 'DE' (${}^{2}\mathrm{S}_k$)
'ABCDE'→'ABC', 'BCD', 'CDE' (${}^{3}\mathrm{S}_k$)

W represents the correlation weights of SMILES attributes. The correlation weights are closely related to the threshold value (T), which classifies SMILES attributes as rare or active. The SMILES attributes are classified as rare when a sufficient number of given SMILES attributes is not included in the training set. For example, if $T = 5$ and in the training set fewer than 5 sought SMILES attributes are found, this SMILES attribute is assigned as rare. Consequently, the correlation weights of the rare SMILES attributes are equal to zero and have no influence on the model. The correlation weights of active SMILES attributes (not rare S_k in the data set) are calculated using the Monte Carlo method optimization. There are two possible systems of optimization. The first scheme assumes the split of all available data into a training set and a test set. Then, the correlation weights are calculated on the basis of the correlation coefficients between DCW(T) and the modeled parameter. The second scheme, named "balance of correlation," assumes the split of available data into a subtraining set, a calibration set, and a test set. Then, the target function for the balanced correlation weight is calculated according to the following equation:

$$\mathrm{CB}=R_{\mathrm{s}}+R_{\mathrm{c}}-\mathrm{ABS}\left(R_{\mathrm{s}}-R_{\mathrm{c}}\right)\times 0.1, \quad (6.2)$$

where R_S and R_C are the correlation coefficients between DCW(T) and the modeled parameter for the subtraining and calibration sets, respectively. It is noteworthy that the calibration set is a preliminary test set. Thus, the split into three sets allows obtaining more balanced correlation weights in comparison to the first scheme, where the training set combines the subtraining and calibration sets into one.

In the recalculations of the first nano-QSAR model, Toropov et al. [65] assumed a low threshold, $T = 2$. In the case of such a simple SMILES description of analyzed metal oxides, only ^1S_k could be used in the model and these are '[', '=', and 'O'. The statistic parameters of all the developed models (six models varied in the random split of compounds into training and test sets) are in the following range: R^2 = 0.74–0.83, $RMSE_C$ = 0.17–0.23, Q^2_{CV} = -, $RMSE_{CV}$ = -, Q^2_{EXT} = 0.83–0.96, and $RMSE_{EXT}$ = 0.14–0.33.

The advantage of the proposed method for the descriptor calculation is undoubtedly simplicity. The method does not require one to build molecular models of analyzed compounds. Effectively, the method skips all the issues related to the appropriate size of the nanocluster, the method of geometry optimization, the level of theory, etc. However, in the case of compounds that do not have SMILES notations with a sufficient number of repeating elements, the SMILES notation may not be an adequate method to distinguish them. Due to the low number of repeating SMILES attributes (rare S_k elements) in an analyzed data set, their correlation weight will be assigned as irrelevant ($W = 0$) and will result in their exclusion from the model. Moreover, the mechanistic interpretation of the model on the basis of the SMILES descriptors is limited. Nevertheless, in response to this issue, Pan et al. [45] upgraded and modified the method of SMILES-based optimal descriptors with additional parameters of nanostructures.

6.3.3 Improved SMILES-Based Optimal Descriptors

Pan et al. [45] proposed the extension of the SMILES-based optimal descriptors to increase the interpretability of results by including physicochemical properties of analyzed structures, that is, molecular weight, cationic charge, mass percentage of metal atoms, size, and aggregation size. Effectively, the improved SMILES-based optimal descriptors are calculated as follows:

$$\mathrm{DCW}(T, N_{\mathrm{epoch}}) = \sum W(^{1}\mathrm{S}_k) + \sum W(^{2}\mathrm{S}_k) + \sum W(^{1}\mathrm{C}_k) + \sum W(^{2}\mathrm{C}_k) \quad (6.3)$$

Here $^1\mathrm{S}_k$ and $^2\mathrm{S}_k$ represent one- and two-element SMILES attributes, $^1\mathrm{C}_k$ and $^2\mathrm{C}_k$ represent the code of k-physicochemical features, and W represents the correlation weight of the SMILES attribute or the physicochemical feature. Then, the procedure of descriptor calculations involves standardization of the X_k feature (X is an improved SMILES-based optimal descriptor, and k can be any feature for which X is calculated) according to the formula

$$\mathrm{Norm}\,(X_k) = \frac{\min X_k + X_k}{\min X_k + \max X_k} \quad (6.4)$$

and differentiation of physicochemical features according to the proposed scale into one class from 1 to 9 (Fig. 6.3):

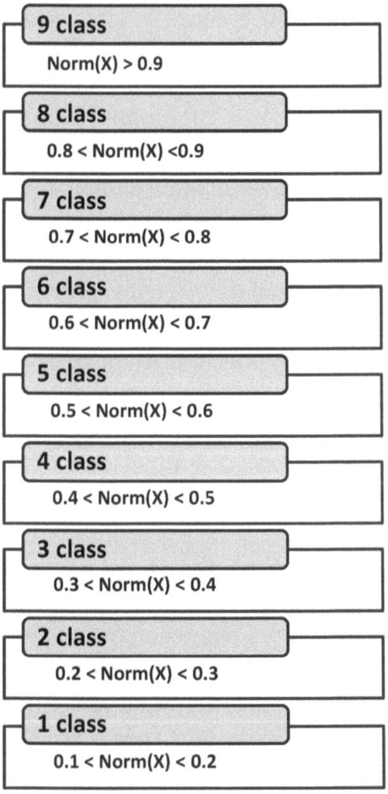

Figure 6.3 Classes of standardized physicochemical features [45].

The construction of ^1C_k and ^2C_k is similar to the construction of ^1S_k and ^2S_k, which means when A, B, C, D, and E represent molecular weight, cationic charge, mass percentage of metal atoms, size, and aggregation size, respectively, the procedure is as follows:

'ABCDE'→'A', 'B', 'C', 'D', 'E' (^1C_k)

'ABCDE'→'AB', 'BC', 'CD', 'DE' (^2C_k)

Then, the optimal descriptors are calculated according to the correlation weights of active physicochemical features by Monte Carlo optimization, where T represents the threshold, which classifies the feature as rare or active, and N_{epoch} represents the number of epochs in the optimization process. In the optimization process, the correlation coefficient (W) of each attribute is determined initially by randomly setting the value 1 ± 0.1. The random value is in the range (0, 1). The order of attributes number, 1, 2, 3, 4, etc., is replaced by a random sequence, 3, 2, 4, 1, etc. Then, the correlation coefficient (R_1) between the modeled parameter and the descriptor of the training set is calculated. Simultaneously, the correlation weight of each attribute is modified according to the following algorithm:

1. $\Delta W_i := 0.001 \times W_i$; Eps $:= 0.0100\, \Delta W_i$;
2. $W_i := W_i + \Delta W_i$;
3. Calculation of the correlation coefficient (R_2) after W_i modification;
4. If $R_2 > R_1$, then $R_1 := R_2$; go to Step 2;
5. $W_i := W_i + \Delta W_i$
6. $\Delta W_i := -0.5 \times \Delta W_i$
7. If the absolute value $\Delta W_i >$ Eps then go to Step 2.

The seven steps are performed for correlation weights of each attribute in the model. The whole process stops when the increase of correlation coefficient is less than 0.001.

Pan et al. [45] implemented physicochemical features and SMILES attributes of nanostructures (size-dependent and size-independent descriptors) in two nano-QSAR models. Effectively, the applied descriptors helped to explain the relationship between the nanostructure characteristic and the estimated toxicity towards *E. coli* cells. Both models based on the DCW(T, N_{epoch}) descriptors obtained with different thresholds and numbers of epochs in Monte Carlo optimization were characterized with better statistics of fit and robustness (R^2 = 0.89–0.98, Q^2_{CV} = 0.84–0.97, s = 0.007–0.179, Q^2_{EXT}

= -, $RMSE_{EXT}$ = -, F = 164–391, and $p < 0.0001$) than the first nano-QSAR model. However, the models were not externally validated; thus, the predictive ability of the models was not verified and the reliability of the results can be disputable.

The great advantage of the proposed upgraded methodology of SMILES-based optimal descriptors is the incorporation of physicochemical parameters into descriptors. Effectively, the ability of interpretation significantly increased. As in the basic version, the improved SMILES-based optimal descriptors do not require advanced computational resources. Still, the method can be applied solely to the data set of structurally similar compounds. However, the proposed method is an example of the consensus in the connection of physicochemical parameters, including experimental measurements, and simplicity in the expression of chemical composition.

6.3.4 Periodic Table Descriptors

Previously, recalculation with the improved SMILES-based descriptors performed by Pan et al. [45] and then Kar et al. [54] proposed the other simple method of descriptor calculation based solely on the periodic table. Kar et al. proposed universal descriptors focused on the variety of electronegativity, valence, ionic radius, atomic number, charge of cation, etc., which are simultaneously independent of the nanoparticle size. Kar et al. assumed that if metal oxide nanopowders are similar in size, they would have also similar-sized agglomerates in water suspension. Thus, a better way to explain the differences in nanomaterials' toxicity is to concentrate on the diversity in their chemical compositions. Kar et al. proposed seven molecular descriptors based on the periodic table details (Table 6.2). Nevertheless, currently there is online software [74] available that calculates 31 elemental descriptors (Table 6.3).

Table 6.2 Periodic table descriptors for metal oxide nanoparticles proposed by Kar et al. [54]

Descriptor	Definition
N_{Metal}	Number of metal atoms
N_{Oxygen}	Number of oxygen atoms

Descriptor	Definition
χ	Metal electronegativity
$\Sigma\chi$	Sum of metal electronegativity of a singular metal oxide
$\Sigma\chi/N_{Oxygen}$	Sum of metal electronegativity of a singular metal oxide weighted by the number of oxygen atoms in the metal oxide
χ_{OX}	Charge of the metal cation related to the proper oxide
MW	Molecular weight

Table 6.3 Elemental descriptors calculated with software available online [74]

Descriptor	Definition
D1_Metals	Total number of metals
D2_SemiMetals	Total number of semimetals
D3_HeteroNonMetals	Total number of nonmetallic heteroatoms (N, O, F, P, S, Cl, Se, Br, I, At) excluding hydrogen
Metals_SumIP	Sum of ionization potential energies (kJ/mol) of metals
SemiMetals_SumIP	Sum of ionization potential energies (kJ/mol) of semimetals
Electrons_ActiveM	Number of electrons of active metal
AtWt_ActiveM	Atomic weight of active metal
MassNo_ActiveM	Mass number of active metal
Neutons_ActiveM	Number of neutrons of active metal
IP_ActivM	Ionization potential energy (kJ/mol) of active metal
X_ActivM	Electronegativity of active metal
VWR_ActivM	Van der Waals radius of active metal
Electrons_Activ_SM	Number of electrons of active metalloid/semimetal

(Continued)

Table 6.3 (*Continued*)

AtWt_Activ_SM	Atomic weight of active metalloid/semimetal
MassNo_Activ_SM	Mass number of active metalloid/semimetal
Neutons_Activ_SM	Number of neutrons of active metalloid/semimetal
IP_Activ_SM	Ionization potential energy of active metalloid/semimetal
X_Activ_SM	Electronegativity of active metalloid/semimetal
VWR_Activ_SM	Van der Waals radius of active metalloid/semimetal
Electrons_Activ_NM	Number of electrons of active nonmetal
AtWt_Activ_NM	Atomic weight of active nonmetal
MassNo_Activ_NM	Mass number of active nonmetal
Neutons_Activ_NM	Number of neutrons of active nonmetal
IP_Activ_NM	Ionization potential energy of active nonmetal
X_Activ_NM	Electronegativity of active nonmetal
VWR_Activ_NM	Van der Waals radius of active nonmetal
SuM_Active_M_SM	Summation of the number of metals and semimetals
SuMElectrons_Active_M_SM	Summation of electrons of active metals and semimetals
SuMAtWt_Active_M_SM	Summation of atomic weights of active metals and semimetals
SuMMassNo_Active_M_SM	Summation of mass numbers of active metals and semimetals
SuMNeutons_Active_M_SM	Summation of neutrons of active metals and semimetals

Kar et al. [54] applied periodic table descriptors to nano-QSAR models developed by two different methods, stepwise multiple linear regression (s-MLR) and partial least squares regression (PLSR), as well as a few random splits of the data into training and test sets. The MLR method helped obtain a monoparametric model with χ_{OX} (R^2 = 0.81–0.90, $RMSE_C$ = 0.16–0.23, Q^2_{CV} = 0.74–0.85, Q^2_{EXT} = 0.72–0.90, and $RMSE_{EXT}$ = 0.15–0.27). On the other hand, the model developed with the PLSR method includes two descriptors: χ_{OX} and χ (R^2 = 0.82, Q^2_{CV} = 0.75, and Q^2_{EXT} = 0.79). Consequently, the mechanistic interpretation of both models is consistent with the previously suggested mechanism and chemical intuition. The cation metal charge (χ_{OX}) exhibits reductive properties resulting from the easy detachment of metal cations from the nanoparticle surface, while the electronegativity (χ) defines the energy required for this process.

The great advantage of the proposed periodic table–based descriptors is the ease of obtaining them, including SMILES input preparation and the short time of calculations. The calculations do not require any advanced theoretical knowledge, unlike the calculations of quantum-chemical descriptors. Nevertheless, even such simple information obtained from the periodic table allows the exploration of divergences of the electron structure of nanomaterials. The only drawback is that the method does not regard the size aspect, and thus it can be applied for nanomaterials of a similar size range.

6.3.5 SiRMS Descriptors

The next recalculation of the first nano-QSAR model was performed by Sizochenko et al. [64], who combined three different types of nanostructure descriptions: simplex representation of molecular structure (SiRMS), liquid drop model (LDM) descriptors (see Section 6.3.6), and metal–ligand binding (MLB) characteristics (see Section 6.3.7).

The SiRMS was previously developed for virtual screening and molecular design of organic compounds [67]. However, its usefulness for nanostructures [40, 64] was also verified. The SiRMS approach assumes encoding molecular structure on the basis of the first level of organization, that is, on the basis of the fragments of fixed composition and topology of the chemical structure [75]. The SiRMS method extends the previously developed SMILES-based approaches, which takes into account mainly different types of atoms. Here, the differentiation of molecules is also performed on the basis of the other characteristics of atoms, such as charge, lipophilicity, electronegativity, Lennard-Jones potentials, and van der Waals interaction [75]. The original chemical structure is converted to symbols related to the appropriate atomic property, that is, electronegativity. Subsequently, the modified structure is fragmentized to get a combination of simplexes, where the maximal simplex can be composed of tetratomic fragments. Effectively, the level of description increases successively from 1D to 4D and any compound is represented by different systems of simplexes.

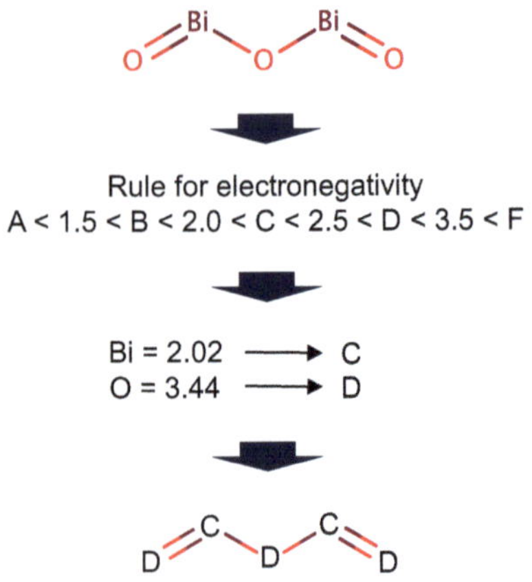

Figure 6.4 The example of Bi_2O_3 conversion into SiRMS representation on the basis of electronegativity. Reproduced from Ref. [64] by permission of The Royal Society of Chemistry.

In the study presented by Sizochenko et al. [64] a nanostructure is represented by a 2D molecular graph. All vertices in a graph symbolize atoms, their type, and proper features, while the connections between the vertices reflect the bonds' nature. The SiRMS approach encodes the molecular structure by organizing discrete groups according to the value range of the analyzed property and assigning atoms to the appropriate group. In the mentioned study, the authors assigned to the vertices appropriate symbols reflecting the value of electronegativity ($A < 1.5 < B < 2.0 < C < 2.5 < D < 3.5 < F$) and Lennard-Jones potential regarding the depth of the potential ($A < 0.01 < B < 0.02 < C < 0.04 < D$) and the distance at which the potential reaches the minimum ($A < 2.5 < B < 3 < C < 3.5 < D < 4 < F$). For example, according to this idea, the representation considering the electronegativity of the analyzed Bi_2O_3 nanoparticle is B_2D_3 because the electronegativity of Bi is equal to 2.02, while the electronegativity of O is 3.44 (Fig. 6.4). Further, the converted structure was fragmentized in accordance with the general rules (Fig. 6.5).

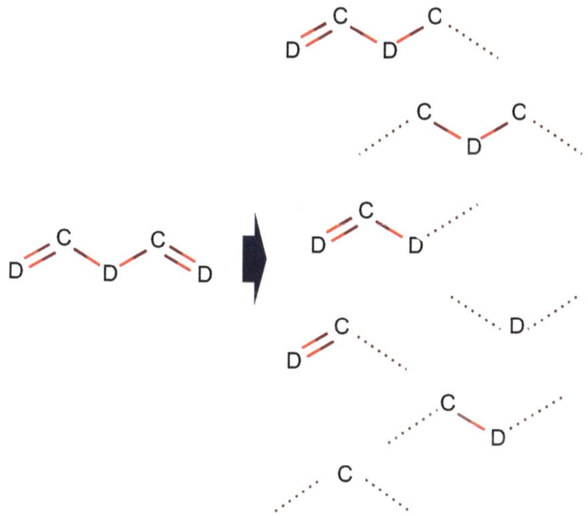

Figure 6.5 The example of fragmentization and simplexes. Reproduced from Ref. [64] by permission of The Royal Society of Chemistry.

In the study discussed here [64], Sizochenko et al. decided to consider molecular fragments of 1 to 3 atoms. The authors combined the simplex representation reflecting the first level of molecular organization with LDM descriptors, which encode the nanoparticle aggregate's characteristics and MLB characteristics. Effectively, in the nano-QSAR model, the contribution of SiRMS-based electronegativity descriptors and SiRMS-based van der Waals descriptors toward the cytotoxic effect on the bacteria *E. coli* was 32% and 7%, respectively, of all descriptors included in the model. Among them, in the developed model were included S_3 (29%), S_1 (7%), and S_2 (3%) SiRMS-derived descriptors. Descriptor S_3 encodes electronegativity of three-atomic fragments [Me]-[O]-[Me], where the smaller electronegativity results in higher toxicity. Descriptor S_1

concerns unbonded two-atomic fragments [Me] ...[Me] and reflects the distance where the potential of van der Waals interactions reaches the minimum. Descriptor S_2 reflects the number of oxygen atoms in the molecule and, consequently, is related to the acid-basic properties of the molecule. A summary of the statistic parameters of the model is contained in the last paragraph of Section 6.3.7 [64].

The SiRMS representation enables characterization of the atomic organization in the molecule. This approach distinguishes nanostructures on the basis of not only types of atoms but also their physicochemical features. The applied 2D approach does not require real geometry of the structure but only a reflection of the chemical composition. In this way, the SiRMS approach disregards the issue of real size and visualization of structural conformation of nanostructures. Hence, the size aspect cannot be investigated and no conclusion can be deduced in this area.

6.3.6 Liquid Drop Model Descriptors

The second type of descriptors used by Sizochenko et al. [64] in the recalculation of the first nano-QSAR model are LDM descriptors. Nanoparticles immersed in solvents display completely different physicochemical properties than their corresponding nanopowders. The interactions between nanoparticles and the medium notably impact the size of agglomerates, surface, surface-volume (SV) ratio, etc. In this regard, Sizochenko et al. [64] decided to include this aspect in the investigations of cytotoxic effect of metal oxide nanoparticles on *E. coli*.

Sizochenko et al. [64] proposed an approach to represent nanostructures in a liquid solution called LDM. In this method, the nanoparticle is illustrated as a spherical drop in which all atoms and molecules are tightly packed. In this way, the density of the entire nanoparticle is equal to the mass density. Analogically, it looks the same in the context of nanoparticles in agglomerates. The minimum radius of interactions between individuals (i.e., molecules in nanoparticles or nanoparticles in agglomerates) is expressed by the Wigner–Seitz radius:

$$r_{\mathrm{w}} = \left(\frac{3M}{4\pi\rho N_{\mathrm{A}}} \right)^{1/3} \tag{6.5}$$

where M is molecular weight, ρ is mass density, and N_A is Avogadro constant.

Due to the assumed spherical shape of the nanoparticle (also, agglomerate), the number of molecules in the analyzed nanoparticle (or the number of nanoparticles in the analyzed agglomerate) can be express as:

$$n = \left(\frac{r_0}{r_w} \right)^3, \tag{6.6}$$

where r_0 is the radius of the nanoparticle (or the agglomerate).

The relationship between r_0 and r_w highlights that a decrease in the particle size results in an increment of its surface area. Simultaneously, it is important to point out that elements located on the surface are affected by different interaction forces than the elements located in the volume (Fig. 6.6).

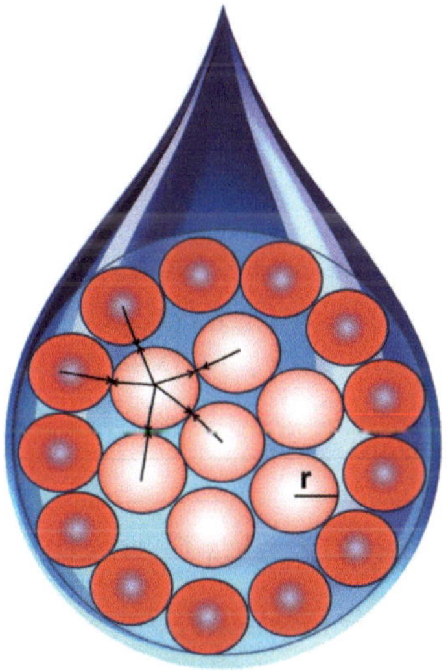

Figure 6.6 Liquid drop model. Red circles, interaction forces between molecules situated on the surface; white circles, interaction forces between molecules situated in the volume. Reproduced from Ref. [64] by permission of The Royal Society of Chemistry.

The number of surface elements can be defined as:

$$F = 4n^{-1/3} \tag{6.7}$$

Consequently, the SV ratio is equal to:

$$SV = \frac{F}{1-F} \tag{6.8}$$

The aggregation parameter is related to the size of the single particle and the size of the agglomerate ratio:

$$AP = \frac{\text{Size of aggregate}}{\text{Size of single particle}} \tag{6.9}$$

Sizochenko et al. used newly developed descriptors in combination with SiRMS-based descriptors (see Section 6.3.5) and MLB characteristics (see Section 6.3.7) to estimate the effective cytotoxic concentration of 17 metal oxide nanoparticles against *E. coli*. LDM-based descriptors constitute 31% of the whole contribution in the model (r_W, 22%; SV, 7%; and ρ, 2%). The great share of r_W indicates that weakly bound molecules on the nanocluster's surface can be responsible for cytotoxic effects. This is also closely connected with the SV ratio and mass density (ρ). A summary of the statistic parameters of the model is in the last paragraph of Section 6.3.7.

The LDM is an example of a nanoparticle's size description. This approach is based on the physical occurrence and enables characterization of the suspended nanoparticles and their agglomerates. The calculations require only knowledge about chemical composition and basic experimentally measured properties, that is, mass density, size of the nanoparticle, and hydrodynamic size. The method does not describe any other feature of a nanoparticle except size.

6.3.7 Metal–Ligand Binding Characteristic

The third type of descriptors used by Sizochenko et al. [64] are the MLB characteristic. This method is focused on the description of the metal ion's affinity toward soft ligands on biomolecules. In the mentioned study, the authors developed two MLB-based descriptors: the covalent index (CI) and cation polarizing power (CPP). The first one encodes the relative significance of covalent interactions in MLB through combinations of electronegativity (χ) and Pauling radius (r):

$$CI = \chi m^2 r \tag{6.10}$$

The second descriptor, CPP, expresses the energy of the electrostatic interactions between metal ion and biochemical ligand through combinations of the ion charge (Z) and Pauling radius (r):

$$CPP = Z^{2/r} \quad (6.11)$$

Among the proposed MLB-based descriptors, only cation polarization power was included in the recalculated nano-QSAR model, but its contribution was as much as 30% of the total descriptors. In this way, CPP strengthened the importance of electrostatic interactions in the cytotoxic effect of analyzed metal oxide nanoparticles, which was also expressed by SiRMS-based descriptors. The final nano-QSAR model combined three different types of descriptors, each carrying out unique information about the nanoparticles' properties, that is, size and electronic properties. So, the model developed with the random forest method has a good prediction ability, confirmed by satisfactory statistical parameters (R^2 = 0.93, $RMSE_C$ = 0.13, Q^2_{CV} = -, $RMSE_{CV}$ = -, Q^2_{EXT} = 0.78, and $RMSE_{EXT}$ = 0.32).

An MLB description can include an important aspect of interactions between metal ions and biochemical ligands. MLB-based descriptors do not require any special software to generate a description, only information available in the periodic table. The method is based on the characteristics of a metal ion's core and does not consider the real size of nanoparticles. Thus, MLB descriptors can be applied solely for nanomaterials with different chemical compositions and similar sizes.

6.3.8 Full-Particle Descriptors

Surface area is one of the most common reasons of the unique features of nanomaterials. For instance, a high aspect ratio between the small size and large surface area is the reason for the easier leaching of atoms from the surface and, consequently, a high surface reactivity. Thus, in order to investigate the surface aspect, Tamm et al. [47] developed the first set of full-particle descriptors to distinguish between atoms from the surface and core atoms of nanoparticles.

The proposed approach is based on the thermodynamically most stable crystal structure of an appropriate metal oxide. The procedure is assumed to replicate the unit cell in three dimensions and then generate a spherical shape by deleting all atoms outside of the assumed diameter of a nanoparticle. However, electroneutrality, potential energies, and coulombic interactions of atoms are also considered. The molecular model divides nanoparticles into two regions, surface and core, where the surface involves atoms from the 1 nm surface layer (Fig. 6.7). Then, the force field method, namely large-scale atomic/molecular massively parallel simulator (LAMMPS software [76]) with 3D periodic boundary conditions, is applied in calculations. It is important to point out that the size of a periodic box has to be sufficiently large to include all atoms within. Similarly, cutoff values of the Buckingham potentials and Wolf summation have to ensure appropriate conditions to include interatomic interactions in the said space. Finally, the presented approach allows one to obtain a set of 35 nanodescriptors about special features of the surface atoms. Among them are descriptors related to size, chemical composition, potential energy, topology, and lattice energy (Table 6.4).

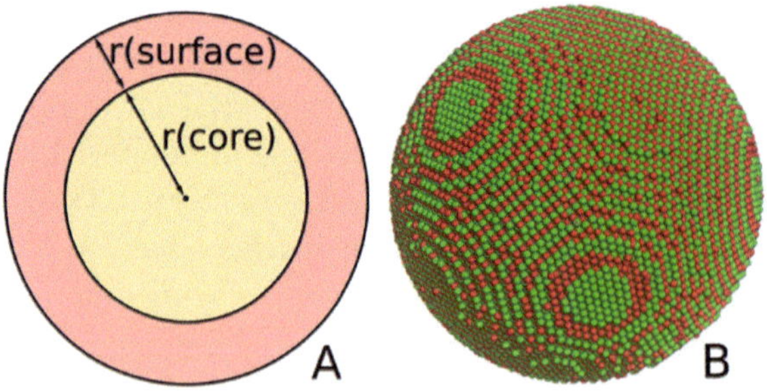

Figure 6.7 The idea for the molecular model of a nanoparticle in the full-particle descriptors. Reproduced from Ref. [47] with permission of The Royal Society of Chemistry.

Table 6.4 Full-particle nanodescriptors

Descriptor	Definition
Geometric	
1	Diameter of the nanoparticle [Å]
2	Surface area of the nanoparticle [Å^2]*
3	Volume of the nanoparticle [Å^3]

Constitutional	
4	Total number of atoms in the nanoparticle*
5	Total number of atoms in the surface region of the nanoparticle*
6	Total number of atoms in the shell region of the nanoparticle*
7	Total number of metal atoms in the nanoparticle*
8	Total number of metal atoms in the surface region of the nanoparticle*
9	Total number of metal atoms in the shell region of the nanoparticle*
10	Total number of oxygen atoms in the nanoparticle*
11	Total number of oxygen atoms in the surface region of the nanoparticle*
12	Total number of oxygen atoms in the shell region of the nanoparticle*
Potential energy related	
13	Average potential energy of atoms in the nanoparticle [eV]
14	Average potential energy of atoms in the surface region of the nanoparticle [eV]
15	Average potential energy of atoms in the shell region of the nanoparticle [eV]
16	Average potential energy of metal atoms in the nanoparticle relative to the infinite crystal [eV]
17	Average potential energy of metal atoms in the surface region of the nanoparticle relative to the infinite crystal [eV]
18	Average potential energy of metal atoms in the shell region of the nanoparticle relative to the infinite crystal [eV]
19	Average potential energy of oxygen atoms in the nanoparticle relative to the infinite crystal [eV]
20	Average potential energy of oxygen atoms in the surface region of the nanoparticle relative to the infinite crystal [eV]

(*Continued*)

Table 6.4 (*Continued*)

21	Average potential energy of oxygen atoms in the shell region of the nanoparticle relative to the infinite crystal [eV]
Topologic	
22	Average coordination number of all atoms in the nanoparticle
23	Average coordination number of atoms in the surface region of the nanoparticle
24	Average coordination number of atoms in the shell region of the nanoparticle
25	Average coordination number of metal atoms in the nanoparticle
26	Average coordination number of metal atoms in the surface region of the nanoparticle
27	Average coordination number of metal atoms in the shell region of the nanoparticle
28	Average coordination number of oxygen atoms in the nanoparticle
29	Average coordination number of oxygen atoms in the surface region of the nanoparticle
30	Average coordination number of oxygen atoms in the shell region of the nanoparticle
Lattice energy related	
31	Lattice energy of the nanoparticle [eV]
32	Difference of the lattice energies of the nanoparticle and the infinite crystal [eV]
33	Lattice energy of the nanoparticle divided by the diameter of the nanoparticle [eV/Å]
34	Lattice energy of the nanoparticle per unit surface area [eV $Å^{-2}$]*
35	Lattice energy of the nanoparticle per unit volume [eV $Å^{-3}$]*

*Logarithmic values

Source: Reproduced from Ref. [47] with permission of The Royal Society of Chemistry.

Due to the high influence of size on nanoparticle features, the contribution of size seems to be crucial in the modeling. However, as previously mentioned, some properties of nanoparticles increase linearly and others nonlinearly to reach the saturation point, which corresponds to the appropriate diameter. Thus, because of differences in the size dependency of nanodescriptors it is impossible to include the size of a nanoparticle as a simple correction parameter in a nano-QSAR/QSPR model. The only way is to investigate each parameter, or descriptor, separately for a relatively narrow range of sizes and then use these functions for extrapolation for sizes outside the function. That kind of investigation was performed for the full-particle descriptors (Fig. 6.8). The authors indicated that the size dependency of descriptors could be positive or negative, and the function of shape and saturation point can be different for different descriptors. However, if the function of size dependence is known, there are opportunities to perform estimations for other sizes.

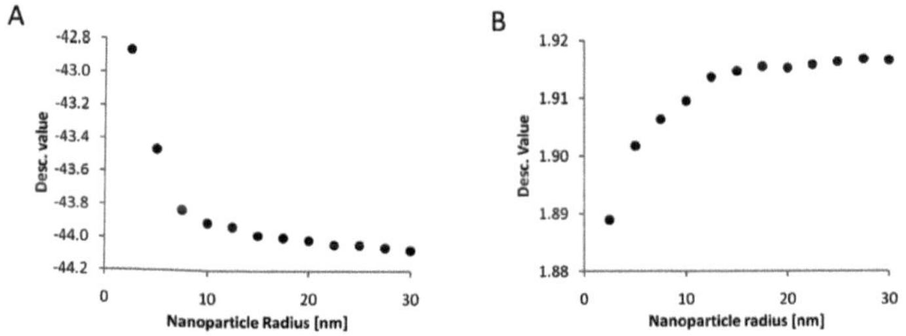

Figure 6.8 Size dependence of random full-particle descriptors: A, average potential energy of metal atoms in Fe_2O_3; and B, average coordination number of all oxygen atoms in SiO_2. Reproduced from Ref. [47] with permission of The Royal Society of Chemistry.

Tamm et al. verified the usability of the full-particle descriptors in the prediction of metal oxide nanoparticles' toxicity toward *E. coli*. The authors obtained a two-parameter nano-QSAR model with satisfactory statistics (R^2 = 0.87 and Q^2_{CV} = 0.81). These descriptors are D_1, the average potential energy of atoms in the shell region of a nanoparticle; and D_2, the average potential energy of oxygen atoms in the surface region of a nanoparticle. Thus, the mechanistic interpretation of the model includes the significant contribution of the chemical composition as well as highlights the role of the oxidation stage of metal ions in the observed toxicity of metal oxide nanoparticles. In other words, the results are consistent with the previous studies provided by other research groups.

Tamm et al. developed a set of theoretical nanodescriptors based solely on the unit cell and Buckingham potentials. Nevertheless, according to the specific size dependence of each descriptor, it is possible to extrapolate these descriptors to nanoparticles of different sizes. Thus, these are one of the already developed size dependence descriptors. By default, full-particle descriptors are addressed to spherical nanoparticles, but the authors indicate that it is also possible to construct specific algorithms for other shapes. Moreover, it is noteworthy to point out that despite the close interlinkage between core and surface, the general intercorrelation between descriptors is low. The calculation of descriptors is technically simple and does not consume computational resources intensively. As a result, full-particle descriptors can be effectively applied to QSAR/QSPR modeling for diverse nanoparticles.

6.4 Other Nanodescriptors

Independently to the first nano-QSAR models/recalculations, new types of nanodescriptors were developed for a variety physicochemical and biological endpoint. Furthermore, the proposed nanodescriptors are based both on experimentally and theoretically generated data. Among them, we can distinguish a diverse level of details in a nanostructure's description. Obviously, it is inherently related to the diverse degree of difficulty in obtaining them as well as the knowledge required for interpretation.

Experimentally measured parameters seem to be the best in explaining nanomaterial features because they directly reflect the specific nature of the nanostructures, that is, size, surface area, and crystallographic structure. However, the application of such descriptors in nano-QSAR/QSPR models has some drawbacks. Firstly, even for simple analysis, sophisticated laboratory equipment and qualified staff are required. Secondly, the analysis can be time consuming and expensive. Thirdly, this approach can be applied solely for existing nanomaterials. Nonetheless, the information extracted from the experimental results can be very useful in the interpretation of modeling results. Consequently, the obtained results can help design new, efficient nanomaterials because, in practice, it is much easier to manipulate physical parameters, for example, surface area, size, and porosity, than any theoretical parameters, like electronic structure, aspect ratio, or some structural indexes. Thus, there are various approaches to including experiments in nano-QSAR/QSPR modeling.

Designing new materials, including nanomaterials, requires knowledge about the physicochemical parameters responsible for the expected properties. Much of the basic knowledge can be obtained from previous experimental investigations. However, in the case of some physicochemical parameters, there is no possibility to investigate them in an experimental way, especially when any similar material does not exist and there is no possibility for any estimation. Fortunately, the solution for these cases is the theoretical modeling of these parameters. The crucial properties of nanomaterials, that is, interactions between nanoparticles and chemical individuals present in the surrounding, surface activity, influence on the electronic structure, etc., can be reflected through molecular modeling. Besides, the obvious advantage is that the opportunity to perform predictions for any theoretically generated nanomaterial and the calculation of theoretical descriptors requires solely a computer and software, which are relatively low in cost in comparison to the experiment.

To complete the review of the latest approaches of nanodescriptors, presented here are a few examples of other experimental and theoretical nanodescriptors developed in the last few years.

6.4.1 Perturbation Approach

The simplest and cheapest approach of introducing experimental data into the nano-QSAR/QSPR modeling is to use available experimental data, already gathered by others. However, the typical QSAR/QSPR approach is assumed to be generated on complete, homogeneous data, which are obtained in the same conditions. Unfortunately, when nanomaterial characteristics are mostly partial or performed in varied conditions, it is hard to include this information in typical modeling. Nevertheless, Kleandrova et al. [43] proposed a way to solve these issues and presented the approach, which merges different kinds of available experimental data independently of the measurement conditions.

Kleandrova et al. [43] proposed the QSAR-perturbation model for the ecotoxicological profile of nanoparticles, which is sensitive to physicochemical properties of nanoparticles, external conditions, and the type of ecotoxicological test. The authors based this on the data set of 85 uncoated nanoparticles with 18 different chemical compositions, namely metals and metal oxides. The ecotoxicity of each of the investigated nanoparticles was measured by at least 1 of 4 ecotoxicity tests (EC_{50}, IC_{50}, LC_{50}, and TC_{50}) toward at least 1 of 20 bioindicators/assay organisms (*Danio magna*, Pseudokirchneriella *subcapitata*, *Danio reiro*, *Raphanus sativus*, etc.). Subsequently, all 85 nanoparticles were classified as ecotoxic [$Tox_i(c)_j = -1$] or nonecotoxic [$Tox_i(c)_j = 1$] in accordance with an arbitrary but rigorous cutoff value of proper ecotoxicity measures ($EC_{50} \geq 168.45$, $IC_{50} \geq 177.62$ for bacteria and algae; $LC_{50} \geq 225.81$, $TC_{50} \geq 334.08$ for crustaceans and fish). Furthermore, all nanoparticles were described by at least 1 of 7 labels regarding shape (spherical, irregular, needle, pyramidal, etc.), at least 1 of 5 descriptions of external conditions in which size measurements were performed (powder, H_2O, H_2O/TMAOH, H_2O/HMT, etc.), and at least 1 of 9 time intervals of exposure (24 h, 48 h, 120 h, 168 h, etc.). As a result, the complete experimental characteristic was used to describe each nanoparticle with a $(c)_j$ parameter as follows:

$$(c)_j = (m_t, a_o, p_s, c_p, t_e), \tag{6.12}$$

where m_t reflects the measure of ecotoxicity, a_o the assay organisms, p_s the nanoparticle shape, c_p the condition of the nanoparticle's size measurements, and t_e the time for which the organisms were in contact with the nanoparticles. Next, the authors characterized all 85 nanoparticles using four descriptors: molar volume (V), electronegativity (E), and polarizability (P) extracted from the online Chemicool [77] periodic table and the experimentally measured size (L). Thus, all nanoparticles could be compared with each other according to the moving average approach as per following equation:

$$\Delta D_i(c_j) = D_i - D_i(c_j)_{avg}, \tag{6.13}$$

where D_i reflects the appropriate descriptor (V, E, P, or L) of i nanoparticles and $D_i(c_j)_{avg}$ represents the average value of this descriptor for a set of n_j nonecotoxic nanoparticles [$\text{Tox}_i(c)_j = 1$] assayed under the same element of condition $(c)_j$. This means that the $\Delta D_i(c_j)$ descriptor takes into account the molecular structure of nanoparticles and the experimental condition of measurements at the same time. Additionally, the procedure of QSAR-perturbation modeling assumes to compare each nanoparticle with other nanoparticles assayed under the same conditions. To accomplish this, nanoparticles in the data set are randomly paired, where one nanoparticle is predicted (taken as the new, output state) and the second nanoparticle is regarded as the reference. In general, it can be described as follows:

$$\text{Tox}_i(c_j)_{\text{nw}} = a_0 \times \text{Tox}_i(c_j)_{\text{rf}} + \Sigma b_j \times \left[\Delta D_i\left(c_j\right)_{\text{nw}} - \Delta D_i(c_j)_{\text{rf}}\right] + d_0 \tag{6.14}$$

In this way, $\text{Tox}_i(c_j)_{\text{nw}}$ (the ecotoxicity of a new nanoparticle) can be estimated on the basis of $\text{Tox}_i(c_j)_{\text{rf}}$ (the ecotoxicity profile of the corresponding nanoparticle, or the reference) and differences in terms of chemical composition, size, shape, assay organisms, time of exposure, and measure of ecotoxicity. $\text{Tox}_i(c_j)_{\text{nw}}$ and $\text{Tox}_i(c_j)_{\text{rf}}$ reflect the relative ecotoxicity dependent on the new or reference state, but together form $\text{Tox}_i(c_j)$. Similarly, $\Delta D_i(c_j)_{\text{nw}}$ and $\Delta D_i(c_j)_{\text{rf}}$ constitute $\Delta D_i(c_j)$. Thus, the previous equation can be reduced to:

$$\text{Tox}_i(c_j)_{\text{nw}} = a_0 \times \text{Tox}_i(c_j)_{\text{rf}} + \Sigma b_j \times \Delta\Delta D_i(c_j) + d_0 \tag{6.15}$$

Here $\Delta\Delta D_i(c_j)$ is the perturbation term, which depends on the differences in physicochemical properties between paired nanoparticles and experimental conditions c_j. Furthermore, a_0, b_j, and d_0 coefficients are calculated using statistical methods, that is, MLR or linear discriminant analysis (LDA).

In the mentioned study, Kleandrova et al. [43] generated 5520 random pairs from 85 metal and metal oxide nanoparticles. The authors applied the LDA method and obtained a nano-QSAR model with seven descriptors (according to the absolute value of contribution): $\Delta\Delta E(a_0)(-6.047)$, $\Delta\Delta E(c_p)(-3.133)$, $\Delta\Delta P(p_s)(-1.706)$, $\Delta\Delta E(t_e)(-1.138)$, $\Delta\Delta V(a_0)(0.765)$, $\Delta\Delta V(m_t)(0.306)$, and $\Delta\Delta L(c_p)$ (0.034). In this way, the model maintains the significance of the role of electronegativity in the cytotoxic effect of nanoparticles. The great contribution of the electronegativity-derived descriptors emphasizes that the presence of strongly electronegative atoms (i.e., oxygen) makes nanoparticles more toxic. These results are consistent with the observations that metal oxide nanoparticles exhibit much higher toxicity than their corresponding metal nanoparticles. Next, the polarizability input reflects the chemical composition and, as a result, the probability of cation formation. The decreasing polarizability, in other words, lower proneness to releasing electrons, results in lower ecotoxic activity. Here also the theory is in accordance with observations that metals are less toxic than metal oxides. Further, molar volume–based descriptors highlight the role of nanoparticle size in ecotoxicity. Ecotoxicity increases due to the diminished molar volume of chemical individuals. Furthermore, these descriptors also indirectly reflect other parameters, that is, the ability to undergo aggregation. Finally, the last descriptor directly reflects the experimental size of nanoparticles and compares nanoparticles measured in the same conditions, that is, as dry powders or in the same aqueous suspensions. This descriptor suggests that increasing the size of nanoparticles leads to higher ecotoxic activity. The detailed interpretations of the applied descriptors are included in Table 6.5. The statistical quality of the developed nano-QSAR-perturbation model was described with small values of Wilks' lambda ($\lambda = 0.297$) and *p*-level ($p < 10^{-16}$) as well as a large value of the chi-square ($\chi^2 = 5014.43$). Moreover, the predicting power was confirmed with high accuracy, sensitivity, and specificity, for both training and validation sets (~99%).

Table 6.5 Perturbation nanodescriptors

Descriptor	Definition
$\mathrm{Tox}_i(c_j)_{rf}$	The binary reflection of the nanoparticle reference's ecotoxicity
$\Delta\Delta E(a_0)$	Perturbation term of electronegativity between the predicted nanoparticle (output state) and the nanoparticle used as a reference in accordance with the assay organism
$\Delta\Delta E(c_p)$	Perturbation term of electronegativity between the predicted nanoparticle (output state) and the nanoparticle used as a reference in accordance with the conditions of the nanoparticle's size measurements

Descriptor	Definition
$\Delta\Delta P(p_s)$	Perturbation term of polarizability between the predicted nanoparticle (output state) and the nanoparticle used as a reference in accordance with the particle shape
$\Delta\Delta E(t_e)$	Perturbation term of electronegativity between the predicted nanoparticle (output state) and the nanoparticle used as a reference in accordance with the time exposure
$\Delta\Delta V(a_0)$	Perturbation term of molar volume between the predicted nanoparticle (output state) and the nanoparticle used as a reference in accordance with the assay organism
$\Delta DV(m_t)$	Perturbation term of molar volume between the predicted nanoparticle (output state) and the nanoparticle used as a reference in accordance with the measure of ecotoxicity
$\Delta\Delta L(c_p)$	Perturbation term of size between the predicted nanoparticle (output state) and the nanoparticle used as a reference in accordance with the conditions of the nanoparticle's size measurements

Source: Reprinted from Ref. [43], Copyright (2014), with permission from Elsevier.

The presented QSAR-perturbation approach is an example of how to compare nanoparticles using theoretically obtained descriptors and available experimental data at the same time. The presented approach is based on the information extracted from the periodic table and simultaneously incorporates available data from experimental measurements and physical characteristic of investigated nanoparticles. To perform the QSAR-perturbation model, any statistical tool can be applied. In the QSAR-perturbation approach the whole procedure of descriptors and $\Delta D_i(c_j)$ calculations, random pairing of a nanoparticle with its reference, $\Delta\Delta D_i(c_j)$ calculations, and finally model creation, is carried out each time from the start. Thus, it is impossible to reuse obtained descriptors in the next nano-QSAR modeling. Nevertheless, the great advantage of the presented approach is the fact that it is quite a simple procedure of modeling and intuitive interpretation of results.

6.4.2 Image Descriptors

The statement that the shape of a nanoparticle can influence its properties became popular when the first detailed description of possible forms of nanostructures came out [78] (Fig. 6.9). Nevertheless, the suggestion to include microscope images as descriptors in nano-QSAR/QSPR modeling was unrealized for a long period of time due to the lack of a concept about how to express what is seen on the images in a numerical way. Over the following years, this challenge was taken up a few times.

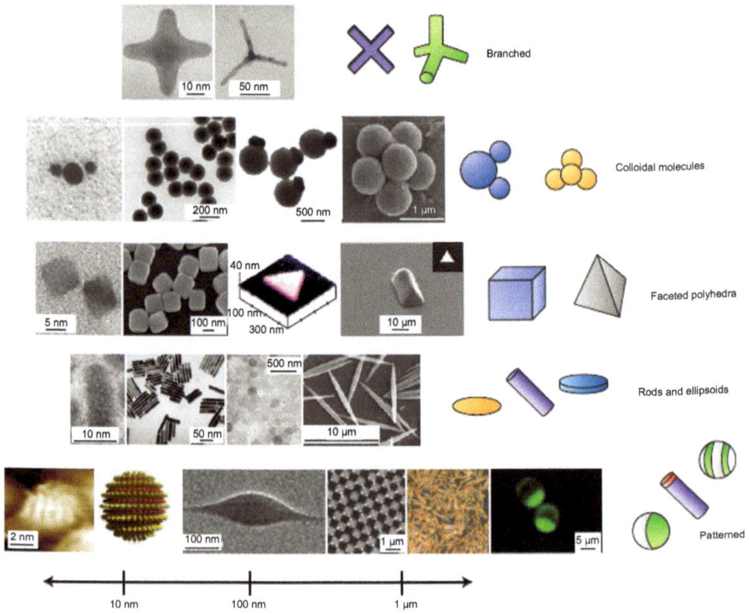

Figure 6.9 Representative building blocks of nanoparticles. Reprinted by permission from Springer Customer Service Centre GmbH: Springer Nature, *Nature Materials*, Ref. [78], Copyright (2007).

First, Puzyn et al. [66] proposed characterizing objects as per the scanning electron microscopy (SEM)/transmission electron microscopy (TEM)/atomic force microscopy (AFM) images using RGB or grayscale. The idea was to convert the image into a numerical matrix in which values reflect individual pixels according to the selected scale (RGB or grayscale). In this way, physical characteristics of nanoparticles, that is, shape or porosity, could be described on the basis of these values. However, at that time, this idea was not applied in nano-QSAR/QSPR modeling but it was an inspiration for other research groups.

The next research in this area was performed by Bigdeli et al. [79]. The authors applied a binary grayscale to investigate TEM images of metal and metal oxide nanoparticles. The developed procedure involved a few steps (Fig. 6.10). The first one was to provide a general code to ensure the maximum accuracy of TEM images and simultaneously minimize inaptness of blurred/noisy images. This was achieved by defining the thresholds/contour algorithm of a proper feature using machine learning techniques. The second step was the conversion of grayscale images to binary scale images. The value of each pixel was assigned on the basis of its neighbors, where 1 corresponded to the objects while 0 to the background. Finally, the algorithm was coded, compiled, and implemented using the self-developed software. Then, the calculation of descriptors for imported images was performed. The authors suggested a set of image-based nanodescriptors describing size, surface area, agglomeration state, curvature, corner count, and shape type.

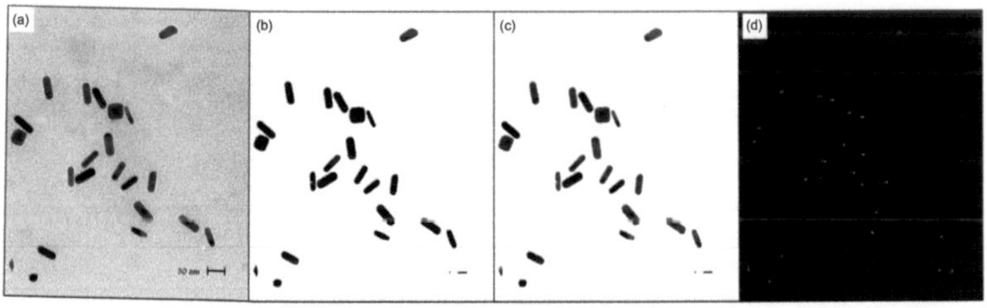

Figure 6.10 Preprocessing of image to descriptors: (a) original image, (b) threshold for background separation, (c) image without a background, and (d) all objects found in the investigated image. Reproduced from Ref. [79] with permission of The Royal Society of Chemistry.

Finally, Bigdeli et al. [80] utilized the newly created descriptors in the nano-QSAR model to determine the factors most responsible in gold nanoparticle exocytosis. The combination of image-based descriptors (shape, surface area, corner count, curvature, aggregation state, etc.), experimental measurements (hydrodynamic size, zeta potential, and wavelength before and after protein coating), and combined descriptors, including charge density and spectral parameters, allowed them to obtain the PLSR QSAR model for gold nanoparticles' exocytosis in macrophages.

Now, the method of image descriptors in nano-QSAR/QSPR modeling is even more improved. Odziomek et al. [81] extended the morphological description of nanoparticles with a set of parameters such as perimeter, major and minor axes, Feret's diameter, circularity, roundness, and solidity. Similarly, as in the above-mentioned procedure [79], the preprocessing involved (Fig. 6.11) (i) filtering to avoid noise derived from artifacts, dust, etc., (ii) thresholding to separate the background on the basis of pixel intensity, and (iii) segmentation, when the most interesting region of nanoparticles is selected to calculate descriptors. Then, the grayscale expresses the intensity of visual signals and each pixel is assigned with a number from 0 to 255 (0 represents no signal, whereas 255 represents complete blackness).

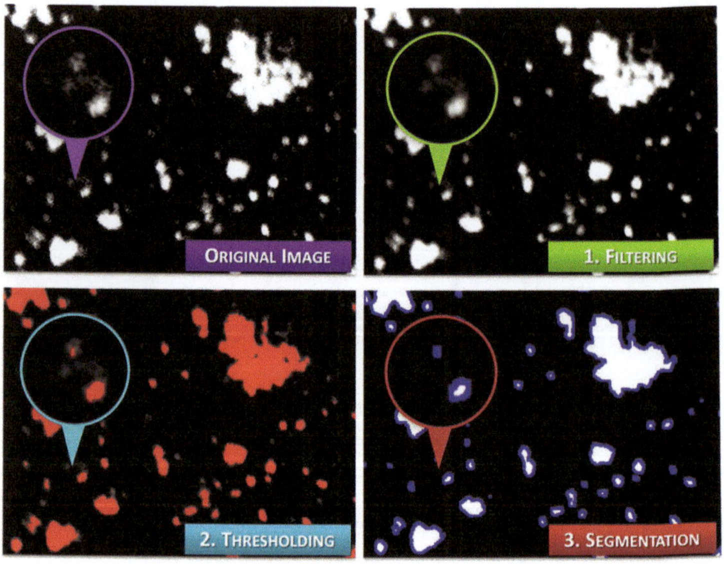

Figure 6.11 Preprocessing of image to descriptors. Reproduced from Ref. [81] with permission from John Wiley and Sons.

This approach was successfully applied by Mikolajczyk et al. [39] in nano-QSPR modeling, where the study presented the influence of particles' sphericity and electronic structure on the zeta potential of metal oxide nanoparticles. However, it is noteworthy to point out that sphericity was reflected as a combination of the volume and surface area extracted from TEM images:

$$\psi = \frac{\pi^{1/3} 6V^{2/3}}{A}, \tag{6.16}$$

where V represents volume and A surface area.

Image-based descriptors have a great potential in designing target nanomaterials. Firstly, they can be useful in the mechanistic interpretation of relationships between nanostructures and their final features. Then, this knowledge can be useful in designing new nanomaterials with strictly determined morphology, because, in practice, it is much easier to manipulate physical parameters than any theoretical index. Finally, the analysis of image-based descriptors performed for newly synthesized nanostructures allows verifying the application potential or human/environmental safety before their implementation in products.

6.4.3 Reusing Toxicity Measurements

Parameters of toxicity—genotoxicity, cytotoxicity, and ecotoxicity—are investigated using organisms at different trophic levels, that is, bacteria, algae, rodents, and mammals. However, due to the current policy of reducing the participation of laboratory animals, they are often replaced by cell lines of these organisms, including human cell lines. Moreover, in order to further reduce time and cost, currently, considerable attention is being given to the development of a strategy for reusing available results of experiments for the prediction of interspecies toxic effects (Fig. 6.12). This strategy aims to use experimental studies performed for organisms at lower trophic levels or even single-cell organisms to extrapolate the probable toxic effects to higher organisms, that is, mammals.

In the context of toxicity modeling for nanomaterials, the first models were proposed by Kar et al. [82] as interspecies nano-QTTR (quantitative toxicity–toxicity relationship) models. The models were about the correlation between cytotoxicity effects of metal oxide nanoparticles for *E. coli* and HaCaT human cell lines, in other words, between prokaryotic and eukaryotic species. The authors used the toxicology endpoint (pEC_{50}) of one species (*E. coli* or HaCaT) as a descriptor to estimate the same toxicology endpoint for the second species (HaCaT or *E. coli*, respectively). The application of such a descriptor allows including information about the mechanisms of toxicity in a specific form. Moreover, combination with quantum mechanical and periodic table descriptors enabled deeper interpretation of these mechanisms.

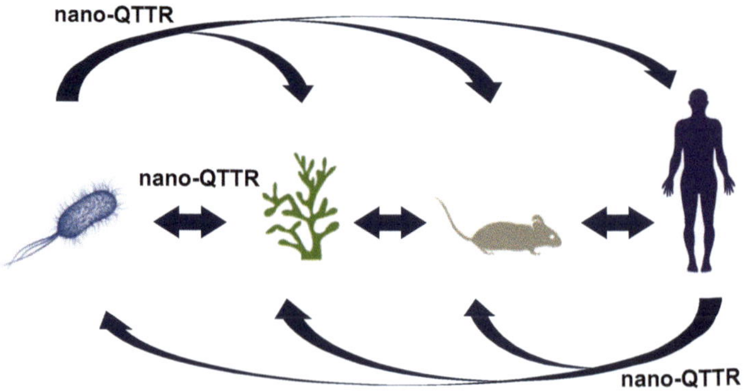

Figure 6.12 Idea of interspecies predictions in nano-QTTR modeling.

In the case of modeling HaCaT cytotoxicity effect, the model combined three descriptors: cytotoxicity measures against *E. coli*, electronic chemical potential (μ), and standard enthalpy of metal oxide nanocluster formation (ΔH_f^c). The electronic chemical potential reflects the escaping tendency of electron transfer from an electron cloud at an equilibrium state, and it is a counterpart of negative Mulliken electronegativity. Knowing that electronegativity describes the attraction of a bonding pair of electrons to the atom, it is obvious that the higher value leads to the higher catalytic activity and consequently higher toxicity of nanoparticles. The second descriptor, ΔH_f^c, corresponds to the standard enthalpy of formation of a metal oxide nanocluster, and its high value suggests that metal oxide nanoparticles with a strongly positive charge will exhibit higher toxicity. Thus, the first theoretical descriptor describes a surface redox activity and transfer of electrons from a valence band to a conduction band during the intercellular process in biological media. And the next descriptor is responsible for the detachment of charge from the surface of the nanoparticle. This means, both descriptors correspond to the process of the oxidative stress induction in the HaCaT human cell line.

In the case of the model for assessing the cytotoxicity effect against *E. coli*, the modeled parameter was explained by using the HaCaT cytotoxicity endpoint, enthalpy of a gaseous cation formation ΔH_{Me+}, and the charge of the metal cation of an adequate metal oxide χ_{OX}. The first theoretical descriptor responds to the lattice energy, and its high value means a high charge of the metal cation. Consequently, energetically, the release of a cation with a smaller charge is preferred than the release of one with a larger charge ($Me^{2+} > Me^{3+} > Me^{4+}$). In turn, the low value of the second descriptor, χ_{OX}, corresponds to the strong reductive properties and the simplicity of the metal cation's detachment from the surface. Effectively, these descriptors consider the reinforcement of the processes that are accountable for the reductive properties of metal oxide nanoparticles and consequently, point out that the low charge of cations is primarily responsible for the cytotoxicity activity against *E. coli*.

Although the toxicity endpoint in nano-QTTR models has no clear contribution in mechanistic interpretation, this kind of descriptor considerably improves the predictive power of the models. The available toxicological data, independently of the bioindicator, can be applied in the predictive models for other organisms. The proposed approach makes easier the identification of relationships between organisms in the ecosystem. Consequently, the construction of interspecies nano-QTTR models could have a significant impact on the improvement of toxicological investigations in human and environmental risk assessments.

6.4.4 Mixture Descriptors

In the above-mentioned approaches, the authors were focused mainly on the pristine metal oxide nanoparticles. However, due to the rapid evolution of nanotechnology, heterogeneous (second generation) nanomaterials were also developed [83, 84]. These are nanomaterials with surface modifications or composed of different types of nanomaterials. So far, nano-QSAR/QSPR methods have not been applied for these kinds of nanomaterials because of the lack of suitable and reliable descriptors dedicated to heterogeneous nanomaterials. Nevertheless, Mikolajczyk et al. [85] attempted to adopt additive descriptors, which were previously applied for conventional organic compounds, for heterogeneous TiO_2-based-nanoparticle-modified noble metals. This is the first approach of nanodescriptors that describes the modified nanoparticles as a whole, not as separate core and surface modifications.

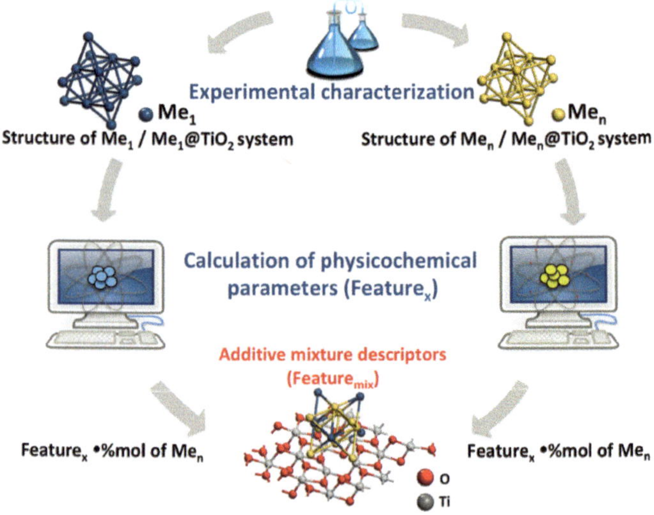

Figure 6.13 The idea of computing mixture descriptors for heterogeneous nanomaterials. Reproduced from Ref. [85] with permission of The Royal Society of Chemistry.

Mikolajczyk et al. applied additive descriptors to predict the cytotoxic potential of a set of 34 TiO_2-based nanoparticles modified by (poly)metallic clusters (Pt, Ag, Au) toward the Chinese hamster ovary cell line. The authors assumed that the properties of heterogeneous nanomaterials are a result of the additive contribution of individual components (Fig. 6.13). Simultaneously, the contribution is proportional to the mole fraction of each component in the mixture. Thus, the mixture property is as follows:

$$c_{\text{mix}} = \sum_{i=1}^{n} a\text{C}_i \tag{6.17}$$

where a is dose/concentration of the particular component C_i.

In the presented additive approach, the authors used QM descriptors calculated for (i) noble metal clusters in the gas phase and (ii) solid stated systems of modified clusters. The authors obtained a set of parameters, such as the energy of the HOMO and LUMO (E_{HOMO} and E_{LUMO}), absolute electronegativity (μ), absolute hardness (η), ionization potential (I), electron affinity (A), total energy of the system (E_{tot}), and adsorption energy of the metal (E_{ads}). The mixture descriptors have the following form:

$$D_{mix} = \%\,mol_{Me_1} \times F_1 + \rightleftharpoons + \%\,mol_{Me_n} \times F_n \tag{6.18}$$

where D_{mix} is a mixture descriptor, $\%\,mol_{Me_n}$ is a weighted contribution of each component in the final mixture feature, and F_n is a quantum mechanical descriptor for each component in the mixture.

Mikolajczyk et al. applied the above-mentioned mixture descriptors in nano-QSAR modeling for a set of TiO_2-based nanoparticles modified with different types and amounts of Ag, Au, and Pt metals. The authors obtained two nano-QSAR models, and each model includes one mixture descriptor. The first model includes the amount of silver precursor used in the preparation of heterogeneous nanomaterial ($\%mol_{Ag}$), and the second model includes the absolute electronegativity of the whole mixture (μ_{mix}). Both nano-QSAR models explain a great amount of variability of the observed cytotoxicity (93% and 94%, respectively, for the first and second models) and are characterized by good statistical parameters (R^2 = 0.93, $RMSE_C$ = 0.11, Q^2_{CV} = 0.92, $RMSE_{CV}$ = 0.13, Q^2_{EXT} = 0.88, $RMSE_{EXT}$ = 0.14; and R^2 = 0.94, $RMSE_C$ = 0.11, Q^2_{CV} = 0.93, $RMSE_{CV}$ = 0.12, Q^2_{EXT} = 0.83, and $RMSE_{EXT}$ = 0.16, respectively, for the first and second models). The mechanistic interpretation of the models indicates how great is the influence of silver on the cytotoxicity in comparison to the other noble metals (Au and Pt), which are not included in the models. Due to the nature of precursor concentration–based descriptor, the first model can practically influence the design of new nanomaterials according to the idea of safe-by-design. In turn, in the second model, the electronegativity-based descriptor highlights the role of electron-donor centers caused by the presence of noble atoms on the TiO_2 surface in the formation of reactive oxygen species.

According to the presented idea, the features of heterogeneous nanomaterials can be predicted on the basis of the features of individual components and their proportional contribution in the mixture. The presented approach of additive nanodescriptors can be helpful in the designing and estimation properties of new modified nanomaterials through a detailed analysis of all the components. Simultaneously, this approach is based on another class of descriptors, that is, QM descriptors. Thus, all kinds of property-related information can be used in this modeling.

6.5 Summary and Future Perspectives

It is a great challenge to develop suitable and reliable descriptors for nanomaterials because many aspects influence a nanomaterial's properties and there are limitations in the form of the sophistication, time, and cost of experiments and calculations. The presented approaches indicate many ways to avoid some of these problems, but none of the experimental or theoretical approaches solve all the issues and each approach has its own limitations (Table 6.6). Nevertheless, these methods constitute a step forward in designing safer nanomaterials with high application potential and the simultaneous ability to reduce laboratory animal use.

The existing approaches to nanodescriptors provide great potential in estimating properties of a wide range of nanomaterials. In the future, nanodescriptors will cover the real sizes and complex chemical compositions of nanomaterials, as well as the external conditions that influence nanomaterial features. However, even now nano-QSAR/QSPR modeling could be much more efficient if access to the tools and know-how of experiments, as well as calculation procedures, is opened. The investigations, performed by both research groups and industry, in the nanomaterials area could be much more fruitful if results of all previously performed experiments and calculations are put in one database. This could help avoid duplication of investigations in the same area. Consequently, the consumption of chemicals, laboratory animals, and computational resources and, importantly, time and cost could be significantly reduced. Even the presence of all tools for descriptor calculations in larger software could notably improve research in the nanomaterial area. Simultaneously, it is noteworthy to point out that despite the fact that some of the developed approaches are very valuable, the guidelines for descriptor generation/reproduction are often not sufficiently clear, especially for new users. Thus, it is important to share detailed and transparent manuals. Another aspect is communication between researchers. There is the need for closer collaboration between experimentalists and theorists so that they could more easily understand the results and then draw appropriate conclusions.

Table 6.6 Summary of presented approaches in nanodescriptors

Kind of descriptors	Advantages	Limitations	Software	Webpages
Quantum-mechanical descriptors	Deep analysis of intrinsic properties, as in any other approach Efficient diversification of similar and dissimilar chemical individuals Appropriate for a wide range of nanomaterials	Size of molecular model Level of theory Advanced calculations Computationally demanding Requirement of background in computational chemistry for performing and interpretation of results Labor intensive	MOPAC Gaussian	http://www.openmopac.net/ http://www.gaussian.com/
SMILES-based optimal descriptors	Simple calculation procedure Fast Open-source software	Size not taken into account Distinction between only chemicals varied in chemical composition Only for structurally similar compounds Physical aspects of materials not taken into account Nonintuitive interpretation	CORAL software	http://www.insilico.eu/CORAL/

(Continued)

Table 6.6 (*Continued*)

Kind of descriptors	Advantages	Limitations	Software	Webpages
Improved SMILES-based optimal descriptors	Simple calculation procedure Fast Distinction of chemicals varied in chemical composition and physical properties Open-source software	Size not taken into account Only for structurally similar compounds	CORAL software	http://www.insilico.eu/CORAL/
Periodic table descriptors	Simple calculation procedure Fast Open-source software	Size not taken into account Distinction between only chemicals varied in chemical composition Physical aspects of materials not taken into account	Elemental-Descriptor 1.0 – (Metal, nonmetal, semimetal)	http://nanobridges.eu/software/
Simplex representation of molecular structure descriptors	Simple calculation procedure Fast Diversification of chemicals varied in chemical composition Consideration of physical properties of materials Open-source software	Size not taken into account Nonintuitive interpretation	SiRMS	https://github.com/DrrDom/sirms/

Kind of descriptors	Advantages	Limitations	Software	Webpages
Liquid drop model descriptors	Simple calculation procedure Fast Consideration of tendency of nanoparticles to aggregate in liquids No requirement of any software	No nanoparticle property other than size taken into account	-	-
Metal–ligand binding characteristic	Consideration of metal-ligand binding, as in no other approach No requirement of any software	No nanoparticle property other than binding characteristic taken into account	-	-
Full-particle descriptors	Consideration of the surface/core aspect of nanoparticles, as in no other approach Consideration of size Opportunity to extrapolate descriptors to nanoparticles of other sizes Open-source software Computationally nondemanding	Currently, algorithm available only for spherical particles Required background in force field modeling	LAMMPS software	http://lammps.sandia.gov

(*Continued*)

Table 6.6 *(Continued)*

Kind of descriptors	Advantages	Limitations	Software	Webpages
Perturbation approach	Simple calculation procedure Simultaneous consideration of all available knowledge about chemicals Opportunity to include other descriptors Intuitive interpretation	No possibility to reuse descriptors for another data set	Any statistical software	-
Image descriptors	Intuitive interpretation	Required SEM/TEM/AFM and other images Complicated procedure of calculation Labor intensive	Self-developed algorithm	-
Reusing of toxicity measurements	Based on available experimental results Improved predicting power of models with classic descriptors Easier identification of interspecies relationships in the ecosystem	No clear contribution to mechanistic interpretation	-	-
Mixture descriptors	Consideration of properties of heterogeneous nanomaterials Simple calculation procedure Opportunity to include other descriptors	Limitations related to the included descriptors	-	-

Acknowledgments

EW and TP acknowledge the funding received from the European Union Horizon 2020 (Grant Agreement No. 646221, NanoREG2). KJ thanks the European Commission for financial support through the PATROLS project (Grant Agreement No. 760813). This work was also supported in part by the National Science Foundation through the ND EPSCoR Award #IIA-1355466 and by the State of North Dakota (BR).

References

1. Radomska, A., Leszczyszyn, J. and Radomski, M. W. (2016). The Nanopharmacology and nanotoxicology of nanomaterials: new opportunities and challenges, *Adv. Clin. Exp. Med.*, **25**, 151–162.
2. Chung, W. H., Hwang, Y. T., Lee, S. H. and Kim, H. S. (2016). Electrical wire explosion process of copper/silver hybrid nano-particle ink and its sintering via flash white light to achieve high electrical conductivity, *Nanotechnology*, **27**, 1–13.
3. Bhagat, S. A. and Singh, M. C. (2016). Development and evaluation of silver nanoparticles and its applications in topical drug delivery systems, *Asian J. Pharm.*, **10**, 16–21.
4. Sehati, N., Dalali, N., Soltanpour, S. and Dorraji, M. S. S. (2015). Application of hollow fiber membrane mediated with titanium dioxide nanowire/reduced graphene oxide nanocomposite in preconcentration of clotrimazole and tylosin, *J. Chromatogr. A*, **1420**, 46–53.
5. Moakhar, R. S., Goh, G. K. L., Dolati, A. and Ghorbani, M. (2015). A novel screen-printed TiO_2 photoelectrochemical sensor for direct determination and reduction of hexavalent chromium, *Electrochem. Commun.*, **61**, 110–113.
6. Khan, M. E., Khan, M. M. and Cho, M. H. (2015). Green synthesis, photocatalytic and photoelectrochemical performance of an Au-Graphene nanocomposite, *RSC Adv.*, **5**, 26897–26904.
7. Katz, L. M., Dewan, K. and Bronaugh, R. L. (2015). Nanotechnology in cosmetics, *Food Chem. Toxicol.*, **85**, 127–137.
8. Amini, S. M., Gilaki, M. and Karchani, M. (2014). Safety of nanotechnology in food industries, *Electron. Physician*, **6**, 962–968.
9. Lem, K. W., Choudhury, A., Lakhani, A. A., Kuyate, P., Haw, J. R., Lee, D. S., Iqbal, Z. and Brumlik, C. J. (2012). Use of nanosilver in consumer products, *Recent Pat. Nanotech.*, **6**, 60–72.

10. Shanaghi, A., Sabour, A. R., Shahrabi, T. and Aliofkhazraee, M. (2009). Corrosion protection of mild steel by applying TiO_2 nanoparticle coating via sol-gel method, *Prot. Met. Phys. Chem.*, **45**, 305–311.

11. Thines, R. K., Mubarak, N. M., Nizamuddin, S., Sahu, J. N., Abdullah, E. C. and Ganesan, P. (2017). Application potential of carbon nanomaterials in water and wastewater treatment: a review, *J. Taiwan Inst. Chem. E*, **72**, 116–133.

12. Kannaiyan, K. and Sadr, R. (2017). The effects of alumina nanoparticles as fuel additives on the spray characteristics of gas-to-liquid jet fuels, *Exp. Therm. Fluid Sci.*, **87**, 93–103.

13. Foster, E. J., Moon, R. J., Agarwal, U. P., Bortner, M. J., Bras, J., Camarero-Espinosa, S., Chan, K. J., Clift, M. J. D., Cranston, E. D., Eichhorn, S. J., Fox, D. M., Hamad, W. Y., Heux, L., Jean, B., Korey, M., Nieh, W., Ong, K. J., Reid, M. S., Renneckar, S., Roberts, R., Shatkin, J. A., Simonsen, J., Stinson-Bagby, K., Wanasekara, N. and Youngblood, J. (2018). Current characterization methods for cellulose nanomaterials, *Chem. Soc. Rev.*, **47**, 2609–2679.

14. Lin, P. C., Lin, S., Wang, P. C. and Sridhar, R. (2014). Techniques for physicochemical characterization of nanomaterials, *Biotechnol. Adv.*, **32**, 711–726.

15. Cho, E. J., Holback, H., Liu, K. C., Abouelmagd, S. A., Park, J. and Yeo, Y. (2013). Nanoparticle characterization: state of the art, challenges, and emerging technologies, *Mol. Pharmaceut.*, **10**, 2093–2110.

16. Baer, D. R., Engelhard, M. H., Johnson, G. E., Laskin, J., Lai, J. F., Mueller, K., Munusamy, P., Thevuthasan, S., Wang, H. F., Washton, N., Elder, A., Baisch, B. L., Karakoti, A., Kuchibhatla, S. V. N. T. and Moon, D. (2013). Surface characterization of nanomaterials and nanoparticles: important needs and challenging opportunities, *J. Vac. Sci. Technol. A*, **31**, 1–34.

17. Abukhdeir, N. M. (2016). Computational characterization of ordered nanostructured surfaces, *Mater. Res. Express*, **3**.

18. Mikolajczyk, A., Gajewicz, A., Rasulev, B., Schaeublin, N., Maurer-Gardner, E., Hussain, S., Leszczynski, J. and Puzyn, T. (2015). Zeta potential for metal oxide nanoparticles: a predictive model developed by a nano-quantitative structure-property relationship approach, *Chem Mater*, **27**, 2400–2407.

19. Laramy, C. R., Brown, K. A., O'Brien, M. N. and Mirkin, C. A. (2015). High-throughput, algorithmic determination of nanoparticle structure from electron microscopy images, *ACS Nano*, **9**, 12488–12495.

20. Pederzoli, F., Tosi, G., Vandelli, M. A., Belletti, D., Forni, F. and Ruozi, B. (2017). Protein corona and nanoparticles: how can we investigate on?, *Wires. Nanomed. Nanobi.*, **9**, 1–23.

21. Munteanu, C. R., Fernandez-Blanco, E., Seoane, J. A., Izquierdo-Novo, P., Rodriguez-Fernandez, J. A., Prieto-Gonzalez, J. M., Rabunal, J. R. and Pazos, A. (2010). Drug discovery and design for complex diseases through QSAR computational methods, *Curr. Pharm. Des.*, **16**, 2640–2655.

22. Jagiello, K., Sosnowska, A., Walker, S., Haranczyk, M., Gajewicz, A., Kawai, T., Suzuki, N., Leszczynski, J. and Puzyn, T. (2014). Direct QSPR: the most efficient way of predicting organic carbon/water partition coefficient (log K (OC)) for polyhalogenated POPs, *Struct. Chem.*, **25**, 997–1004.

23. Odziomek, K., Gajewicz, A., Haranczyk, M. and Puzyn, T. (2013). Reliability of environmental fate modeling results for POPs based on various methods of determining the air/water partition coefficient (log K-AW), *Atmos. Environ.*, **73**, 177–184.

24. Puzyn, T., Mostrg, A., Suzuki, N. and Falandysz, J. (2008). QSPR-based estimation of the atmospheric persistence for chloronaphthalene congeners, *Atmos. Environ.*, **42**, 6627–6636.

25. Barycki, M., Sosnowska, A., Gajewicz, A., Bobrowski, M., Wilenska, D., Skurski, P., Gieldon, A., Czaplewski, C., Uhl, S., Laux, E., Journot, T., Jeandupeux, L., Keppner, H. and Puzyn, T. (2016). Temperature-dependent structure-property modeling of viscosity for ionic liquids, *Fluid Phase Equilib.*, **427**, 9–17.

26. Das, R. N. and Roy, K. (2013). Advances in QSPR/QSTR models of ionic liquids for the design of greener solvents of the future, *Mol. Diversity*, **17**, pp. 151–196.

27. Sosnowska, A., Barycki, M., Zaborowska, M., Rybinska, A. and Puzyn, T. (2014). Towards designing environmentally safe ionic liquids: the influence of the cation structure, *Green Chem.*, **16**, 4749–4757.

28. Rajeswari, R. and Jothilakshmi, R. (2016). Predicting toxicity of nanomaterials using QSAR methods: a review, *Mater. Focus*, **5**, 335–341.

29. Wyrzykowska, E., Mikolajczyk, A., Sikorska, C. and Puzyn, T. (2016). Development of a novel in silico model of zeta potential for metal oxide nanoparticles: a nano-QSPR approach, *Nanotechnology*, **27**, 1–8.

30. Mikolajczyk, A., Gajewicz, A., Mulkiewcz, E., Rasulev, B., Marchelek, M., Hirano, S., Zaleska-Medynska, A. and Puzyn, T. (2018). Nano-QSAR

modeling for ecosafe design of heterogeneous TiO_2-based nano-photocatalysts, *Environ. Sci.: Nano*, 1150–1160.

31. Praetorius, A., Labille, J., Scheringer, M., Thill, A., Hungerbuhler, K. and Bottero, J. Y. (2014). Heteroaggregation of titanium dioxide nanoparticles with model natural colloids under environmentally relevant conditions, *Environ. Sci. Technol.*, **48**, 10690–10698.
32. Hotze, E. M., Phenrat, T. and Lowry, G. V. (2010). Nanoparticle aggregation: challenges to understanding transport and reactivity in the environment, *J. Environ. Qual.*, **39**, 1909–1924.
33. Liu, H. X., Lu, J., Liu, D. M. and Cui, F. Y. (2018). Brownian dynamic study of the aggregation process of TiO_2 nanoparticles in aqueous suspensions, *Environ. Eng. Sci.*, 1–9.
34. Shevlin, D., O'Brien, N. and Cummins, E. (2018). Silver engineered nanoparticles in freshwater systems - likely fate and behaviour through natural attenuation processes, *Sci. Total. Environ.*, **621**, 1033–1046.
35. Hu, Y., Chen, X. J., Yang, K. and Lin, D. H. (2018). Distinct toxicity of silver nanoparticles and silver nitrate to Daphnia magna in M4 medium and surface water, *Sci. Total Environ.*, **618**, 838–846.
36. Jagiello, K., Chomicz, B., Avramopoulos, A., Gajewicz, A., Mikolajczyk, A., Bonifassi, P., Papadopoulos, M. G., Leszczynski, J. and Puzyn, T. (2017). Size-dependent electronic properties of nanomaterials: how this novel class of nanodescriptors supposed to be calculated?, *Struct. Chem.*, **28**, 635–643.
37. Abbas, Z., Labbez, C., Nordholm, S. and Ahlberg, E. (2008). Size-dependent surface charging of nanoparticles, **112**, 5715–5723.
38. Gajewicz, A., Puzyn, T., Bakhtiyor, R., Leszczynska, D. and Leszczynski, J. (2011). Metal oxide nanoparticle: size-dependence of quantum-mechanical properties, *Nanosci. Nanotechnol.-Asia*, **1**, 53–58.
39. Mikolajczyk, A., Gajewicz, A., Rasulev, B., Schaeublin, N., Maurer-Gardner, E., Hussain, S., Leszczynski, J. and Puzyn, T. (2015). Zeta potential for metal oxide nanoparticles: a predictive model developed by a nano-quantitative structure–property relationship approach, *Chem. Mater.*, **27**, 2400–2407.
40. Sizochenko, N., Jagiello, K., Leszczynski, J. and Puzyn, T. (2015). How the „liquid drop" approach could be efficiently applied for quantitative structure-property relationship modeling of nanofluids, *J. Phys. Chem. C*, **119**, 25542–25547.
41. Puzyn, T., Rasulev, B., Gajewicz, A., Hu, X., Dasari, T. P., Michalkova, A., Hwang, H. M., Toropov, A., Leszczynska, D. and Leszczynski, J.

(2011). Using nano-QSAR to predict the cytotoxicity of metal oxide nanoparticles, *Nat. Nanotechnol.*, **6**, 175–178.

42. Gajewicz, A., Schaeublin, N., Rasulev, B., Hussain, S., Leszczynska, D., Puzyn, T. and Leszczynski, J. (2015). Towards understanding mechanisms governing cytotoxicity of metal oxides nanoparticles: hints from nano-QSAR studies, *Nanotoxicology*, **9**, 313–325.
43. Kleandrova, V. V., Luan, F., Gonzalez-Diaz, H., Ruso, J. M., Melo, A., Speck-Planche, A. and Cordeiro, M. (2014). Computational ecotoxicology: simultaneous prediction of ecotoxic effects of nanoparticles under different experimental conditions, *Environ. Int.*, **73**, 288–294.
44. Luan, F., Kleandrova, V. V., Gonzalez-Diaz, H., Ruso, J. M., Melo, A., Speck-Planche, A. and Cordeiro, M. (2014). Computer-aided nanotoxicology: assessing cytotoxicity of nanoparticles under diverse experimental conditions by using a novel QSTR-perturbation approach, *Nanoscale*, **6**, 10623–10630.
45. Pan, Y., Li, T., Cheng, J., Telesca, D., Zink, J. I. and Jiang, J. C. (2016). Nano-QSAR modeling for predicting the cytotoxicity of metal oxide nanoparticles using novel descriptors, *RSC Adv.*, **6**, 25766–25775.
46. Roy, K., Kar, S. and Das, R. N. (2015). *A Primer on QSAR/QSPR Modeling: Fundamental Concepts* (Springer).
47. Tamm, K., Sikk, L., Burk, J., Rallo, R., Pokhrel, S., Madler, L., Scott-Fordsmand, J. J., Burk, P. and Tamm, T. (2016). Parametrization of nanoparticles: development of full-particle nanodescriptors, *Nanoscale*, **8**, 16243–16250.
48. Roy, K., Kar, S. and Das, R. N. (2015). *Understanding the Basics of QSAR for Applications in Pharmaceutical Sciences and Risk Assessment* (Elsevier Academic Press, Amsterdam).
49. Natarajan, R. and Nirdosh, I. (2008). Quantitative structure-activity relationship (QSAR) approach for the selection of chelating mineral collectors, *Miner. Eng.*, **21**, 1038–1043.
50. Tokarski, J. S., Hopfinger, A. J., Hobbs, J. D., Ford, D. M. and Faulon, J. L. M. (1997). Molecular modelling of polymers 17. Simulation and QSPR analyses of transport behavior in amorphous polymeric materials, *Comput. Theor. Polym. S.*, **7**, 199–214.
51. Puzyn, T., Jeliazkova, N., Sarimveis, H., Robinson, R. L. M., Lobaskin, V., Rallo, R., Richarz, A. N., Gajewicz, A., Papadopulos, M. G., Hastings, J., Cronin, M. T. D., Benfenati, E. and Fernandez, A. (2018). Perspectives from the NanoSafety Modelling Cluster on the validation criteria for (Q)SAR models used in nanotechnology, *Food Chem. Toxicol.*, **112**, 478–494.

52. Winkler, D. A., Mombelli, E., Pietroiusti, A., Tran, L., Worth, A., Fadeel, B. and McCall, M. J. (2013). Applying quantitative structure-activity relationship approaches to nanotoxicology: Current status and future potential, *Toxicology*, **313**, 15–23.

53. Toropov, A. A., Achary, P. G. R. and Toropova, A. P. (2016). Quasi-SMILES and nano-QFPR: the predictive model for zeta potentials of metal oxide nanoparticles, *Chem. Phys. Lett.*, **660**, 107–110.

54. Kar, S., Gajewicz, A., Puzyn, T., Roy, K. and Leszczynski, J. (2014). Periodic table-based descriptors to encode cytotoxicity profile of metal oxide nanoparticles: a mechanistic QSTR approach, *Ecotox. Environ. Saf.*, **107**, 162–169.

55. Chen, G. C., Vijver, M. G., Xiao, Y. L. and Peijnenburg, W. J. G. M. (2017). A review of recent advances towards the development of (quantitative) structure-activity relationships for metallic nanomaterials, *Materials*, **10**, 1–30.

56. Shao, C. Y., Chen, S. Z., Su, B. H., Tseng, Y. F. J., Esposito, E. X. and Hopfinger, A. J. (2013). Dependence of QSAR models on the selection of trial descriptor sets: a demonstration using nanotoxicity endpoints of decorated nanotubes, *J. Chem. Inf. Model.*, **53**, 142–158.

57. Kar, S., Gajewicz, A., Puzyn, T. and Roy, K. (2014). Nano-quantitative structure-activity relationship modeling using easily computable and interpretable descriptors for uptake of magnetofluorescent engineered nanoparticles in pancreatic cancer cells, *Toxicol. in Vitro*, **28**, 600–606.

58. Bouhaik, I. S., Leroy, P., Ollivier, P., Azaroual, M. and Mercury, L. (2013). Influence of surface conductivity on the apparent zeta potential of TiO_2 nanoparticles: application to the modeling of their aggregation kinetics, *J. Colloid Interface Sci.*, **406**, 75–85.

59. Holmberg, J. P., Ahlberg, E., Bergenholtz, J., Hassellov, M. and Abbas, Z. (2013). Surface charge and interfacial potential of titanium dioxide nanoparticles: experimental and theoretical investigations, *J. Colloid Interface Sci.*, **407**, 168–176.

60. Behera, M. and Giri, G. (2014). Green synthesis and characterization of cuprous oxide nanoparticles in presence of a bio-surfactant, *Mater. Sci.-Poland*, **32**, 702–708.

61. Dahle, J. T., Livi, K. and Arai, Y. (2015). Effects of pH and phosphate on CeO_2 nanoparticle dissolution, *Chemosphere*, **119**, 1365–1371.

62. Jiang, J. K., Oberdorster, G. and Biswas, P. (2009). Characterization of size, surface charge, and agglomeration state of nanoparticle dispersions for toxicological studies, *J. Nanoparticle Res.*, **11**, 77–89.

63. Llorente, I., Fajardo, S. and Bastidas, J. M. (2014). Applications of electrokinetic phenomena in materials science, *J. Solid State Electrochem.*, **18**, 293–307.

64. Sizochenko, N., Rasulev, B., Gajewicz, A., Kuz'min, V., Puzyn, T. and Leszczynski, J. (2014). From basic physics to mechanisms of toxicity: the "liquid drop" approach applied to develop predictive classification models for toxicity of metal oxide nanoparticles, *Nanoscale*, **6**, 13986–13993.

65. Toropov, A. A., Toropova, A. P., Benfenati, E., Gini, G., Puzyn, T., Leszczynska, D. and Leszczynski, J. (2012). Novel application of the CORAL software to model cytotoxicity of metal oxide nanoparticles to bacteria Escherichia coli, *Chemosphere*, **89**, 1098–1102.

66. Puzyn, T., Leszczynska, D. and Leszczynski, J. (2009). Toward the development of "nano-QSARs": advances and challenges, *Small*, **5**, 2494–2509.

67. Puzyn, T., Leszczynski, J. and Cronin, M. T. (2010). *Recent Advances in QSAR Studies: Methods and Applications* (Springer).

68. Linkov, I., Steevens, J., Adlakha-Hutcheon, G., Bennett, E., Chappell, M., Colvin, V., Davis, J. M., Davis, T., Elder, A., Hansen, S., Hakkinen, P. B., Hussain, S. M., Karkan, D., Korenstein, R., Lynch, I., Metcalfe, C., Ramadan, A. B. and Satterstrom, F. K. (2009). Emerging methods and tools for environmental risk assessment, decision-making, and policy for nanomaterials: summary of NATO Advanced Research Workshop, *J. Nanoparticle Res.*, **11**, 513–527.

69. Heinlaan, M., Ivask, A., Blinova, I., Dubourguier, H. C. and Kahru, A. (2008). Toxicity of nanosized and bulk ZnO, CuO and TiO_2 to bacteria Vibrio fischeri and crustaceans Daphnia magna and Thamnocephalus platyurus, *Chemosphere*, **71**, 1308–1316.

70. Adams, L. K., Lyon, D. Y. and Alvarez, P. J. J. (2006). Comparative ecotoxicity of nanoscale TiO_2, SiO_2, and ZnO water suspensions, *Water Res.*, **40**, 3527–3532.

71. Toropov, A. A., Toropova, A. P., Benfenati, E., Leszczynska, D. and Leszczynski, J. (2010). SMILES-based optimal descriptors: QSAR analysis of fullerene-based HIV-1 PR inhibitors by means of balance of correlations, *J. Comput. Chem.*, **31**, 381–392.

72. Toropov, A. A., Toropova, A. P., Benfenati, E., Leszczynska, D. and Leszczynski, J. (2010). InChI-based optimal descriptors: QSAR analysis of fullerene[C60]-based HIV-1 PR inhibitors by correlation balance, *Eur. J. Med. Chem.*, **45**, 1387–1394.

73. CORAL Software, http://www.insilico.eu/CORAL/

74. Elemental-Descriptor 1.0, http://nanobridges.eu/software/
75. Kuz'min, V., Artemenko, E., Muratov, E. N., Polischuk, P. G., Ognichenko, L. N., Liahovsky, A. I., Hromov, A. I. and Varlamova, E. V. (2010). Challenges and advances in computational chemistry and physics, in *Recent Advances in QSAR Studies: Methods and Applications* **8**, Puzyn, T., Leszczynski, J. and Cronin, M. T., eds. (Springer), pp. 127–176.
76. Plimpton, S. (1995). Fast parallel algorithms for short-range molecular-dynamics, *J. Comput. Phys.*, **117**, 1–19.
77. Periodic table, https://www.chemicool.com/
78. Glotzer, S. C. and Solomon, M. J. (2007). Anisotropy of building blocks and their assembly into complex structures, *Nat. Mater.*, **6**, 557–562.
79. Bigdeli, A., Hormozi-Nezhad, M. R., Jalali-Heravi, M., Abedini, M. R. and Sharif-Bakhtiar, F. (2014). Towards defining new nano-descriptors: extracting morphological features from transmission electron microscopy images, *RSC Adv.*, **4**, 60135–60143.
80. Bigdeli, A., Hormozi-Nezhad, M. R. and Parastar, H. (2015). Using nano-QSAR to determine the most responsible factor(s) in gold nanoparticle exocytosis, *RSC Adv.*, **5**, 57030–57037.
81. Odziomek, K., Ushizima, D., Oberbek, P., Kurzydlowski, K. J., Puzyn, T. and Haranczyk, M. (2017). Scanning electron microscopy image representativeness: morphological data on nanoparticles, *J. Microsc.*, **265**, 34–50.
82. Kar, S., Gajewicz, A., Roy, K., Leszczynski, J. and Puzyn, T. (2016). Extrapolating between toxicity endpoints of metal oxide nanoparticles: predicting toxicity to Escherichia coli and human keratinocyte cell line (HaCaT) with nano-QTTR, *Ecotoxicol. Environ. Saf.*, **126**, 238–244.
83. Hou, Y. H., Liu, Y. P., Gao, R. Q., Li, Q. J., Guo, H. Z., Goswami, A., Zboril, R., Gawande, M. B. and Zou, X. X. (2017). Ag@CoXP core-shell heterogeneous nanoparticles as efficient oxygen evolution reaction catalysts, *ACS Catal.*, **7**, 7038–7042.
84. Zheng, Z. J., Li, F. Z., Liu, J., Pastore, R., Raos, G., Wu, Y. P. and Zhang, L. Q. (2018). Effects of chemically heterogeneous nanoparticles on polymer dynamics: insights from molecular dynamics simulations, *Soft Matter*, **14**, 1219–1226.
85. Mikolajczyk, A., Gajewicz, A., Mulkiewicz, E., Rasulev, B., Marchelek, M., Diak, M., Hirano, S., Zaleska-Medynska, A. and Puzyn, T. (2018). Nano-QSAR modeling for ecosafe design of heterogeneous TiO_2-based nano-photocatalysts, *Environ. Sci. Nano*, **5**, 1150–1160.

Chapter 7

Nano-QSAR for Environmental Hazard Assessment: Turning Challenges into Opportunities

Willie Peijnenburg,[a,b] **Guangchao Chen,**[a,b] **and Martina Vijver**[a]
[a]*Institute of Environmental Sciences, University Leiden, PO Box 9518, 2300 RA Leiden, The Netherlands*
[b]*Center for Safety of Substances and Products, Dutch National Institute of Public Health and the Environment, PO Box 1, 3720 BA Bilthoven, The Netherlands*
willie.peijnenburg@rivm.nl

The virtually exponential increase of nanotechnology has raised concerns on the risks posed by engineered nanomaterials (ENMs). Recent studies on the hazards of ENMs showed that these materials are capable of inducing adverse effects in the biota, and a comprehensive assessment of the hazards of ENMs is thus urgently needed. Such an assessment is, however, severely hindered by considerations of time, financial burden, and ethics. Collecting the necessary information in as efficient a manner as possible is thus a key challenge. In this context, the extension of the conventional QSAR approach to nanotoxicology (i.e., nano-QSAR) is a possible

Computational Nanotoxicology: Challenges and Perspectives
Edited by Agnieszka Gajewicz and Tomasz Puzyn

ISBN 978-981-4800-64-8 (Hardcover), 978-0-429-34137-3 (eBook)
www.jennystanford.com

opportunity. Various attempts have been made so far to meet the challenge of correlating physicochemical characteristics of ENMs to biological effects elicited by ENMs. These efforts have highlighted the potential applicability of QSAR in the nanotoxicity field as it aids in prioritizing information on nanotoxicity and in rationalizing the risk assessment of ENMs. This chapter summarizes and discusses the current knowledge on nano-QSAR for metal-based ENMs with regard to (i) sources of data, (ii) existing nano-QSARs, (iii) mechanistic interpretation, and (iv) an outlook on the opportunity for further development of this frontier. The basic aim is to present key advances in relevant nanomodeling studies and to stimulate future research efforts in the quickly developing field of computational hazard assessment of ENMs.

7.1 Introduction

7.1.1 General

Nanotechnology basically deals with controlling the structure of matter at the nanoscale (1–100 nm) with respect to one or more external dimensions in order to produce materials with novel and hitherto unexploited properties [1, 2]. Nanomaterials are tailored to the needs of inimitable characteristics (e.g., electromagnetic, catalytic, optical, and thermal properties) that are often not observed in their bulk counterparts [3, 4]. As such, nanomaterials have been designed and engineered for a broad spectrum of applications. An online database called the "Nanotechnology Consumer Products Inventory" has listed 8 general categories (including 37 subcategories) of nano-enabled products, namely appliances, automotive, cross cutting, electronics and computers, food and beverage, goods for children, health and fitness, and home and garden [5, 6]. By early December 2016, this inventory contained in total 1827 consumer products on the market from 715 companies in 33 countries, which were manufacturer-identified as incorporating engineered nanomaterials (ENMs). The controllable production and widespread commercial applications of ENMs have shown the immense promise of nanomaterials to benefit the world economy and quality of life. According to the European Commission, direct employment in the

European Union (EU) involving nanotechnology is estimated to comprise up to 400,000 jobs [7]. The global production rates for ENMs involved in structural applications (e.g., catalysts, films and coatings, composites) have been predicted to increase from about 103 tons per year in 2010 to an order of 104–105 tons per year by 2020 [8, 9]. By 2020, the global market for nanotechnology is likely to continue to grow at double-digit rates (around 17% annually) for the coming decade and reach a global value of $75 billion in 2020 [10].

7.1.2 Safety Concerns

The rapid development of nanotechnology and extensive use of ENMs for industrial and commercial applications have caused safety concerns [11, 12]. The ongoing production of ENMs of all types certainly increases the likelihood of the release of ENMs into the environment. A proposed life cycle assessment of ENMs showed that the release of ENMs into the environment can be traced back to the stages of ENM production; incorporation of ENMs into products; and consumption, recycling, and disposal of ENM-containing products [13]. As estimated, every year about 189,200 tons of ENMs are released into landfills; 69,200 tons of ENMs are released into water bodies; 51,600 tons of ENMs are released into soil; and 8100 tons of ENMs are released into the air [14]. It is beyond doubt that the ongoing release of ENMs into the environment will result in higher exposure of humans and ecosystems to ENMs. As illustrated in Fig. 7.1, the exposure of humans to ENMs may occur via a number of exposure routes:

- Dermal contact, for instance, by applying personal care products incorporating ENMs such as TiO_2 and ZnO [15]
- Inhalation and subsequent deposition in the lungs [16]
- Ingestion of ENMs added to food or of unwarranted ENMs that leach out from package materials into food [17, 18]

In addition, therapeutic and medical applications of ENMs can also result in direct uptake into the human body, even though this option is still seen as underdeveloped [8]. Theoretically, the proposed exposure routes of ENMs also apply for environmental organisms just as for humans. Given the diversity of organisms, exposure of environmental species to ENMs seems to be more complicated and additional exposure and uptake routes do exist. For instance, the gill has been identified as being the principal site of the uptake of Cu and Ag ENMs for fishes and other gill-keeping species [19, 20]. Plants could interact with ENMs adsorbed on the soil and sediments via their roots [21], whereas ENMs deposited and aggregated on the leaves or other aerial parts of plants are able to penetrate through stomatal pathways [22, 23]. Finally, uptake of ENMs via bacterial cell membranes has been shown to occur [24, 25].

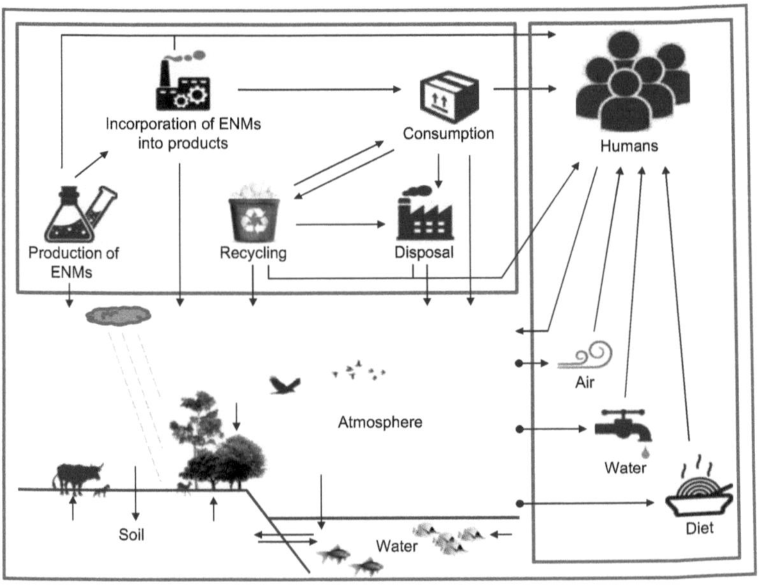

Figure 7.1 Possible exposure routes of humans and the environment to ENMs (adapted from Dowling et al. [9] and Gottschalk et al. [13]).

7.1.3 Toxicity of ENMs

The uptake of nanomaterials may lead to adverse effects, and there is consensus across a variety of studies regarding the occurrence of damage triggered by ENMs at the cellular level [26]. A comprehensive study by Shaw et al. [27] evidenced the effects of ENMs on the cellular viability and physiology of different mammal cell lines. Gajewicz et al. [28] also reported the impacts of metal oxide ENMs on cell viability (human keratinocyte cells), which confirmed the observations of other independent reports [29–31]. Experimental assays of nanotoxicity have also been generally performed on various trophic levels of organisms, such as algae (e.g., *Pseudokirchneriella subcapitata*), bacteria (e.g., *Escherichia coli*), crustaceans (e.g., *Daphnia magna* and *Daphnia pulex*), fish (e.g., *Danio rerio* and *Oryzias latipes*), nematodes (e.g., *Caenorhabditis elegans*), plants (e.g., *Lemna minor*), protozoa (e.g., *Tetrahymena thermophila*), yeast (e.g., *Saccharomyces cerevisiae*), and mammals (e.g., *Rattus*) [32–34].

The exposure of ENMs to different species may lead to the occurrence of, for example, mortality; immobilization; malformation; inflammatory response; and inhibition of cell viability, growth, luminescence, reproduction, feeding, and fertilization [33]. For instance, exposure of algae (e.g., *Chlamydomonas reinhardtii*, *Pseudokirchneriella subcapitata*, and *Scenedesmus obliquus*) to ENMs may result in the inhibition of growth and the loss of cell viability [35–37]. Bacteria that were exposed to ENMs exhibited mortality (e.g., *E. coli*), luminescence inhibition (e.g., *E. coli*, *Vibrio fischeri*, and *Pseudomonas putida*), and growth inhibition (e.g., *E. coli*) [38–42]. Exposure of crustaceans, such as *Daphnia magna*, *Ceriodaphnia dubia*, and *Daphnia pulex*, to ENMs was found to cause mortality; immobilization; and inhibition of growth, feeding, and reproduction [43–46]. ENMs tested on fish are likely to induce mortality, growth inhibition, delay of hatching, and developmental malformation [47–50]. The introduction of ENMs to nematodes, protozoa, and yeast may cause mortality, growth inhibition, inhibition of cell viability, reproduction inhibition, and immobilization [51–58]. For rats, in vivo experiments have also evidenced the harmful effects of ENMs, such as hepatotoxicity and nephrotoxicity after oral gavage [59].

ENMs may pose effects via a single pathway or via combinations of a few possible pathways. ENMs can, for instance, induce the generation of reactive oxygen species (ROS) or induce direct steric hindrance or interferences with important reaction sites [60]. ENMs are also considered to be able to act as vectors for transporting other toxic chemicals into cells, a phenomenon that is described as the Trojan horse mechanism [61]. Nanotoxicity could as well occur due to the shedding of ions from ENM crystals. This process is generally believed to be one of the important pathways of induction of toxicity for soluble metallic nanoparticles (NPs) [62]. The released ions are able to interact with subcellular structures, initiating cellular damages, or stimulate ROS formation, which in turn has been reported to induce oxidative stress, resulting in the disturbance of cellular physiological redox-regulated functions [63, 64]. The released ions can also promote the production of ROS [65].

7.1.4 Environmental Risk Assessment and Safe-by-Design Development of ENMs

The concerns raised regarding potential adverse effects of NMs released in the environment have induced the development of environmental risk assessment strategies that include nano-specific considerations of fate and adverse effects. To ensure nanosafety and to optimally benefit from nanotechnology, two strategies stand out:

- To design and produce ENMs that are safe and environmentally benign while with desired properties. This strategy aims at minimizing the potential risks of ENMs from the very beginning of the development of an ENM application. It is referred to as the safe-by-design development of ENMs and relates to ex ante safety assessment.
- To assess the risks of existing and newly introduced ENMs. This strategy is aimed at controlling relevant risks during the stages of manufacture, use, and disposal of ENMs with prospective risk assessment.

To answer the question whether an ENM is environmentally safe, a series of key steps, including hazard assessment, exposure assessment, and risk characterization, is required, whereas measures are to be taken based on the established conclusion [66]. As suggested by the European Chemicals Agency (ECHA), the process of risk assessment of ENMs (as of any chemical to be regulated within the EU) begins with the identification and assessment of ENM hazard, provided the ENMs are subject to registration under the Registration, Evaluation, Authorisation and Restriction of Chemicals (REACH) regulation, that is, when reaching an annual production or import of at least 10 tons [67]. The hazard assessment of ENMs may provide important feedback for safe-by-design approaches for ENMs with regard to key characteristics of ENMs governing relevant toxicity pathways, upon which modifications toward designing safer materials could be determined [68].

As described in Fig. 7.2, hazard assessment for human health and the environment comprises evaluation of information, classification and labeling, and identification of predicted no-effect concentrations (PNECs) or derived no-effect levels (DNELs). The gathering and evaluation of relevant physicochemical, (eco)toxicological information of ENMs is certainly fundamental to the assessment of ENM hazard as the very first step. It involves the retrieval and sharing of existing data, consideration of needed information, identification of information gaps, and the generation of new data or preparation of a proposal for a tailored testing strategy [67]. On the basis of the first step, it will be determined whether or not the ENM is meeting the criteria for any of the hazard classes or categories proposed by ECHA, that is, the step of classification and labeling. Once an ENM is categorized in at least one of the listed classes, derivation of the hazard threshold levels of the ENM for human health and the environment, for example, PNECs and DNELs, is required in light of a qualitative risk characterization for relevant ENMs.

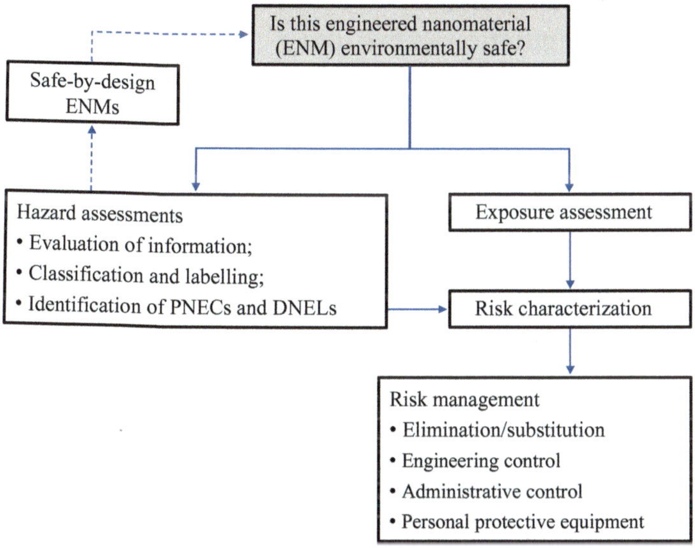

Figure 7.2 Schematic explanation of the safe handling of engineered nanomaterials. PNEC: predicted no effect concentration; DNEL: derived no effect level. The figure is adapted from the ECHA guidance [67].

Gathering and evaluation of hazard information of ENMs is, therefore, essential. In a survey provided by the NanoSafety Cluster Database working group, a total number of 38 online ENM databases developed under various projects were listed. These were provided with the names of databases and website addresses [69]. Another online inventory, named StatNano, was established in 2010 for access to up-to-date information and statistics in nano-based science, technology, and industry [70]. This website also provides a so-called Nanotechnology Products Database (established in January 2016) for the analysis and characterization of nanotechnology-based consumer products [71]. Undoubtedly, those databases and platforms are of significant importance as the first step in gathering, evaluation, and processing of information regarding the hazard of ENMs. As a follow-up, it is crucial to develop comprehensive databases containing reliable and sufficient information on ENM characterization, experimental conditions, and toxicity for ENM hazard assessment.

7.1.5 Handling Nanosafety with the Aid of Computational Toxicology

A large number of ENMs have been carefully tested on various species and cell lines [26, 33]. Yet, given the substantial number of existing, nontested ENMs and the enormous growth of nanotechnology, testing every single type of ENM to support the comprehensive evaluation of ENM safety is expensive and time consuming and, thus, virtually impossible. Testing of all hitherto nontested ENMs and of all newly developed ENMs also conflicts with the 3Rs principle of refining, reducing, and replacing animal use in toxicity testing [72]. Thus, researchers have been seeking to develop alternatives for testing assays of conventional chemicals and of ENMs. One of the helpful tools within this search for testing alternatives is computational toxicology. Computational toxicology is defined as a discipline that integrates information from various sources in order to develop computer-based models for the improved interpretation and prediction of effects of chemicals [73]. A few typical tools in this field are, for example, quantitative structure–activity relationships (QSARs), identification of structural alerts, read-across extrapolations, and dose- and time-response modeling.

The QSAR method enables the correlation of chemical characteristics with experimental toxicity data and thus enables one to encode existing knowledge into predictive models. To build a

QSAR model, the measured or calculated descriptors characterizing key structures of chemicals and the toxicity endpoints reflecting the chemical biological effects are required. The role of QSARs in predictive toxicology is:

- To provide efficient and inexpensive screening tools for the evaluation of chemical hazards
- To assist in the categorization and labeling of chemicals on the basis of their hazard effects
- To help interpret the underlying mechanisms of toxicity of substances [74]

QSARs have already been successfully used as very helpful tools for conventional chemicals in relating structural characteristics to chemical properties and biological effects in order to fill data gaps [75–77]. According to REACH, data derived from QSARs may support the waiving of laboratory testing or serve as a trigger for proposing further testing; when certain required conditions are met, QSAR results could be used instead of testing data for registration under REACH [78].

The limited data availability on ENM hazards necessitates the need of extending conventional QSAR approaches to nanotoxicology, that is, nano-QSARs. For the hazard assessment of ENMs, nano-QSARs could be potentially used to generate nontesting data during the gathering of information in the first step or to assist in the second step of classification and labeling of ENMs by directly categorizing ENMs into different hazard classes. The descriptors in nano-QSARs may also be helpful for understanding related mechanisms and identifying key factors affecting ENM toxicity, which also provides guidance for the modification of ENM characteristics for the safe-by-design development of ENMs. To date, attempts have already been made to correlate the characteristics of ENMs to their biological responses [79–81]. These studies showed the tantalizing possibility that the QSAR method may indeed be feasible and useful in predicting the biological activity profiles of novel ENMs. However, it meanwhile also revealed that nano-QSAR is still in its infancy and further challenges in this field need to be overcome. One issue standing out on this background relates to the comprehensive representation of ENM structures. As known, ENMs often exist as populations of materials varying in structural characteristics, for example, composites, sizes, shapes, and functional groups. The structural ambiguousness of ENMs makes it difficult for experimentalists to provide precise information on ENM characterization, which consequently hinders the calculation of representative descriptors for ENMs. Another issue of importance in this context concerns the dynamics of ENMs in media. ENMs often strongly interact with constituents in the medium and undergo dramatic changes in their surface properties and dissolution and aggregation behavior [82]. These changes consequently alter the mobility and bioavailability and, ultimately, toxicity of ENMs. Therefore, in some cases the toxicity information on ENMs can be poorly correlated to their characteristics if the dynamics of the ENMs in the media are not considered.

Overall, it is to be concluded that the use of computational toxicology in support of the hazard assessment of ENMs is a field of research that needs further development. This chapter aims to explore the contribution of computational toxicology in the safe handling of metal-based ENMs. This is done by evaluating the currently existing literature data on metal-based ENMs for the use of computational toxicology in light of the safety assessment of ENMs, by identifying data gaps and by reviewing nano-QSARs developed so far for the prediction and categorization of ENM hazards.

7.2 Inventory of Existing Toxicity Data of Metal-Based ENMs

7.2.1 Need for Reliable Experimental Data

The QSAR approach offers a rapid way of filling data gaps caused by the limited availability, or the absence, of experimental information. Attempts have also been made to use experimentally obtained data to link the physicochemical characteristics of ENMs to their cellular uptake, cytotoxicity, and ecotoxicity [3, 28, 29, 60, 83–95]. To improve the development of nano-QSARs, with the ultimate goal of employing nano-QSARs as alternative in silico screening methods in toxicity testing, it is essential that all published nanotoxicity data are properly summarized and organized. Meanwhile, ENM-related regulatory frameworks also require the gathering of nanotoxicity information to enable the optimal use of the existing scientific sources. For instance, according to EU Directive 93/67/EEC, it is preferable that the classification of chemical hazards to aquatic organisms is based on toxicity data from at least three standard test organisms (i.e., algae, crustacea, and fish), with hazard initially determined by the lowest median L(E)C50 value (the chemical concentration found to cause 50% death or effect of interest) of the species tested: if the L(E)C50 value is <1 mg/L, the compound is considered very toxic to aquatic organisms; if the L(E)C50 value is 1–10 mg/L, the compound is considered toxic to aquatic organisms; if the L(E)C50 value is 10–100 mg/L, the compound is considered harmful to aquatic organisms; if the L(E)C50 value is >100 mg/L, the compound is not classified as being toxic or harmful [66, 96]. Blaise et al. [97] and Sanderson et al. [98] have subsequently extended this classification scheme by adding one more category: L(E)C50 value <0.1 mg/L corresponds to compounds that are extremely toxic to aquatic organisms [96, 99]. Two kinds of data are essential for developing predictive nano-QSAR models: data that characterize the physicochemical properties of groups of pristine ENMs and data that describe the relevant biological effects of ENMs on test organisms (toxicity endpoints), including a detailed description of the experimental conditions or of the test protocols used.

7.2.2 Overview of Experimental Data

As a first step toward the development of nano-QSARs, and on the basis of a Web of Science™ Core Collection bibliometric data search using the advanced search facilities, we established an inventory of toxicity data of metal-based ENMs that are widely used in a variety of applications. The inventory included information on ENM characterization as far as provided in the study reports. The organisms were selected on the basis of the work of Ivask et al. [96], Bondarenko et al. [26], and Kahru and Dubourguier [99] and are restricted to algae, yeast, bacteria, protozoa, nematodes, crustacea, and fish. The evaluation of the toxicity data obtained was performed in light of nano-QSAR development and was based on the characteristics of pristine ENMs. The toxicity endpoints reported in the literature, and hence included in the database, consisted of the lethal concentration (LC), the effect concentration (EC, or IC when the effect refers to inhibition), the lowest-observed-effect concentration (LOEC), the no-observed-effect concentration (NOEC), the minimum bactericidal concentration, and the minimum inhibitory concentration (MIC; more commonly used in antimicrobial assays). For QSAR modeling purposes, further analysis focused on the number of different toxicity endpoints, the type of biological effects induced by the ENMs, data availability (i.e., the amount of accessible toxicity data), and also the characterization of the ENMs provided.

The search yielded a total of 910 papers, of which 134 dealt with algae, 102 dealt with yeast, 406 papers were on bacteria, 17 were on protozoa, 43 were on nematodes, 193 were on crustacea, and 245 were on fish. With regard to the ENMs studied, the majority of the studies focused on nanosilver toxicity (383 studies), followed by 239 on titanium, 130 on copper, and 137 on zinc ENMs. Thereupon, 45 studies dealt specifically with quantum dots. A general overview of the information obtained is provided in Fig. 7.3. Deleting organisms with less than six data points yielded a database with 886 records.

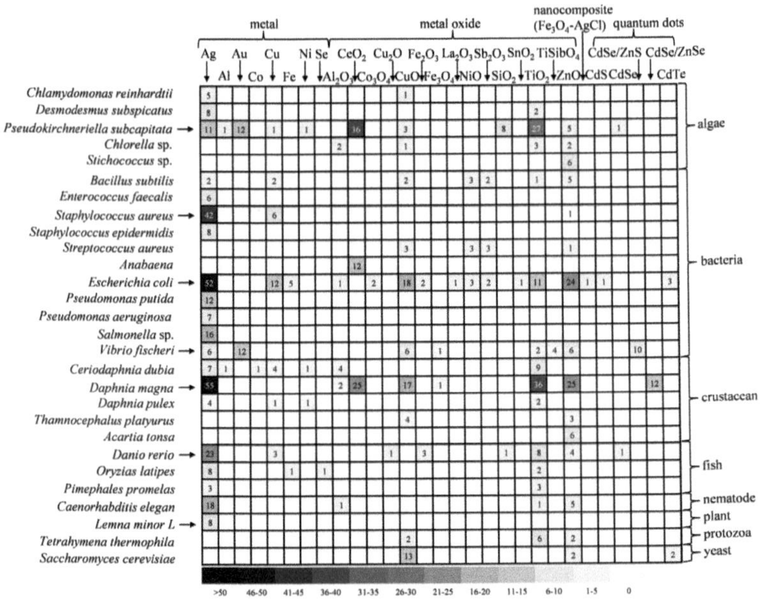

Figure 7.3 Overview of toxicity data on metal-based ENMs and test species and number of records of each endpoint tested. Organisms with less than six records of toxicity are not included.

Ag ENMs were most widely studied, with a total of 332 records of endpoints and with a particular focus on two bacteria, *Staphylococcus aureus* and *E. coli*, and one crustacean, *Daphnia magna*. Meanwhile, more than average attention was also paid to TiO_2 (126 records) and ZnO (109) ENMs. As for the test organisms most often used, *Daphnia magna*, *E. coli*, and *Pseudokirchneriella subcapitata* (an alga) were the dominant species in the database, with 173, 139, and 106 toxicity records, respectively. They were followed by *Staphylococcus aureus* (49 toxicity records), the bacterium *V. fischeri* (47 records), and the zebrafish (*Danio rerio*, 44 records). Given the numbers of available records, these data are potentially useful for nano-QSAR modeling, but care should be taken regarding data consistency. If we take the endpoints in *Daphnia magna* as an example, the 191 records retrieved on this water flea consisted of 65 values for LC_{50}, 69 values for EC_{50}, 20 values for NOEC, 14 values for LOEC, and 23 others (e.g., LC10, LC20, EC10, and EC20). In addition, there was further variation with regard to the various biological effects that were assessed and the duration of the exposure of the organisms to the NPs.

The biological effects commonly investigated included immobilization, mortality, growth inhibition, cell viability inhibition, malformation, luminescence inhibition, and reproduction inhibition. Growth inhibition and mortality are generally applied in ecotoxicity assays; the other endpoints are used as appropriate on different groups of organisms. This is not surprising as the standard test organisms in OECD guidelines, algae, bacteria, crustacea, and fish, are paid relatively more research attention. Immobilization, for instance, is an important factor that is often used to characterize the effects of ENMs on crustacea. Inhibition of reproduction is another commonly studied endpoint. Meanwhile, luminescence inhibition is examined only with bacteria. With regard to the issue of data availability of toxicity endpoints for nano-QSAR modeling, the EC_{50} for growth inhibition of algae accounts for 93 records. Concerning bacteria, 110 MIC data records and 86 EC_{50} values for inhibition of luminescence were retrieved. For crustacea, 108 LC_{50} values and 61 EC_{50} values for immobilization were found. For fish, 44 LC_{50} (mortality) values were obtained. Besides the test species used and the biological effects and toxicity endpoints measured, the diversity of metal-based ENMs in a data set is also of major importance for the development of nano-QSARs. In this context, two issues stand out: first, regardless of the number of records available in a data set with the same toxicity endpoint, the data set should cover different ENMs in order to be potentially modeled against ENM properties; second, the ENMs of interest should share a degree of structural similarity in order to be grouped and described in terms of descriptors suitable for modeling. A diversity analysis of the metal-based ENMs was performed on the toxicity data of the six species mentioned above: *Daphnia magna* (173 records), *E. coli* (139 records), *Pseudokirchneriella subcapitata* (106 records), *Staphylococcus aureus* (49 records), *V. fischeri* (47 records), and *Danio rerio* (44 records). LC_{50} values for *Daphnia magna*, *E. coli*, and *Danio rerio* are reported for eight ENMs. Meanwhile, growth inhibition to *Pseudokirchneriella subcapitata* (EC_{50}) was reported for 10 ENMs. These endpoints can be considered for building local nano-QSAR models for a specific endpoint and a specific organism.

7.2.3 Suitability of Experimental Data for QSAR Modeling

In addition to making data available for well-specified toxicity endpoints, ENM characterization in the form of measured and/ or calculated ENM properties also plays an essential role in the development of nano-QSARs. Figure 7.4a shows the frequency distribution of the measured ENM properties in the data retrieved. The ENM properties analyzed included size, zeta potential, shape, surface charge and surface area, crystallinity, purity, and coating. The results show that the size (primary) of metal-based ENMs was generally provided (847 records), followed by the zeta potential (316 records), the surface area (224 records), coating (117 records), and purity (87 records). Only a limited number of studies offered information about the shape (67 records), crystallinity (57 records), and surface charge (3 records). It is worth noting that some of the data on these properties are hardly suited for the purpose of nano-QSAR studies. For instance, ENM size is on occasion given as a range between 20 and 60 nm, or <100 nm [44, 100]. Unfortunately, this is not precise enough for developing models. Thus, for the purpose of nano-QSAR development, data availability on measured ENM properties is even more limited than reported. As shown in Fig. 7.4b, after an analysis of the number of measured properties for a certain

ENM in a publication, it is clear that relatively few ENM features are usually reported. Most of the published toxicity endpoints contain one or two ENM properties, and only 5.3% of the assembled records contain more than three ENM properties. Thus, according to our analysis of the availability of ENM properties, the development of nano-QSAR models simply on basis of the reported experimentally determined descriptors would, at this time, be a challenging task.

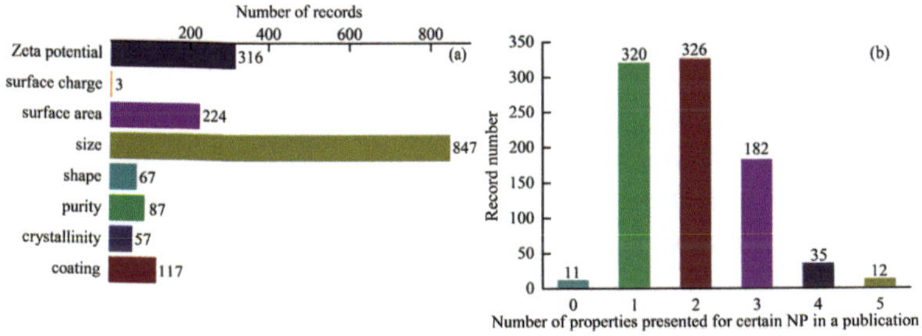

Figure 7.4 Overview of basic particle properties reported in ecotoxicity studies of ENMs, showing the number of records (Fig. 7.4a) for each property and the number of properties studied for a given ENM (Fig. 7.4b).

The QSAR paradigm dictates that it is possible to predict properties like toxicological effects directly from the physicochemical properties of the entities of interest [101]. A specific characteristic of metal-based ENMs is, however, that when in contact with artificial and natural aqueous media, they often interact strongly with constituents in the medium [102] and undergo dramatic changes in their surface properties [103] as well as in their aggregation and dissolution behavior [104, 105]. These changes affect ENM mobility and bioavailability and, ultimately, toxicity to organisms [106–109]. It should be acknowledged that these interactions are dynamic in nature and are often kinetically rather than thermodynamically controlled, as is usually the case for (conventional) nonparticulate chemicals. Therefore, toxicity information can be poorly modeled in case information on physicochemical properties is available solely for pristine ENMs. Relationships developed between toxicity endpoints and the characteristics of pristine ENMs without considering the dynamic transformations of ENMs in the media will in these cases most likely result in models of low statistical significance, predictability, and relevance. In such a context, better interpretation of the dynamic processes influencing ENMs in aqueous environments is highly required for modeling and predicting the biological effects of ENMs.

Overall, our results show that the existing data currently cannot be used to the extent that would be needed to efficiently develop predictive toxicity models for metal-based ENMs, keeping the OECD criteria for development of QSARs suited for regulatory purposes in mind [110]. When performing in-depth quantitative analysis of the experimental data, data consistency is shown to play a vital role in this respect as numerous data gaps were identified when comparing species and ENMs tested. An obvious recommendation at this stage is, therefore, that further testing be performed on additional key species and ENMs in order to more accurately assess the impacts of metal-based ENMs on ecosystems and to develop widely applicable global and local nano-QSARs. To obtain data that are acceptable for use in modeling applications, it should be emphasized that experimental protocols used in future data generation must be based on consistent experimental conditions or on generally accepted and widely applied guidelines (e.g., OECD or US Environmental Protection Agency guidelines). Currently, QSAR approaches have limited potential when used for predicting ENM toxicity on the basis of the characteristics of pristine ENMs. However, taking account of the issues raised above will certainly facilitate future developments in this emerging area of research.

7.3 Recent Advances toward the Development of QSARs for Metallic ENMs

7.3.1 Representation of the Intrinsic Properties of ENMs

In silico methods are now routinely used to estimate physicochemical properties, biological activity, or toxicological effects of a wide range of chemical substances. Apparently, in the case of substances such as ENMs with an unclear molecular structure, the standard QSAR approach cannot be used. Apart from the issues related to data availability described above, a key issue is related to the selection of proper ENM-specific parameters that can be linked to a specific endpoint. Frequently, a few parameters, such as particle size, surface area, and surface characteristics, have been envisaged to be the major properties discriminating ENM-specific fate and effect properties from the corresponding properties of dissolved, nonparticulate, molecules [111]. There are most likely other parameters that play an equally important role in driving the properties and effects of ENMs. For example, ENMs of the small nanometer size are able to cross biological membrane barriers and may reach different body organs that are otherwise protected against the entry of larger materials [111]. However, it is also known that ENMs are not always more detrimental compared to their corresponding larger forms [112]. This points out that there are properties other than size and surface area that play an important role in determining the effects and interactions of NPs in biological systems. As pointed out by Oksel [113], surface coatings are probably very important in this respect. In this regard, QSAR modeling approaches have the potential to answer the question of which properties of ENMs determine toxicity, and these approaches can therefore help to elucidate the key parameters that control ecotoxic effects of ENMs. In such cases, the QSAR models can be built up using some additional information, for example, available physicochemical parameters, solubility, partition coefficients between different solvents, technological conditions, and parameters selected for manufacturing various nanomaterials. Thus, there is a need to develop publicly and freely available models that can relate reliable and consistently determined effect data to the intrinsic properties of ENMs.

Proper description of the intrinsic properties of ENMs in light of QSAR development is still in its infancy, and development of ENM-specific descriptors is clearly a key research need. Apart from cross sections of the physicochemical parameters already discussed, efforts have been made to incorporate theoretical molecular descriptors into modeling. Thousands of descriptors of varying nature have been made available for characterizing nonparticulate organic and inorganic chemicals and for representing molecular structures. Most of them are, however, inapplicable to ENMs or need adaptation to allow for their use for characterization of ENMs. A typical example of descriptors used nowadays in QSAR modeling of metallic ENMs is given in Table 7.1. This example refers to a set of quantum-mechanical and image descriptors successfully used to model the cytotoxicity of a set of 18 metal oxide ENMs [28].

Table 7.1 Overview of quantum-mechanical and image descriptors of 18 metal oxide ENMs, as retrieved from the study by Gajewicz et al. [28]

Quantum-mechanical descriptors	Image descriptors
• Standard enthalpy of formation of metal oxide nanoclusters (ΔH_f^c)	• Volume (V)
• Total energy (TE)	• Surface diameter (d_S)
• Electronic energy (EE)	• Equivalent volume diameter (d_V)
• Core–core repulsion energy (Core)	• Equivalent volume / surface (d_{Sauter})
• Solvent-accessible surface (SAS)	• Area (A)
• Energy of the highest occupied molecular orbital (HOMO)	• Porosity (P_x)
• Energy of the lowest unoccupied molecular orbital (LUMO)	• Porosity (P_y)
• Chemical hardness (η)	• Sphericity (Ψ)
• Total softness (S)	• Circularity (f_{circ})
• HOMO-LUMO energy gap (E_g)	• Anisotropy ratio (AR_x)
• Electronic chemical potential (μ)	• Anisotropy ratio (AR_y)
• Valence band (E_v)	
• Conduction band (E_c)	
• Mulliken's electronegativity (χ^c)	
• Parr and Pople's absolute hardness (Hard)	
• Schuurmann MO shift alpha (Shift)	
• Polarizability derived from the heat of formation (A_{hof})	
• Polarizability derived from the dipole moment (A_d)	

7.3.2 Overview of Models and Modeling Approaches

7.3.2.1 General considerations

Facing the strong need for extending the conventional QSAR approach to nanotoxicology, some attempts have been made to link ENMs' biological effects with the characteristics of ENMs. Table 7.2 offers an overview of the research achievements so far. The underlying literature search was performed by means of an advanced search in the Web of Science^{TM} Core Collection database. The search was manually supplemented with relevant publications not included in the search records. As can be deduced from Table 7.2, various nano-QSARs were constructed on the basis of a variety of modeling techniques, such as linear and nonlinear regression, support vector machine (SVM), artificial neural networks (ANN), and *k* nearest neighbor (*k*-NN). Distinct biological responses induced by ENMs, such as cellular uptake and cytotoxicity in different cell lines, and the ecotoxicity of ENMs were addressed. The studies provided in Table 7.2 highlight the potential of QSAR methods to be adopted as a tool in predicting nanotoxicity.

To provide an overview of recent key advances in the field of nano-QSAR, the state-of-the-art aspects of the field are discussed as related to the following:

- Sources of data for modeling, including the number and type of ENMs involved, targets or target species in tests, toxicity endpoints, characteristics of ENMs provided by nanomodelers, and accessibility of this information. For the relatively large data sets, methods of ENM characterization are summarized given the absence of calculated descriptors for the whole data sets.
- Existing nano-QSARs. Presentation of the models is based on the aspects of toxicity data utilized, descriptor calculation and consequent elimination of less significant descriptors used, modeling techniques, actual expression of the nano-QSARs (if applicable), predictive performance, and last but not least, characterization of applicability domains of the models.
- Mechanistic interpretation. On the basis of the developed nano-QSARs, the contributions of the various descriptors to the models are summarized. Thereupon, this contribution is

discussed in light of interpreting the underlying mechanisms of toxicity and cellular uptake of ENMs.

- An outlook on further development of nano-QSARs identifying major research gaps in the field, as related to data scarcity, integration of relevant existing information, and evaluation of data quality.

Table 7.2 Overview of the peer-reviewed literature on nano-QSARs for metallic ENMs[a]

Description	Reference
The model predicts the oxidative stress potential of metal oxide ENMs on the basis of the bandgap energy.	Burello and Worth [114]
A final consensus model was developed on the basis of the top five candidate models constructed by naive Bayes, logistic regression, *k* nearest neighbor, and support vector machine, predicting the cellular uptake of 105 ENMs (single-metal core) by PaCa2 (pancreatic cancer cells).	Chau and Yap [115]
Global classification models were developed to predict the ecotoxicity of metallic ENMs to different species; classification models were also built for *Danio rerio*, *Daphnia magna*, *Pseudokirchneriella subcapitata*, and *Staphylococcus aureus*.	Chen et al. [116]
(i) Cytotoxicity of 31 ENMs to vascular smooth muscle cells was modeled on the basis of multiple linear regression and Bayesian regularized artificial neural network, and (ii) cellular uptake of 108 ENMs was modeled in human umbilical vein endothelial cells (HUVECs) and PaCa2 cells using multiple linear regression with the expectation maximization method.	Epa et al. [83]
Generated models predict the cytotoxicity of 44 ENMs with diverse metal cores using the SVM method and the cellular uptake of 109 ENMs in PaCa2 cells using the *k*-NN method.	Fourches et al. [85]

Description	Reference
The multiple linear regression method combined with a genetic algorithm was applied to describe the toxicity of 18 metal oxide ENMs to the human keratinocyte cell line (HaCaT).	Gajewicz et al. [28]
The cellular uptake of 109 magnetofluorescent ENMs in PaCa2 cells was modeled using multiple linear regression and the multilayered perceptron neural network; descriptor selection was performed by combining the self-organizing map and stepwise multiple linear regression.	Ghorbanzadeh et al. [86]
Using the toxicity data set of 17 metal oxide ENMs to *E. coli*, models were built with the multiple linear regression and partial least-squares methods.	Kar et al. [93]
A model was developed establishing the cellular uptake of 109 magnetofluorescent ENMs in PaCa2 cells.	Kar et al. [87]
A perturbation model was introduced for the prediction of ecotoxicity and cytotoxicity of ENMs; molar volume, electronegativity, polarizability, size of the particles, hydrophobicity, and polar surface area are included in the model.	Kleandrova et al. [3]
Classification models (logistic regression) were developed to predict the cytotoxicity of nine ENMs to the transformed bronchial epithelial cells (BEAS-2B).	Liu et al. [29]
A nano-SAR was developed classifying 44 iron-based ENMs into bioactive or inactive, using a naive Bayesian classifier based on four descriptors: primary size, spin-lattice and spin-spin relaxivities, and zeta potentials.	Liu et al. [117]
The SVM nano-SAR model was constructed on the basis of the cytotoxicity data of 24 metal oxide ENMs to BEAS-2B cells and murine myeloid (RAW 264.7) cells.	Liu et al. [88]

(*Continued*)

Table 7.2 (*Continued*)

Description	Reference
The perturbation model was presented predicting the cytotoxicity of ENMs against several mammalian cell lines; the influence of molar volume, polarizability, and size of the particles was indicated.	Luan et al. [89]
A quantitative model was developed on the basis of the toxicity data of 16 metal oxide ENMs to *E.* coli using enthalpy of formation of a gaseous cation (ΔH_{Me+}) and polarization force (Z/r). The toxicity of 35 other metal oxide ENMs was predicted and depicted in the periodic table.	Mu et al. [118]
Models were constructed to predict (i) the toxicity of 17 metal oxide ENMs to *E. coli* and (ii) the cytotoxicity in HaCaT cells of 18 different metal oxide ENMs. The factors of molecular weight, cationic charge, mass percentage of metal elements, and individual and aggregation sizes were discussed.	Pan et al. [119]
The cytotoxicity of TiO_2 and ZnO ENMs was modeled by LMR and C4.5 algorithms.	Papa et al. [120]
The toxicity and photoinduced toxicity of 17 metal oxide ENMs to *E. coli* was assessed using a self-written least-squares fitting program.	Pathakoti et al. [94]
The cytotoxicity of 17 metal oxide ENMs to *E. coli* was predicted with only one descriptor: enthalpy of formation of a gaseous cation having the same oxidation state as in the metal oxide structure.	Puzyn et al. [60]
Predictive models were built on the basis of (i) cytotoxicity of different ENMs (with diverse metal cores) in four cell lines (endothelial and smooth muscle cells, monocytes, and hepatocytes), (ii) cellular uptake of 109 ENMs in PaCa2 cells, and (iii) cytotoxicity of 17 different metal oxide ENMs to *E. coli*.	Singh and Gupta [90]

Description	Reference
On the basis of random forest regression, predictive classification models were developed for the (i) toxicity of 17 metal oxide ENMs to *E. coli* and (ii) cytotoxicity of 18 metal oxide ENMs to HaCaT cells.	Sizochenko et al. [91]
Structure–activity relationship models (random forest) were introduced for toxicity of 24 metal oxide ENMs toward BEAS-2B and RAW 264.7 cell lines.	Sizochenko et al. [121]
The toxicity of 17 metal oxide ENMs to *E. coli* was estimated by employing the SMILES-based optimal descriptors.	Toropov et al. [95]
Cellular uptake of 109 ENMs with the same core but different surface modifiers in the PaCa2 cells was modeled on the basis of SMILES-based optimal descriptors.	Toropov et al. [92]
A classification model was built for 24 metal oxide ENMs on the basis of the dissolution of metals and energy of the conduction band (E_c).	Zhang et al. [30]

[a]Apart from the references obtained, a general description of each model is given.

7.3.2.2 Sources of data actually used for modeling

As a data-driven approach, the field of nano-QSAR relies highly on generating or assembling qualified experimental data. To integrate the existing information obtained from the various data sets that were successfully used in nano-QSARs, and therefore to aid further studies of nanomodeling, the underlying experimental data in the nano-QSARs mentioned in Table 7.2 were analyzed. As can be seen in Table 7.3, research attention was mainly on the cellular uptake of ENMs by different cell lines, on cytotoxicity, and on the toxicity of ENMs to *E. coli*. Despite the numerous nano-related tests that are being carried out, it is to be concluded that only a few data sets were generally used as the data source for the nano-QSARs developed so far. The most widely applied data in QSAR-like studies (Table 7.3) are from Weissleder et al. [122], Puzyn et al. [60], and Shaw et al. [27].

These experimental data sets are presented and further discussed below in the order of cellular uptake, cytotoxicity in cell lines, and toxicity to *E. coli* in terms of the following aspects (when available): types and numbers of ENMs, targets or target species, toxicity endpoints, characteristics of the ENMs provided, and accessibility of relevant information.

Table 7.3 Summary of experimental data of ENMs used in nano-QSAR studies

Nano-QSAR	Data set used	Number of ENMs	Core of ENMs	Test organism
Gajewicz et al. [28] Pan et al. [119] Sizochenko et al. [91]	Gajewicz et al. [28]	18	Metal oxide	HaCaT cells
Pathakoti et al. [94]	Pathakoti et al. [94]	17	Metal oxide	*E. coli*
Kar et al. [93] Pan et al. [119] Mu et al. [118] Puzyn et al. [60] Singh and Gupta [90] Sizochenko et al. [91] Toropov et al. [95]	Puzyn et al. [60]	17	Metal oxide	*E. coli*
Epa et al. [83] Fourches et al. [85] Liu et al. [117] Singh and Gupta [90]	Shaw et al. [27]	50	Metal oxide and quantum dots	Endothelial cells, vascular smooth muscle cells, human HepG2 cells, RAW 264.7 cells

Nano-QSAR	Data set used	Number of ENMs	Core of ENMs	Test organism
Chau and Yap [115] Epa et al. [83] Fourches et al. [85] Ghorbanzadeh et al. [86] Kar et al. [87] Singh and Gupta [90] Toropov et al. [92]	Weissleder et al. [122]	146	Metal oxide	PaCa2 (pancreatic cancer cells)
Liu et al. [88] Zhang et al. [30] Sizochenko et al. [121]	Zhang et al. [30]	24	Metal oxide	BEAS-2B cells, RAW 264.7 cells
Liu et al. [29]	Liu et al. [29]	9	Metal oxide	BEAS-2B cells
Papa et al. [120]	Sayes and Ivanov [123]	24 TiO_2, 18 ZnO	TiO_2, ZnO ENMs	Rat L2 lung epithelial cells, rat lung alveolar macrophages

7.3.2.2.1 *Cellular uptake assays*

Weissleder et al. [122] modified the surface of monocrystalline magnetic ENMs (3 nm core of $(Fe_2O_3)_n(Fe_3O_4)_m$) with 146 various small molecules (modifiers) and created a library of 146 water-soluble, magnetic, and fluorescent ENMs. ENMs were made magnetofluorescent by adding the fluorescein isothiocyanate to the ENM surfaces. Uptake of these ENMs by five cell lines was screened afterward. The cell lines used include pancreatic cancer cells (PaCa2), a macrophage cell line (U937), resting primary human macrophages, activated primary human macrophages, and human umbilical vein

Table 7.4 Overview of reported information on the data on cellular uptake of ENMs

Reference	Method of ENM characterization	Data accessibility	ENM number	Other information
Weissleder et al. [122]			146	Molecular weight and structures
Chau and Yap [115]	Calculation of 679 1D and 2D chemical descriptors of modifiers using PaDEL-Descriptor (v2.8)	Values of PaCa2 cellular uptake were available (unit: number of ENMs per cell).	109	SMILES (simplified molecular input line entry system)
Epa et al. [83]	691 molecular descriptors of modifiers from DRAGON (v5.5), ADRIANA (v2.2) and an in-house modeling software package		108	List of modifiers
Fourches et al. [85]	MOE descriptors for modifiers, including physical properties, surface areas, atom and bond counts, Kier & Hall connectivity indices, kappa shape indices, adjacency and distance matrix descriptors, pharmacophore feature descriptors, and molecular charges	Values of PaCa2 cellular uptake were available (log10[ENM]/cell pM).	109	SMILES

Reference	Method of ENM characterization	Data accessibility	ENM number	Other information
Ghorbanzadeh et al. [86]	Hyperchem program (v7) for constructing the molecular structure of modifiers; geometry, optimized with the Austin Model 1 (AM1) semiempirical method; DRAGON for descriptor calculation	Values of PaCa2 cellular uptake were available (log10[ENM]/cell pM).	109	List of modifiers and SMILES
Kar et al. [87]	Calculation of a pool of 307 descriptors using Cerius 2 (v4.10), DRAGON 6, and PaDEL-Descriptor (v2.11)	Values of PaCa2 cellular uptake were available (log10[ENM]/cell pM).	109	List of modifiers
Singh and Gupta [90]	Calculation of 174 molecular descriptors for the modifiers (topological, electronic, geometrical, and constitutional) using Chemistry Development Kit (CDK v1.0.3)		109	List of modifiers, chemical structures, and SMILES
Toropov et al. [92]	Use of SMILES-based optimal descriptors		109	SMILES, correlation weights (CWs) of SMILES attributes (SA)

Source: [122]

endothelial cells (HUVECs). A diversity of cellular uptake of various functionalized ENMs and a high dependence of ENM uptake on the composition of their surface were observed especially in the PaCa2 cells [115, 124]). Data on PaCa2 cellular uptake of ENMs can be retrieved from the studies of Fourches et al. [85, 124] and also from the studies of Chau and Yap [115], Ghorbanzadeh et al. [86], and Kar et al. [87]. In the absence of data on calculated descriptors for the whole data set, methods of characterizing ENMs in previous studies are presented in Table 7.4. An analysis of the methods reported in literature shows that emphasis in ENM characterization was so far largely put on the characteristics of ENM surface modifiers, given the conclusion of Weissleder et al. [122] that the PaCa2 cellular uptake of ENMs depends highly on the surface modification of the ENMs. Descriptor calculation of the modifiers was performed within different software packages (e.g., PaDEL-Descriptor, DRAGON, and ADRIANA), each providing various molecular descriptors.

7.3.2.2.2 Cytotoxicity in cell lines

One of the most widely used cell line–based toxicity data for ENMs is from the work of Shaw et al. [27]. In their study, four cell-based assays were performed on the basis of four cell types at four different doses. The four types of cells, namely endothelial cells (human aorta), vascular smooth muscle cells (human coronary artery), hepatocytes (human HepG2 cells), and murine RAW 264.7 leukemic monocyte/macrophage cells, were employed to assess the cytotoxicity of 50 ENMs (iron-based ENMs, pseudo-caged ENMs, and quantum dots). The four cell-based assays were mitochondrial membrane potential, adenosine triphosphate (ATP) content, apoptosis, and reducing equivalents assays. The ENMs were characterized by their coating, surface modification, size, the spin-lattice (R1) and spin-spin (R2) relaxivities, and zeta potential. Experimental values were expressed in units of standard deviations of the distribution assessed when cells were only treated with PBS (Z score). Fourches et al. [85] afterward transformed the 64 features (4 assays × 4 cell lines × 4 doses) of 48 iron-based ENMs into 1 by calculating their arithmetic mean (Z_{mean}), which enabled binary classification studies on the basis of this data set.

Gajewicz et al. [28] tested the cytotoxicity of 18 metal oxide ENMs to the human keratinocyte cell line (HaCaT) by characterizing cell viability and subsequently expressing the impact of the ENMs in terms of LC_{50} (concentration of the ENMs that leads to 50% fatality). Experimental data are accessible in the original publication. As shown in Table 7.1, 18 quantum-mechanical and 11 image descriptors were calculated for modeling purposes. Information on the (aggregation) size for this data set was provided by Sizochenko et al. [91] as shown in Table 7.5.

By measuring the plasma-membrane leakage via propidium iodide uptake in transformed bronchial epithelial cells (BEAS-2B), Liu et al. [29] studied the cytotoxicity of a variety of ENMs: Al_2O_3, CeO_2, Co_3O_4, TiO_2, ZnO, CuO, SiO_2, Fe_3O_4, and WO_3. The cytotoxicity was expressed in terms of percentage of membrane-damaged cells. Descriptors calculated were the number of metal and oxygen atoms (N_{Metal} and N_{Oxygen}), atomic mass of the ENM metal (m_{Me}), molecular weight of the metal oxide (m_{MeO}), group and period of the ENM metal (G_{Me} and P_{Me}), atomization energy of the metal oxide (E_{MeO}), ENM primary size (d), zeta potential, and isoelectric point (IEP).

Another data set that was provided by Zhang et al. [30] contains information on the toxicity of 24 oxide ENMs. The toxicity was expressed in terms of $\log EC_{50}$, in which EC_{50} represents the effective concentration that causes 50% response. The lactate dehydrogenase (LDH), 3-(4,5-dimethylthiazol-2-yl)-5-(3-carboxymethoxyphenyl)-2-(4-sulfophenyl)-2H-tetrazolium, and ATP assays were implemented to assess the nanotoxicity to BEAS-2B and rat alveolar macrophage cells (RAW264.7) in the study. Information on the crystalline structure of the ENMs (crystal system, space group, and unit cell parameters), primary and hydrodynamic sizes of metal oxide ENMs, and parameters for calculating ENM band energies (conduction and valence band, bandgap energy, absolute electronegativities, and point of zero charge and zeta potential) were also provided by these authors. Liu et al. [88] built a nano-SAR model based on these data, along with a summary of the calculated physicochemical properties of the ENMs. Information on 13 descriptors was provided, including the ENM primary size (d), energy of the conduction band (E_C), energy of the valence band (E_V), metal oxide atomization energy (E_{Amz}), metal oxide electronegativity (χ_{MeO}), metal oxide sublimation enthalpy (ΔH_{sub}), metal oxide ionization energy (ΔH_{IE}), metal oxide

Table 7.5 Toxicity data of metal oxide ENMs to *E. coli* along with the corresponding ENM characterization, as reported by Puzyn et al. [60] and Pathakoti et al. [94]

Reference	Endpoint or descriptor[a]	Al_2O_3	Bi_2O_3	CoO	Cr_2O_3	CuO	Fe_2O_3	In_2O_3	La_2O_3
Puzyn et al. [60]	$\log 1/EC_{50}$ (mol/L)	2.49	2.82	3.51	2.51	3.2	2.29	2.81	2.87
	HoF (kcal/mol)	−8244.39	−1966.06	−8799.74	−2829.39	−954.75	−1051.01	−3088.47	N/A[b]
	TE (au[c])	−31,466.24	−36,107.9	−17,006.65	−20,103.79	−45,632.03	−6971.23	−40,744.58	N/A
	EE (au)	−630,309.4	−695,662.6	−298,812.5	−307,815.1	−874,569.2	−44,000.25	−872,314.9	N/A
	Core (au)	598,843.2	659,554.7	281,805.8	287,711.4	828,937.2	37,029	831,570.3	N/A
	CA (A^2)	1109.25	1550.69	1072.12	658.55	639.23	243.1	1313.54	N/A
	CV (A^3)	2259.8	4106.88	1548.39	1161.13	1108.01	319.14	3094.57	N/A
	HOMO (eV)	−4.88	−4.11	−10.48	−6.9	−6.11	−7.13	−8.16	N/A
	LUMO (eV)	−0.29	−1.4	−8.28	−0.49	−2.25	−0.68	−3.37	N/A
	GAP (eV)	−4.59	−2.71	−2.2	−6.41	−3.85	−6.45	−4.79	N/A
	ΔH_{Clust} (kcal/mol)	−8016.75	−1600.96	−8318.03	−2264.36	−759.32	−140.11	−3189.89	N/A
	ΔH_{Me+} (kcal/mol)	1187.83	1137.4	601.8	1268.7	706.25	1408.29	1271.13	1017.22
	ΔH_L (kcal/mol)	−3694.7	−3199.02	−933.06	−3645.23	−991.61	−3589.06	−3449.01	−2969.34
Kar et al. [93]	χ	1.61	2.02	1.88	1.66	1.9	1.83	1.78	1.1

Reference	Endpoint or descriptor[a]	Al_2O_3	Bi_2O_3	CoO	Cr_2O_3	CuO	Fe_2O_3	In_2O_3	La_2O_3
	$\sum\chi$	3.22	4.04	1.88	3.32	1.9	3.66	3.56	2.2
	$\sum\chi/n_O$	1.073	1.347	1.88	1.107	1.9	1.22	1.187	0.733
	MW	101.96	465.96	74.93	151.98	79.546	159.6	277.62	325.8
	N_{Metal}	2	2	1	2	1	2	2	2
	N_{Oxygen}	3	3	1	3	1	3	3	3
	χ_{ox}	3	3	2	3	2	3	3	3
Singh and Gupta [90]; Toropov et al. [95]	SMILES notation	O = [Al]O[Al] = O	O = [Bi]O[Bi] = O	[Co] = O	O = [Cr] O[Cr] = O	[Cu] = O	O = [Fe] O[Fe] = O	O = [In]O[In] = O	O = [La] O[La] = O
Sizochenko et al. [91]	Size (nm)	44	90	100	60	N/A	32	30	46
	Aggregation size (nm)	372	2029	257	617	N/A	298	224	673
Pathakoti et al. [94]	Toxicity under darkness, log1/EC_{50} (mol/L)	2.42	3.55	3.13	2.06	4.24	2.4	2.83	4.96
	Toxicity under sunlight exposure, log1/EC_{50} (mol/L)	2.75	4.02	3.33	2.06	5.71	2.54	3.48	5.56

Table 7.5 (*Continued*)

Reference	Endpoint or descriptor[a]	Al_2O_3	Bi_2O_3	CoO	Cr_2O_3	CuO	Fe_2O_3	In_2O_3	La_2O_3
	Particle size (vendor) (nm)	<50	90–210	<100	<100	<50	<50	<100	<100
	Particle size TEM (nm)	55 ± 17	144 ± 7	55 ± 13	47 ± 27	28 ± 7	68 ± 20	60 ± 14	65 ± 19
	Hydrodynamic size (nm)	330	4084	262	426	285	>6000	308	508
	Zeta potential (mV) (H_2O)	30.3 ± 1.3	−(16.5 ± 0.8)	17.5 ± 1.5	−(12 ± 1.3)	24.4 ± 0.6	−(6.3 ± 1)	22.6 ± 0.4	−(3.6 ± 1.1)
	Zeta potential (mV) (KCl)	25.3 ± 1.1	−(4.9 ± 0.1)	26 ± 0.5	23.3 ± 1	19.1 ± 0.3	−(19.5 ± 1.9)	28.7 ± 0.4	22.3 ± 1.7
	Surface area (m^2/g)	37	N/A	>8	N/A	33	36	28	20
	HHOMO (au)	−0.28332	−0.25282	−0.22125	−0.24513	−0.23584	−0.28333	−0.26458	−0.18716
	LZELEHHO (au)	0.21053	0.18445	0.16853	0.1985	0.17822	0.17472	0.196	0.12053
	LUMOA (au)	−0.13774	−0.11608	−0.1158	−0.15186	−0.1206	−0.06611	−0.12741	−0.05389
	LUMOB (au)	−0.13774	−0.11608	−0.13073	−0.11721	−0.11888	−0.1625	−0.12741	−0.05389
	ALZLUMO (au)	−0.13774	−0.11608	−0.12327	−0.13454	−0.11974	−0.11431	−0.12741	−0.05389
	C_p ($J\ mol^{-1}\ K^{-1}$)	79.04	113.51	55.23	118.74	42.3	103.85	92	108.78
	MHOMOA (au)	−0.21849	−0.31904	−0.23238	−0.22219	−0.28868	−0.22944	−0.20161	−0.18806
	MLUMOA (au)	0.01695	0.11366	0.0355	0.02728	0.03649	0.03114	0.01047	0.0152
	QMELECT (au)	0.10077	0.10269	0.09844	0.09746	0.1261	0.09915	0.09557	0.08643

Table 7.5 (a) Toxicity data of metal oxide ENMs to *E. coli* along with the corresponding ENM characterization, as reported by Puzyn et al. [60] and Pathakoti et al. [94]

Reference	Endpoint or descriptor[a]	NiO	Sb_2O_3	SiO_2	SnO_2	TiO_2	V_2O_3	Y_2O_3	ZnO	ZrO_2
Puzyn et al. [60]	$\log 1/EC_{50}$ (mol/L)	3.45	2.64	2.2	2.01	1.74	3.14	2.87	3.45	2.15
	HoF (kcal/mol)	63.89	−2140.84	−4117.99	−2611.4	−9826.23	−3192.8	−11,486.06	−5307	−9834.79
	TE (au[c])	−28,053.23	−18,039.21	−21,060.12	−41,962.21	−31,518.11	−26,082.85	−30,633.6	−23,157.5	−23,405.1
	EE (au)	−432,596.3	−221,601.9	−321,878.6	−874,369	−576,824.1	−441,765.9	−511,018.9	−379,004.7	−358,169.2
	Core (au)	404,543.1	203,562.7	300,818.5	832,406.7	545,306	415,683.1	480,385.3	355,847.2	334,764.1
	CA (A^2)	658.97	974.52	752.62	1734.21	1099.97	1130.15	1804.54	854.96	1055.34
	CV (A^3)	1087.54	1797.3	1466.75	3958.85	2340.2	2425.65	5400.97	1848.79	2403.07
	HOMO (eV)	−5.75	−8.3	−7.12	−6.14	−10.33	−3.54	−1.28	−10.76	−6.19
	LUMO (eV)	−1.03	−1.03	−3.89	−2.29	−2.86	0.64	1.2	−6.89	−4.54
	GAP (eV)	−4.73	−7.27	−3.23	−3.85	−7.47	−4.17	−2.48	3.87	−1.65
	ΔH_{Clust} (kcal/mol)	325.13	−1525.61	−3295.09	−2090.85	−8730.58	−3157.19	−11,485.2	−5357.21	−8956
	ΔH_{Me+} (kcal/mol)	596.7	1233.06	1686.38	1717.32	1575.73	1097.73	837.15	662.44	1357.66

(Continued)

Table 7.5 (a) (*Continued*)

Reference	Endpoint or descriptor[a]	NiO	Sb_2O_3	SiO_2	SnO_2	TiO_2	V_2O_3	Y_2O_3	ZnO	ZrO_2
	ΔH_L (kcal/mol)	−964.84	−3280.99	−3157.91	−2820.92	−2895.72	−3555.36	−3110.59	−971.3	−2640.71
Kar et al. [93]	χ	1.91	2.05	1.9	1.96	1.54	1.63	1.22	1.65	1.33
	$\sum\chi$	1.91	4.1	1.9	1.96	1.54	3.26	2.44	1.65	1.33
	$\sum\chi/n_O$	1.910	1.367	0.95	0.98	0.77	1.087	0.813	1.650	0.665
	MW	74.69	291.52	60.08	150.7	79.86	149.88	225.82	81.38	123.2
	N_{Metal}	1	2	1	1	1	2	2	1	1
	N_{Oxygen}	1	3	2	2	2	3	3	1	2
	χ_{ox}	2	3	4	4	4	3	3	2	4
Singh and Gupta [90]; Toropov et al. [95]	SMILES notation	[Ni] = O	O = [Sb] O[Sb] = O	O = [Si] = O	O = [Sn] = O	O = [Ti] = O	O = [V]O[V] = O	O = [Y]O[Y] = O	O = [Zn]	O = [Zr] = O
Sizochenko et al. [91]	Size (nm)	30	20	150	15	46	15	38	71	47
	Aggregation size (nm)	291	223	640	810	265	1307	1223	189	661

Reference	Endpoint or descriptor[a]	NiO	Sb_2O_3	SiO_2	SnO_2	TiO_2	V_2O_3	Y_2O_3	ZnO	ZrO_2
Pathakoti et al. [94]	Toxicity under darkness, log1/EC_{50} (mol/L)	3.79	3.12	2.54	2.53	2.14	3.48	5.79	5.8	2.58
	Toxicity under sunlight exposure, log1/EC_{50} (mol/L)	3.87	3.66	2.92	3.24	4.68	3.78	5.84	6.23	3.04
	Particle size (vendor) (nm)	<50	90–210	10–20	<100	<100	N/A	<50	<100	<100
	Particle size TEM (nm)	14 ± 9	84 ± 23	20 ± 5	15 ± 4	42 ± 9	N/A	38 ± 9	71 ± 17	27 ± 6
	Hydrodynamic size (nm)	399	619	1230	3971	748	307	357	1614	2337
	Zeta potential (mV) (H_2O)	26 ± 0.4	−20.7 ± 1.3	−29.8 ± 1.9	−21.1 ± 0.4	−(10.7 ± 2.5)	−(27.9 ± 0.9)	16.3 ± 0.9	−(20.9 ± 0.5)	−(6.9 ± 0.5)
	Zeta potential (mV) (KCl)	26.8 ± 1.2	−(12.7 ± 0.4)	−(33.7 ± 1.6)	−(16.7 ± 0.2)	−(2.2 ± 0.4)	−(32.6 ± 0.5)	17.9 ± 1	−(24.9 ± 0.3)	4 ± 2.7
	Surface area (m^2/g)	80	N/A	N/A	18.6	36	N/A	31	15	22
	HHOMO (au)	−0.24118	−0.26243	−0.34322	−0.30533	−0.26454	−0.21916	−0.18944	−0.22819	−0.24301

(*Continued*)

Table 7.5 (a) *(Continued)*

Reference	Endpoint or descriptor[a]	NiO	Sb_2O_3	SiO_2	SnO_2	TiO_2	V_2O_3	Y_2O_3	ZnO	ZrO_2
	LZELEHHO (au)	0.18042	0.17398	0.24497	0.22422	0.19486	0.17387	0.12854	0.13198	0.18424
	LUMOA (au)	−0.11966	−0.08552	−0.14672	−0.14311	−0.12517	−0.12857	−0.06763	−0.03576	−0.12546
	LUMOB (au)	−0.11378	−0.08552	−0.14672	−0.14311	−0.12517	−0.1058	−0.06763	−0.13869	−0.12546
	ALZLUMO (au)	−0.11672	−0.08552	−0.14672	−0.14311	−0.12517	−0.11719	−0.06763	−0.08723	−0.12546
	C_p (J mol^{-1} K^{-1})	44.31	101.63	44.43	52.59	55.48	103.22	102.51	40.25	56.19
	MHOMOA (au)	−0.23629	−0.33428	−0.30135	−0.26703	−0.23244	−0.24678	−0.21085	−0.29251	−0.23175
	MLUMOA (au)	0.03517	0.13016	−0.00653	−0.01678	0.02141	0.024	0.01766	0.04335	0.01644
	QMELECT (au)	0.10056	0.10206	0.15394	0.14191	0.10552	0.11139	0.0966	0.12458	0.10766

[a]EC_{50}, the effective concentration that causes 50% response; HoF, the standard heat of formation of the oxide cluster; TE, total energy of the oxide cluster; EE, electronic energy of the oxide cluster; Core, core-core repulsion energy of the oxide cluster; CA, area of the oxide cluster calculated on the basis of COSMO; CV, volume of the oxide cluster calculated on the basis of COSMO; HOMO, energy of the highest occupied molecular orbital of the oxide cluster; LUMO, energy of the lowest unoccupied molecular orbital of the oxide cluster; GAP, energy difference between HOMO and LUMO energies; ΔH_{Clust}, enthalpy of detachment of metal cations Me^{n+} from the cluster surface; ΔH_{Me+}–, enthalpy of formation of a gaseous cation; ΔH_L, lattice energy of the oxide

[b]N/A, data not available

[c]au, atomic units

standard molar enthalpy of formation (ΔH_{sf}), metal oxide lattice enthalpy (ΔH_{Lat}), first molar ionization energy of metal ($\Delta H_{IE,1+}$), ionic index of metal cation (Z^2/r), IEP, and zeta potential in water at pH 7.4.

7.3.2.2.3 *Toxicity to E. coli*

Puzyn et al. [60] tested the toxicity of 10 metal oxide ENMs to an *E. coli* (Migula) Castellani & Chalmers (ATCC#25254) strain. Metal oxide ENMs covered in the test are Bi_2O_3, CoO, Cr_2O_3, In_2O_3, NiO, Sb_2O_3, SiO_2, V_2O_3, Y_2O_3, and ZrO_2. Meanwhile, results of another seven metal oxide ENMs tested with the same protocol, namely Al_2O_3, CuO, Fe_2O_3, La_2O_3, SnO_2, TiO_2, and ZnO ENMs, were taken from a previous study [40] and a data set consisting of 17 metal oxide ENMs was built. Toxicity to *E. coli* was expressed in terms of the logarithm of the molar $1/EC_{50}$. Data are shown in Table 7.5. As shown in Table 7.5, Kar et al. [93] calculated seven molecular descriptors in their modeling study of the same data set: metal electronegativity (χ), sum of metal electronegativity for individual metal oxide ($\sum\chi$), sum of metal electronegativity for individual metal oxides divided by the number of oxygen atoms present in a particular metal oxide ($\sum\chi/n_O$), N_{Metal}, N_{Oxygen}, the charge of the metal cation corresponding to a given oxide (χ_{ox}), and molecular weight (MW). Two studies [90, 95] provided 2D structural information of the same set of ENMs in the form of simplified molecular input line entry system (SMILES). Information on ENM size and aggregation size can also be found in Sizochenko's study [91]. In addition, 12 electronic descriptors were provided (structural parameters of the ENMs were given by Puzyn et al. [60], including the standard heat of formation (HoF) of the oxide cluster, total energy (TE) of the oxide cluster, electronic energy (EE) of the oxide cluster, core-core repulsion energy of the oxide cluster (Core), area of the oxide cluster calculated on the basis of COSMO (CA), volume of the oxide cluster calculated on the basis of COSMO (CV), energy of the highest occupied molecular orbital (HOMO) of the oxide cluster, energy of the lowest unoccupied molecular orbital (LUMO) of the oxide cluster, energy difference between HOMO and LUMO energies (GAP), enthalpy of detachment of metal cations Me^{n+} from the cluster surface (ΔH_{Clust}), enthalpy of formation of a gaseous cation (ΔH_{Me+}), and lattice energy of the oxide (ΔH_L). Mu et al. [118] also presented data of 26 computational descriptors for this data set.

Using the same types of 17 ENMs as in Puzyn's study [60], Pathakoti et al. [94] examined the nanotoxicity to the ATCC#25254 strain under dark conditions and sunlight exposure for 30 min. The toxicity of ENMs was expressed by the logarithmic values of LC_{50}. Information was provided regarding the ENM size (by suppliers), transmission electron microscopy (TEM) particle size, hydrodynamic size, zeta potential in water and in KCl solution, and surface area. Moreover, six electronic descriptors for metal oxides and three for metal atoms were calculated: the larger (less negative) of the HOMO energies of the alpha spin and beta spin orbitals (HHOMO), the alpha and beta LUMO energies (LUMOA and LUMOB, respectively), the absolute electronegativity of the metal oxide calculated from HHOMO and LUMOA (LZELEHHO), the average of LUMOA and LUMOB (ALZLUMO), molar heat capacity of the metal oxide at 298.15 K (C_p), the alpha HOMO and LUMO energies of metal atoms (MHOMOA and MLUMOA, respectively), and the absolute electronegativity of the metal atom calculated from MHOMOA and MLUMOA (QMELECT).

7.3.2.3 Existing nano-QSARs

Suitable modeling tools are capable of extracting meaningful relationships between nanostructures and the nanotoxicity, thus yielding predictive models. Widely employed methods for deriving nano-QSARs are linear and logistic regressions, together with the approaches of SVMs, ANNs, and the *k*-NN. Details on the workflows for model development and the resulting equations (if applicable) are summarized below, including the number of ENMs, predictive performances, descriptor calculation, and selection. Descriptors used in the developed models or identified factors by relevant studies are summarized in Table 7.6 for further discussion.

7.3.2.3.1 *Linear regression models*

7.3.2.3.1.1 *Cellular uptake*

Epa et al. [83] reparametrized linear models for the cellular uptake of 108 ENMs (87 in training set, 21 in test set) in PaCa2 and HUVECs [122]. A method called mixed linear regression (MLR) with expectation maximization (MLREM) sparse feature reduction

Table 7.6 Overview of computational descriptors or factors discussed in nano-QSAR studies, including information on the original data set for modeling[a]

Reference	Descriptor or identified factor by developed models	Data set
Epa et al. [83]	Number of 10-membered rings (n_{R10}), molecular asphericity (ASP), d COMMA2 value/weighted by atomic masses (DISPm), Qzz COMMA2 value/ weighted by atomic masses (QZZm), number of secondary amides, aliphatic (n_{RCONHR}), number of (thio-) carbamates, aromatic(n_{ArOCON}), CH3X (C-005), number of circuits (n_{CIR}), number of N atoms (n_N), average molecular span *R* (SPAM), Qyy COMMA2 value/weighted by atomic polarizabilities (QYYp), number of total secondary C sp^3 (n_{Cs}), number of aromatic hydroxyls (n_{ArOH}), H attached to C0(sp^3) with 2X attached to next C (H-053), and =O (O-058)	Weissleder et al. [122]
Fourches et al. [85]	Surface area "owned" with SlogP weight −10 to −0.4 (SlogP_VSA0), surface area "owned" with SlogP weight −0.4 to −0.2 (SlogP_VSA1), surface area "owned" with SlogP weight −0.2 to 0 (SlogP_VSA2), surface area "owned" with SlogP weight −0.15 to −0.2 (SlogP_VSA5), van der Waals surface area surface area ofhydrogen bond donors (vsa_don), van der Waals surface area of nondonor/ acceptor atoms (vsa_other), van der Waals surface area surface area ofbasic atoms (vsa_base), sum of the van der Waals surface area of atoms whose PEOE partial charge is positive, divided by the total surface area (PEOE_VSA_FPOS), van der Waals surface area where atomic partial charge $0.05 < q < 0.1$ (PEOE_VSA+1), number of double bonds, and aromatic bonds not considered (b_double)	
Ghorbanzadeh et al., [86]	Number of donor atoms for H bonds (nH_{Don}), Geary autocorrelation of lag 1 weighted by van der Waals volume (GATS1v), 3D-MoRSE-signal 29/unweighted (Mor29u), *D* total accessibility index/weighted by Sanderson electronegativity (D_e), 3D-MoRSE-signal 14/unweighted (Mor14u), and mean electrotopological state (M_s)	

(*Continued*)

Table 7.6 (*Continued*)

Reference	Descriptor or identified factor by developed models	Data set
Kar et al. [87]	Hydrophobicity of the N atom in primary aliphatic amine (Al-NH_2) fragment (A_{type}-N-66), hydrophobicity of the N atom in a secondary aliphatic amine (Al_2-NH) fragment (A_{type}-N-67), measure of electronic features of the molecule relative to molecular size ($\sum\beta'$), relative positive charge surface area (Jurs–RPCS), all-path Wiener index (W_{ap}), and number of aliphatic nitro groups ($n$$RNO_2$)	
Singh and Gupta [90]	Weighted partial negative surface area-3 (WNSA-3), weighted partial positive area-2 (WPSA-2), Chi simple path descriptor of order 5 (SP-5), Chi valence path descriptor of order 4 (VP-4), moment of inertia along X/Z axis (MOMI-XZ), logarithmic form of octanol-water partition coefficient predicted by atomic method (XlogP), number of rotatable bonds (n_{RotB}), number of hydrogen bond donors (nHB_{Don}), Chi valence path cluster of order 6 (VPC-6), ionization potential (IP), and number of hydrogen acceptors (nHB_{Acc})	
Chau and Yap [115]	Number of CH_2 groups; primary, secondary, and tertiary nitrogen; halogens (fluorine, bromine, and iodine), sulfur atoms, fused rings, and hydrogen bonding	
Gajewicz et al. [28]	Enthalpy of formation of metal oxide nanocluster representing a fragment of the surface (ΔH_f^c), and Mulliken's electronegativity of the nanocluster (χ^c)	Gajewicz et al. [28]
Sizochenko et al. [91]	Unbonded two-atomic fragments [Me]· · ·[Me] (S_1), Wigner–Seitz radius of an oxide's molecule (r_w), mass density (ρ), covalent index of the metal ion (CI), SiRMS-derived number of oxygen's atoms in a molecule (S_2), and aggregation parameter (AP)	
Pan et al. [119]	Molecular weight, cationic charge, mass percentage of metal elements, individual size, and aggregation size	
Pathakoti et al. [94]	Absolute electronegativity of the metal atom (QMELECT), absolute electronegativity of the metal oxide (LZELEHHO), literature molar heat capacity of the metal oxide at 298.15 K (C_p), and average of the alpha and beta LUMO energies of the metal oxide (ALZLUMO)	Pathakoti et al. [94]

Reference	Descriptor or identified factor by developed models	Data set
Puzyn et al. [60]	Enthalpy of formation of a gaseous cation having the same oxidation state as that in the metal oxide structure (ΔH_{Me+})	Puzyn et al. [60]
Kar et al. [93]	Charge of the metal cation corresponding to a given oxide (χ_{ox}) and metal electronegativity (χ)	
Singh and Gupta [90]	Oxygen percent, molar refractivity, and polar surface area	
Sizochenko et al. [91]	Unbonded two-atomic fragments [Me]· ·[Me] (S_1), Wigner–Seitz radius of an oxide's molecule (r_w), mass density (ρ), cation polarizing power, SiRMS-derived number of oxygen atoms in a molecule (S_2), triatomic fragments [Me]-[O]-[Me] (S_3), and proportion of surface molecules (SV)	
Mu et al. [118]	Enthalpy of formation of a gaseous cation having the same oxidation state as that in the metal oxide structure (ΔH_{Me+}), and polarization force (Z/r)	
Pan et al. [119]	Molecular weight, cationic charge, mass percentage of metal elements, individual size, and aggregation size	
Epa et al. [83]	Core material (I_{Fe3O4}), surface coating ($I_{dextran}$), and surface charge ($I_{surf.chg}$)	Shaw et al. [27]
Fourches et al. [85]; Liu et al. [117]; Singh and Gupta [90]	Size, R1 relaxivity, R2 relaxivity, and zeta potential	
Zhang et al. [30]	Conduction band energy (E_c) and solubility of metals	Zhang et al. [30]
Sizochenko et al. [121]	Mass density, molecular weight, aligned electronegativity, covalent index, cation polarizing power, Wigner–Seitz radius, surface area, surface-area-to-volume ratio, aggregation parameter, two-atomic descriptor of van der Waals interactions, triatomic descriptor of atomic charges, tetra-atomic descriptor of atomic charges, and size in DMEM	
Liu et al. [88]	Ionic index of metal cation (Z^2/r), ENM conduction band energy (E_c), metal oxide ionization energy (ΔH_{IE}), metal oxide electronegativity (χ_{MeO}), atomization energy of metal oxide (E_{Amz}), primary size (d), and atomic mass of ENM metal (m_{Me})	

(*Continued*)

Table 7.6 (*Continued*)

Reference	Descriptor or identified factor by developed models	Data set
Papa et al. [120]	Size of ENMs (X0), size in water (X1), size in phosphate buffered saline (X2), concentration (X4), and zeta potential (X5)	Sayes and Ivanov [123]
Liu et al. [29]	Size of ENM (d), volume concentration (θ_v), period of the ENM metal in the periodic table (P_{Me}), and atomization energy of the metal oxide (E_{MeO})	
Chen et al. [116]	Molecular polarizability, accessible surface area, and solubility	Others
Luan et al. [89]; Kleandrova et al. [3]	Molar volume, polarizability, size of ENMs, electronegativity, hydrophobicity, and polar surface area of surface coatings	

[a]Name of the descriptors in original publications are given in the parenthesis (whenever available).

was employed to optimize the descriptor set from a pool of 691 descriptors. DRAGON (v5.5), ADRIANA (v2.2), and an in-house modeling software package were used for descriptor calculation. The best performing models used 19 descriptors for PaCa2 cells ($R^2_{training}$ = 0.76, R^2_{test} = 0.79, standard error of estimate [SEE] = 0.19, and standard error of prediction [SEP] = 0.24) and 11 for HUVECs ($R^2_{training}$ = 0.74, R^2_{test} = 0.63, SEE = 0.34, and SEP = 0.36).

A PLS model predicting the cellular uptake (log10[ENM]/cell pM) of 109 magnetofluorescent ENMs in PaCa2 cells [122] was constructed by Kar et al. [87]. In this study, a total of 307 descriptors were calculated using the Cerius 2 (v4.10), DRAGON (v6), and PaDEL-Descriptor (v2.11), which were afterward filtered by the genetic function approximation. Finally, six molecular descriptors appeared in the developed model:

$$\begin{aligned}&\log_{10}[\text{NP}]/\text{cell} = 3.335 + \\&(0.774\times <1-A_{\text{type}}-N-66> -(0.222\times A_{\text{type}}-N-67) \\&+(7.36\times <0.6-\Sigma\beta'>)-(0.101\times \text{Jurs-RPCS}) \\&-(0.00002\times W_{\text{ap}})-(0.462\times n\text{RNO}_2),\end{aligned} \tag{7.1}$$

with $n_{training}$ = 89, LV = 5, R^2 = 0.806, Q^2_{LOO} = 0.758, $Q^2_{Leave\text{-}10\%\text{-}out}$ = 0.634, $Q^2_{Leave\text{-}25\%\text{-}out}$ = 0.648, SEE = 0.2, $\overline{r^2_{m(LOO)Scaled}}$ = 0.665,

$\Delta r^2_{m(LOO)Scaled} = 0.113$, $n_{test} = 20$, $Q^2_{F1} = R^2_{pred} = 0.879$, $SEP = 0.12$, $Q^2_{F2} = 0.868$, $\overline{r^2_{m(test)Scaled}} = 0.793$, $\Delta r^2_{m(test)Scaled} = 0.115$, $\overline{r^2_{m(overall)Scaled}} = 0.679$, and $\Delta r^2_{m(overall)Scaled} = 0.116$.

In the model, the descriptors A_{type}-N-66 and A_{type}-N-67 are the hydrophobicity of the N atom in, respectively, a primary and a secondary aliphatic amine (Al-NH_2 and Al_2-NH, respectively), $\sum\beta'$ characterizes the measure of electronic features of the molecule relative to molecular size, Jurs–RPCS stands for the relative positive charge surface area, W_{ap} represents for the all-path Wiener index, and $n\mathrm{RNO_2}$ is the number of aliphatic nitro groups. The leverage and distance to model in X-space approaches [125, 126] were applied to check the model's domain of applicability.

Using the data from Weissleder et al. [122], Ghorbanzadeh et al. [86] proposed a predictive model of cellular uptake (log10[ENM]/cell pM) on the basis of a multilayered perceptron neural network technique. A self-organizing map (SOM) strategy was employed combined with stepwise MLR to promote the feature reduction. This procedure provided six most informative descriptors, namely number of donor atoms (N and O) for H bonds ($n\mathrm{H_{Don}}$), Geary autocorrelation of lag 1 weighted by van der Waals volume (GATS1v), 3D-MoRSE-signal 29/unweighted (Mor29u), D total accessibility index/weighted by Sanderson electronegativity (D_e), 3D-MoRSE-signal 14/unweighted (Mor14u), as well as the mean electrotopological state (M_s). The linear model has the form

$$\frac{\log_{10}[\mathrm{NP}]}{\mathrm{cell}} = 70 - 0.130 \times n\mathrm{H_{Don}} + 0.412 \times \mathrm{GATS1v} \\ -0.398 \times \mathrm{Mor29u} + 1.243 \times D_e - 0.163 \times \mathrm{Mor14u} + 0.045 \times M_s. \tag{7.2}$$

The model gave a correlation coefficient (R) of 0.782 for the training set (root-mean-square error [RMSE] = 0.369) and 0.755 for the validation set (RMSE = 0.357).

7.3.2.3.1.2 *Cytotoxicity*

On the basis of the apoptosis assay of smooth muscle cells from Shaw et al. [27], Epa et al. [83] developed a model consisting of three descriptors for the core material (I_{Fe3O4}), surface coating ($I_{dextran}$), and surface charge ($I_{surf.chg}$) of ENMs. The descriptors are considered to have a value of 1 when the condition is present and 0 when the condition is absent. For instance, I_{Fe3O4} is set to be 1 for the ENM with an Fe_2O_3 core and 0 when the ENM core is Fe_3O_4; $I_{dextran}$ is equal to 1 in the case of a dextran coating and 0 for others; and surface functionality is encoded as 1 (basic), −1 (acidic), or 0 (neutral). Smooth muscle apoptosis (SMA) was used as the endpoint in the constructed model:

$$\begin{aligned} SMA = 2.26(\pm 0.72) - 10.73(\pm 1.05) \times I_{Fe_2O_3} \\ -5.57(\pm 0.98) \times I_{dextran} - 3.53(\pm 0.54) \times I_{surf.chg}, \end{aligned} \tag{7.3}$$

with $n = 31$, $R^2_{training} = 0.81$, $R^2_{test} = 0.86$, $SEE = 3.6$, and $SEP = 3.3$.

Papa et al. [120] reported three MLR models predicting the potential of ZnO and TiO_2 ENMs to induce the release of LDH in rat lung cells. Data was retrieved from the study by Sayes and Ivanov [123] that provided values of five descriptors: engineered size (X0), size in water (X1), size in phosphate buffered saline (X2), concentration (X4), and zeta potential (X5). The first linear model combined information on both TiO_2 and ZnO ENMs (all together 31 ENMs):

$$LDH_{(TiO_2+ZnO)} = 0.66 + 0.003X4 + 0.005X0 - 4.46E-5X2, \tag{7.4}$$

with $R^2 = 0.82$, $Q^2_{loo} = 0.76$, $Q^2_{lmo30\%} = 0.74$, $r^2_{YS} = 0.1$, $s = 0.11$, and $F = 40$.

The Williams plot for the applicability domain of the model was depicted in the original publication. Besides, linear models were also built separately for TiO_2 (22 ENMs) and ZnO ENMs (15 ENMs):

$$LDH_{(TiO_2)} = 0.599 + 0.003X4 + 0.004X0, \tag{7.5}$$

with $R^2 = 0.84$, $Q^2_{loo} = 0.79$, $Q^2_{lmo30\%} = 0.78$, $r^2_{YS} = 0.1$, $s = 0.12$, and $F = 48$, and

$$LDH_{(ZnO)} = 1.041 + 0.001X1 - 0.001X2 + 0.001X4, \tag{7.6}$$

with $R^2 = 0.91$, $Q^2_{loo} = 0.8$, $Q^2_{lmo30\%} = 0.76$, $r^2_{YS} = 0.22$, $s = 0.08$, and $F = 35$.

Another approach explicitly and completely based on MLR is reported by Gajewicz et al. [28]. In this case, the cytotoxicity of 18 metal oxide ENMs to the HaCaT cell line was modeled. A total of 27 descriptors were calculated, which included 16 quantum-mechanical descriptors and 11 image descriptors derived from TEM images. For calculating the quantum-mechanical descriptors, the molecular geometry was optimized at the level of the semiempirical

PM6 method encoded in MOPAC 2009 [127, 128]. Information on the size, size distribution, shape, porosity, and surface area of ENMs was extracted on the basis of TEM images to generate the 11 image descriptors. Two descriptors were afterward selected by the genetic algorithm (GA), that is, ΔH_f^c and χ^c. The model can be expressed as

$$\log(\mathrm{LC}_{50})^{-1} = 2.47(\pm 0.05) + 0.24(\pm 0.05) \times \Delta H_f^{\ c} + 0.39(\pm 0.05) \times \chi^{c}\ \mathrm{v}, \tag{7.7}$$

with $F = 44.6$, $p = lx10^{-4}$, $n = 18$, $R^2 = 0.93$, $\mathrm{RMSE_C} = 0.12$, $Q^2_{\mathrm{CV}} = 0.86$, $\mathrm{RMSE_{CV}} = 0.16$, $Q^2_{\mathrm{EXT}} = 0.83$, and $\mathrm{RMSE_P} = 0.13$.

Here ΔH_f^c is the enthalpy of formation of the metal oxide nanocluster representing a fragment of the surface and χ^c represents Mulliken's electronegativity of the cluster.

Using the data set reported by Gajewicz et al. [28], Pan et al. [119] developed two predictive models incorporating the so-called improved SMILES-based optimal descriptors. The models predicting the cytotoxicity of metal oxide ENMs to HaCaT cells have the form

$$\log\left(\frac{1}{\mathrm{LC}_{50}}\right) = -0.2909\,(\pm 0.0664) + 0.1038\,(\pm 0.0027) \times \mathrm{DCW}\,(1,3), \tag{7.8}$$

with $n = 13$, $R^2 = 0.9606$, $Q^2_{\mathrm{LMO}} = 0.9393$, $s = 0.008$, $F = 268$, and $p < 0.0001$,

and

$$\log\left(\frac{1}{\mathrm{LC}_{50}}\right) = 0.0012\,(\pm 0.0048) + 0.0778\,(\pm 0.0001) \times \mathrm{DCW}\,(1,3), \tag{7.9}$$

with $n = 12$, $R^2 = 0.9997$, $Q^2_{\mathrm{LMO}} = 0.9996$, $s = 0.007$, $F = 1273$, and $p < 0.0001$.

The number 1 in DCW(1,3) is the coefficient for classification of features into two classes (noise and active), and the number 3 in DCW(1,3) is the number of epochs of the Monte Carlo optimization. The characteristics of ENMs involved in the models are MW, cationic charge, mass percentage of metal elements, individual size, and aggregation size of ENMs.

Besides, in the study by Liu et al. [88] a linear regression model was developed for 24 metal oxide ENMs on the basis of a recently reported data set [30]. Three descriptors were involved in the model, namely E_c, ΔH_{IE}, and χ_{MeO}. The model was reported to give an accuracy of 89% for the samples.

7.3.2.3.1.3 *Toxicity to E. coli*

Puzyn et al. [60] originally built a data set for the toxicity of 17 metal oxide ENMs to *E. coli*. On the basis of the data, a simple and statistically significant nano-QSAR model was obtained that used a single descriptor ΔH_{Me+}:

$$\log(1/\mathrm{EC}_{50}) = 2.59 - 0.5 \times \Delta H_{\mathrm{Me+}}, \tag{7.10}$$

with $R^2 = 0.85$, $\mathrm{RMSE_C} = 0.2$, $Q^2_{\mathrm{CV}} = 0.77$, $\mathrm{RMSE_{CV}} = 0.24$, $Q^2_{\mathrm{EXT}} = 0.83$, and $\mathrm{RMSE_P} = 0.19$.

Calculation of a pool of 12 variables (Table 7.5) was executed using the PM6 method as implemented in MOPAC 2009. GA was applied for selecting the most informative descriptors. PLS Toolbox and the Statistics Toolbox for MATLAB were utilized for model development. The leverage approach and Williams plot were employed to visualize model applicability domain.

Working on the data set from Puzyn et al. [60], Kar et al. [93] built a stepwise MLR model as well as a PLS model. Seven descriptors were used for model construction, namely χ, $\Sigma\chi$, $\Sigma\chi/n_{\mathrm{O}}$, N_{Metal}, N_{Oxygen}, χ_{ox}, and MW (Table 7.5). For the MLR model, feature reduction was accomplished by the "stepping criteria" (*F*), and only the descriptor χ_{ox} was seen in the model:

$$\log(1/\mathrm{EC}_{50}) = 4.781 - (8.380 \times \chi_{\mathrm{ox}}), \tag{7.11}$$

with $n = 17$, $R^2 = 0.84$, $R^2_{\mathrm{adj}} = 0.83$, $Q^2_{\mathrm{LOO}} = 0.81$, $Q^2_{\mathrm{L\text{-}10percent\text{-}OUT}} = 0.82$, $Q^2_{\mathrm{L\text{-}20percent\text{-}OUT}} = 0.83$, $Q^2_{\mathrm{L\text{-}25percent\text{-}OUT}} = 0.8$, ${}^{c}R^2_{\mathrm{P}} = 0.82$.

Meanwhile the developed PLS model contained two descriptors χ_{ox} and χ and has the form

$$\log(1/\mathrm{EC}_{50}) = 4.401 - (1.324 \times \chi_{\mathrm{ox}}) + (0.176 \times \chi), \tag{7.12}$$

with $n = 17$, $LV = 1$, $R^2 = 0.82$, $Q^2_{\mathrm{LOO}} = 0.75$, $Q^2_{\mathrm{L\text{-}10percent\text{-}OUT}} = 0.76$, $Q^2_{\mathrm{L\text{-}20percent\text{-}OUT}} = 0.74$, $Q^2_{\mathrm{L\text{-}25percent\text{-}OUT}} = 0.76$, ${}^{c}R^2_{P} = 0.79$.

Characterization of the applicability domain of the model was performed by the leverage approach [125].

Mu et al. [118] also reported MLR models building on the data from Puzyn et al. [60]. Calculation of descriptors was performed using PM6 methods within MOPAC 2012 software package.

Approaches of Pearson and pairwise correlations and clustering and principal component analysis were incorporated to obtain optimal structure descriptors for modeling. Among the developed models, a simple but statistically nano-QSAR has the form:

$$\log\left(\frac{1}{\mathrm{EC}_{50}}\right)=(4.412\pm0.165)$$
$$+(-0.121\pm0.068)Z/r$$
$$+(-0.001\pm2.57\times10^{-4})\Delta H_{\mathrm{Me+}}, \quad (7.13)$$

where Z is the ionic charge and r is the Pauling ionic radius. The statistical indicators of the model are $R^2 = 0.8793$, RMSE = 0.442, $F = 55.654$, and $p = 4.23 \times 10^{-7}$.

The leverage approach and Williams plots were used for the characterization of the applicability domain of the model. On the basis of the developed model, toxic potencies of 35 other metal oxide ENMs were predicted and visualized in a periodic table. Other models using different descriptors were also described in the study.

Pan et al. [119] also built in silico models using data from Puzyn et al. [60]. The reported models on the basis of the improved SMILES-based optimal descriptors can be expressed as

$$\log\left(\frac{1}{\mathrm{LC}_{50}}\right)=0.0321(\pm0.1443)+0.2658$$
$$(\pm0.0141)\times\mathrm{DCW}(6,11), \quad (7.14)$$

with $n = 10$, $R^2 = 0.8891$, $Q^2_{\mathrm{LMO}} = 0.8378$, $s = 0.179$, $F = 164$, and $p < 0.0001$, and

$$\log\left(\frac{1}{\mathrm{LC}_{50}}\right)=-0.0076(\pm0.0306)$$
$$+0.1420(\pm0.0020)\times\mathrm{DCW}(6,17), \quad (7.15)$$

with $n = 9$, $R^2 = 0.9824$, $Q^2_{\mathrm{LMO}} = 0.9745$, $s = 0.007$, $F = 391$, and $p < 0.0001$.

The characteristics of ENMs involved in the models are MW, cationic charge, mass percentage of metal elements, individual size, and aggregation size of ENMs.

7.3.2.3.2 *Logistic regression models*

Liu et al. [29] constructed logistic regression models to classify the effect of nine metal oxide ENMs on BEAS-2B cells into toxic (T) and nontoxic (N). The model with the best classification performance is

$$\ln\left(\frac{P(NP \in T)}{P(NP \in N)}\right) = 3600.6 + 103.5 \times d + 9.5 \times \theta_v + 97.6 \times P_{Me} - 58.5 \times E_{MeO} \tag{7.16}$$

where $P(NP \in T)$ and $P(NP \in N)$ are the probabilities of an ENM being classified as toxic or nontoxic, respectively, d is the ENM size, θ_v is the volume concentration derived from the mass concentration of ENMs, P_{Me} is the period of the ENM metal in the periodic table, and E_{MeO} is the atomization energy of the metal oxide. The model applicability domain was depicted by the principal component analysis.

Liu et al. [29] developed two nano-SAR models on the basis of the logistic regression and quadratic logistic regression methods. The data set of Zhang et al. [30] was chosen. This data set covered data on the toxicity of 24 metal oxide ENMs to BEAS-2B and RAW264.7 cell lines as described above. The quadratic logistic regression model was shown to achieve an accuracy of 89.97% with only two descriptors, E_C and Z^2/r. Meanwhile a marginally better predictability of 90.09% for the logistic regression model was obtained. The molecular descriptors that were included in the logistic regression model were E_C, E_{Amz}, and d.

Logistic regression models were also built by Liu et al. [117] on the basis of an integration of multiparametric bioactivity assays of 44 iron oxide ENMs [27]. The concept of "hit" (significant bioactivity, signal-to-noise ratio > 1.645) was utilized in the study, and the number of hits served as the bioactivity class definition (identifying an ENM as bioactive or inactive), enabling nano-SAR development. Clustering analysis via SOM was also considered besides the number of hits as an alternative to define a class. ENM descriptors included the primary size, the zeta potential, and the R1 and R2. Results showed that the logistic regression model based on class definition of H5 (five hits) possesses the best predictability of 79%, using ENM size and R2 as descriptors. The class definition H6 also enabled the construction of a simple logistic regression model (R1 as the sole descriptor) with 78.2% accuracy.

7.3.2.3.3 *Support vector machine models*

An SVM classification model has been developed by Fourches et al. [85] using the experimental data of 44 ENMs from Shaw et al. [27]. ENM size, R1, R2, and zeta potential were used as input descriptors, and an arbitrary threshold at Z_{mean} = −0.4 was applied to enable a binary classification. Three clusters of ENMs were identified after assigning a hierarchical clustering procedure. It was found that all monocrystalline iron oxide ENMs were in cluster II and all quantum dots appeared in cluster I. Results of the classification confirmed the good predictability of the clustering-based nano-SARs (fivefold external cross-validation) in cluster II:

Cluster I: n = 13, sensitivity = 0.5, and specificity = 0.8
Cluster II: n = 18, sensitivity = 0.78, and specificity = 0.78
Cluster III: n = 13, sensitivity = 0.7, and specificity = 0.4

Here sensitivity = (number of true positives)/(total number of true positives) and specificity = (number of true negatives)/(total number of true negatives) for the binary classification problems. Another SVM nano-SAR classifying 23 metal oxide ENMs as toxic or nontoxic was built by Liu et al. [88] on the basis of measured toxicological responses in BEAS-2B cells and murine myeloid RAW 264.7 cells following an established protocol [30]. An SOM-based consensus clustering was employed and afterward identified three ENM clusters. Clusters II and III contained ENMs being reported as toxic and thus were grouped into a single cluster of ENMs classified as having a positive response. ENMs in cluster I were labeled as nontoxic. A pool of 30 descriptors was initially considered, including information on the fundamental metal oxide, energies or enthalpies of metal oxide, ENM size, zeta potential and IEP, and ENM energy. Descriptor selection was accomplished by the evaluation of models derived from all possible descriptor combinations. The SVM algorithm successfully correlated the cytotoxicity of ENMs with ENM conduction band energy (E_C) and ionic index of metal cation (Z^2/r). The penalty factor and the kernel width of the SVM model were determined to be 128 and 2, respectively. The discriminant function of the SVM model was given by

$$f(\mathbf{x}) = \sum_{i=1}^{6} \alpha_i e^{-2[(x_{i,1}-x_1)^2+(x_{i,2}-x_2)^2]} + b, \quad (7.17)$$

where **x** refers to the ENM identified by the normalized descriptor vector $[Z^2/r, E_C]$ (i.e., x_1, x_2) and $x_{i,1}$ and $x_{i,2}$ stand for the normalized first and second descriptors identified as support vectors, respectively. The values of α_i (i = 1–6) were represented by ZnO (82.342), Ni_2O_3 (128), Mn_2O_3 (83.696), NiO (−70.471), CeO_2 (−95.566), and Fe_2O_3 (128), with b being −10.888. The model was reported to give a predictive accuracy (obtained via 0.632 estimator) of 93.74%. The model applicability domain was characterized by a probabilistic approach [129].

7.3.2.3.4 *Artificial neural network models*

On the basis of the experimental results of Shaw et al. [27], a Bayesian regularized ANN model was constructed predicting the smooth muscle cells' apoptosis triggered by 31 ENMs [83]. The model statistics were as follows: n = 31, $R^2_{training}$ = 0.8, R^2_{test} = 0.9, SEE = 2.8, and SEP = 2.9. Meanwhile an ANN nano-SAR was also built in the study, modeling the cellular uptake in HUVECs and PaCa2 cells [122]:

Cellular uptake in HUVECs: $R^2_{training}$ = 0.7, SEE = 0.3, R^2_{test} = 0.66, SEP = 0.33, and # descriptors = 11

Cellular uptake in PaCa2 cells: $R^2_{training}$ = 0.77, SEE = 0.15, R^2_{test} = 0.54, SEP = 0.28, and # descriptors = 19

Besides the above-mentioned MLR model developed by Ghorbanzadeh et al. [86], another nano-SAR on the basis of a multilayered perceptron neural network technique was also introduced in their study. The SOM strategy combined with stepwise MLR selected six most informative descriptors, namely $n\mathrm{H}_{\mathrm{Don}}$, GATS1v, Mor29u, D_e, Mor14u, and M_s. The derived model gave a performance in terms of values of R^2 of 0.934 for the training set, 0.945 for the internal test set, and 0.943 for the external test set. The calculated RMSE values are 0.146, 0.121, and 0.214 for training, internal test sets, and external test sets, respectively, while the corresponding values of F are 531, 142, and 65, respectively. The applicability domain of the model was firstly evaluated by the approach based on ranges of individual descriptors. A Williams plot was subsequently put into use for visualizing the domain of applicability.

7.3.2.3.5 *k-nearest-neighbor models*

A classification model employing the *k*-NN approach was developed in the study by Fourches et al. [85]. The model was proposed to predict the cellular uptakes of 109 ENMs in PaCa2 cells [122]. Coefficients of correlation R_{abs}^2 were shown to range from 0.65 to 0.8 for the external sets and from 0.67 to 0.9, taking into account the applicability domain that was defined by the Euclidean distance approach. In the study, the descriptors were identified that most frequently occurred in the models (one- to fivefold cross-validations) with the highest prediction accuracy. The top 10 descriptors ranked by averaged frequency were reported to be SlogP_VSA1, SlogP_VSA2, SlogP_VSA5, b_double, SlogP_VSA0, PEOE_VSA+1, van der Waals surface area surface area of hydrogen bond donors (vsa_don), van der Waals surface area of nondonor/-acceptor atoms (vsa_other), van der Waals surface area surface area of basic atoms (vsa_base), and PEOE_VSA_FPOS. SlogP_VSA0 and SlogP_VSA1, along with other descriptors with relatively low frequencies, such as GCUT_SLOGP_0 and BCUT_SLOGP_0, are considered to be generally related to lipophilicity. For instance, PaCa2 uptake of ENMs was observed to be positively correlated with the enrichment of lipophilic compounds on ENM surfaces (value of GCUT_SLOP_0). Other discriminated factors affecting the PaCa2 uptake of ENMs were found to be molecular refractivity, the specific van der Waals surface area, and electrostatic properties. The applicability domain of the model was characterized by the Euclidean distance.

An attempt to predict the cytotoxicity of 44 iron oxide ENMs based on *k*-NN was also reported by Liu et al. [117]. As described above, different numbers of hits were discussed in the study for introducing class definitions besides the clustering analysis via SOM. The results showed that a *k*-NN model using SOM-based consensus clustering gave the best predictive performance of 74.9% accuracy. Three descriptors, ENM size, R1, and R2, were obtained in this model. Meanwhile, H4 class definition was also deemed to be a suitable choice that enabled the development of a *k*-NN model correctly predicting 74.3% of the samples.

7.3.2.3.6 *Other models*

Chau and Yap [115] attempted to correlate the cellular uptake in PaCa2 with the calculated parameters from PaDEL-Descriptor (v2.8).

By lowering the threshold value of significant uptake into PaCa2, 56 ENMs with cellular uptake of more than 5000 ENMs per cell [122] were considered as a positive class and the other 49 were defined as the negative class. On the basis of the four modeling techniques of naive Bayesian classifier (NBC), logistic regression, *k*-NN, and SVM, 2100 candidate models were developed, while only 102 of them were qualified according to the selection criteria. To build a final consensus nano-SAR model, the top five candidate models were chosen: three *k*-NN, one SVM, and one NBC. The consensus model gave a good predictive performance with a sensitivity of 98.2% and a specificity of 76.6% for the data set. Descriptors that commonly appeared in the candidate models were the number of CH_2 groups; primary, secondary, and tertiary nitrogens; halogens (fluorine, bromine, and iodine); sulfur atoms; fused rings; and hydrogen bonding. Most of the descriptors that contributed to the model were interpreted as related to the lipophilicity (e.g., number of lipophilic groups). Other factors, such as hydrogen bonding between nitrogen and hydrogen and bonding between sulfur and various halogen atoms, were also found to affect the cellular uptake of ENMs. This is in agreement with the study by Fourches et al. [85].

Chen et al. [116] reported several nano-SARs for the categorization of ENM hazards across different biota. The toxicity data were retrieved from the database of Chen et al. [130] and the online chemical modeling environment platform [131]. Approaches of functional tree, C4.5 decision tree, random tree, and Simple CART were employed for model development. Global nano-SARs across species using LC_{50} data were shown to correctly predict more than 70% of the samples in training (320 ENMs) and test sets (80 ENMs) based on functional tree, C4.5 decision tree, and random tree methods. The species-specific nano-SARs were also derived for *Danio rerio*, *Daphnia magna*, *Pseudokirchneriella subcapitata*, and *Staphylococcus aureus* with good predictivity. It could be concluded from the models developed that molecular polarizability, accessible surface area, and solubility are key factors affecting the biological activity of metallic ENMs.

Moreover, Zhang et al. [30] reported a regression tree model using the metal dissolution of metal oxide ENMs and energy of conduction band to predict the toxicity potential of 24 metal oxide ENMs. With the data from Zhang et al. [30], Sizochenko et al. [121] developed causal inference nano-SARs for BEAS-2B and RAW 264.7 cell lines (24 metal oxide ENMs for each cell line) with high predictivity. Luan et al. [89] and Kleandrova et al. [3] developed the novel QSTR-perturbation (quantitative structure–toxicity relationship) models assessing the cytotoxicity and ecotoxicity of various types of ENMs. The factors of molar volume, polarizability, size of ENMs, electronegativity, and the hydrophobicity and polar surface area of surface coatings were indicated by the reported models. Singh and Gupta [90] previously performed three cases of nano-QSAR study for metallic ENMs on the basis of the data sets generated by Puzyn et al. [60], Shaw et al. [27], and Weissleder et al. [122]. In this study, classification and regression models were constructed predicting various biological effects of the ENMs by an ensemble learning-based strategy called stochastic gradient boosting and bagging algorithms. Results showed that the developed models are of robustness and no over-fitting of data was present in all case studies. Besides, attempts to link the information of ENM structures to corresponding biological effects were also made using other modeling techniques, such as Monte Carlo method [92, 95], NBC and linear discriminate analysis [88, 117], random forest regression [91, 121], and self-written least-squares fitting program [94].

7.3.2.4 Interpretation of mechanisms of ENM biological activities using the models developed

To enable fast and inexpensive, high-throughput prediction of diverse biological effects caused by ENMs, reliable nano-QSARs should be based on mechanistic knowledge [110]. Only when information on the underlying mechanisms is incorporated in modeling, proper and reliable extrapolation toward untested ENMs or organisms can be performed. The descriptors introduced in the aforementioned nano-QSAR studies are shown to be able to provide vital structural information on the factors affecting ENMs' cellular uptake and toxicity. Therefore, information on these descriptors as summarized in Table 7.6 is linked here to the current understanding of the mechanisms of nanotoxicity.

7.3.2.4.1 *Cellular uptake of ENMs*

Once entering the medium, ENMs may undergo various extra- and intracellular physical-chemical reactions, such as dissolution, ion release, ROS generation, interaction with subcellular structures (e.g., cellular membrane and mitochondrion), and internalization into the cells (Fig. 7.2). Cellular uptake of ENMs is always seen as an important process of ENMs' internalization and subsequently initiating the ENM contact-mediated or dissolved ion-associated intracellular reactions. As hypothesized, ENMs are conventionally transported into cells through endocytosis, a form of active transport in which cells take in ENMs by engulfing them [132]. Possible endocytotic processes proposed include phagocytosis, macropinocytosis, caveolae-dependent and clathrin-mediated endocytosis, and non-clathrin-, non-caveolae-mediated endocytosis [132, 133]. Besides, other responses of cellular membranes to adsorption of ENMs were also shown to exist. On the basis of a dissipative particle dynamics simulation study, Yue and Zhang [134] concluded that surface adhesion, membrane penetration, and even ENM-induced membrane rupture could occur upon ENM attachment to cellular membranes. Lin et al. [135] and Xia et al. [136] also demonstrated that ENMs could access cellular interior through direct membrane penetration.

In these internalization processes, surface properties of ENMs are essential for the ENM-biomolecule interactions and are deemed to be able to alter the cellular uptake pathways. In the experiment of Weissleder et al. [122], a diversity of cellular uptake processes was observed, especially for the PaCa2 cells. These authors consequently concluded that the translocation process is highly dependent on the surface modification of the ENMs. The studies showed that the lipophilicity of the surface molecules is an important discriminating factor that determines the chemical ability of the ENMs to interact with the lipid core of membranes [150]. Fourches et al. [85] reported that four descriptors (out of the top 10 with the highest averaged frequency) SlogP_VSA0, SlogP_VSA1, SlogP_VSA2, and SlogP_VSA5 are intimately correlated with molecular lipophilicity of surface compounds. The ENMs with a higher PaCa2 cellular uptake are generally highly enriched for lipophilic surface modification (higher descriptor value). This is consistent with the results of Epa et al. [83] in which C-005 (associated with hydrophobicity) was observed as a factor affecting ENMs' cellular uptake. Further confirmation was obtained by the appearance of A_{type}-N-66 and A_{type}-N-67 [87] in a

PLS model and the number of lipophilic groups (CH_2, fused rings) in the consensus model of Chau and Yap [115]. Also, the hydrogen bonding capacity of surface modifiers was explained to be one of the driving factors of ENMs' membrane penetrability [115]. An ANN model predicting cellular uptake of ENMs by HUVECs was reported to include the descriptors n_{RCONHR} and n_{ArOCON}, which characterize molecular hydrogen bonding capacity [83]. In the same study, n_N, n_{ArOH}, H-053, and O-058 were also found in the MLR model and were likewise interpreted as affecting the capability of H bonding. In other nano-QSARs, descriptors considered to correlate with this factor include nH_{Don} [86], WPSA-2, nHB_{Don}, and nHB_{Acc} [90]. Hence, these informative descriptors found in the developed nano-QSARs confirmed again the previous experimental observations of the lipophilicity of surface compounds being of significant importance for the cellular uptake of ENMs.

Additionally, shape, size, and flexibility of the surface compounds play an important role in determining ENMs' passive transport across biological membranes. For instance, descriptors (not exclusively) characterizing molecular branching were constantly observed in the models such as n_{R10}, n_{CIR}, n_{Cs} [83], W_{ap} [87], GATS1v, Mor29u, Mor14u [86], SP-5, VP-4, and VPC-6 [90]. Mor29u, Mor14u, SP-5, VP-4, and VPC-6 meanwhile also contain information of the molecular 3D structures (e.g., mass, size, flexibility, and overall shape). Other relevant descriptors are ASP, DISPm, QZZm, QYYp, SPAM [83], $\sum\beta'$ [87], D_e [86], MOMI-XZ, and n_{RotB} [90]. Moreover, impacts on ENMs' cellular uptake were also reported to derive from the molecular reactive surface and electronegativity. Molecular reactive surface–related descriptors in the nano-QSARs are vsa_don, vsa_other, vsa_base, PEOE_VSA_FPOS, PEOE_VSA+1 [85], Jurs–RPCS [87], and weighted partial negative surface area-3 (WNSA-3) [90]. Descriptors associated with molecular electronegativity were observed to be $nRNO_2$ [87]; primary, secondary, and tertiary N; and halogens [115]. It is not surprising that these factors of ENM surface modifiers may influence the ENM-biosurface interactions and pose effects on the cellular uptake of ENMs, independently or cooperatively. Shape, size, or flexibility of surface modifiers of ENMs would affect the interactions between these molecules and the molecular sites of biosurfaces, change the conformation of binding complexes, and ultimately mediate the subsequent ENM-biosurface reactions in which the nature of the reactive surface and electronegativity also play a role.

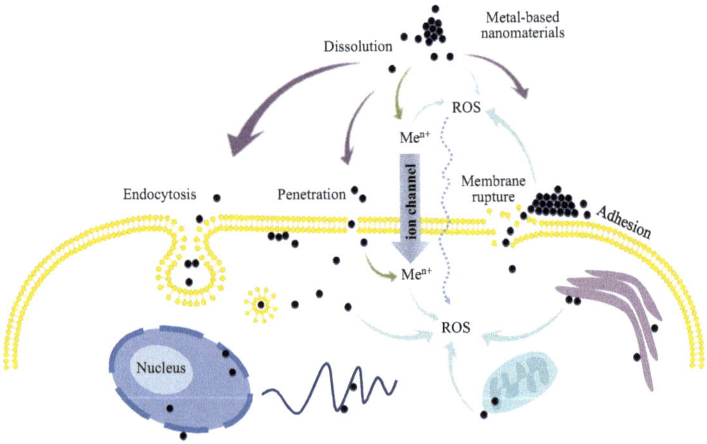

Figure 7.5 Overview of hypotheses associated with the responses of cellular membranes following interactions with ENMs. It is assumed that endocytosis, penetration, adhesion of ENMs upon the cellular membrane, and cellular membrane rupture occur. Cellular membrane rupture is also considered to lead to the internalization of ENMs via the damage sites. Scenarios of relevant ion release from ENMs, ROS generation, and ENM-induced interactions are also depicted.

As can be seen from Fig. 7.5, internalization of ENMs into cells is generally considered as a crucial biological process triggering nanotoxicity. However, the adsorption of ENMs on cellular membranes may also affect cellular membrane integrity and lead to the formation of detects through the membranes [134–137]. This could probably result in the direct internalization of ENMs through the damaged sites of membranes and the release of intracellular components that causes cell death. Notably, extracellular release of ions and formation of ROS are also considered to be factors affecting the toxicity of ENMs in some cases. Von Moos and Slaveykova [65] reported that intracellular ROS generation can be stimulated by the presence of extracellular ROS as a response. The released ions and derived ROS may as well interact with cellular membranes and dependently and/or independently influence ENMs' cellular uptake. This gives a possible explanation for the presence of the descriptor ionization potential (IP) in the nano-SAR of Singh and Gupta [90] and may also be able to explain why the electronegativity-related descriptors $nRNO_2$ [87]; primary, secondary, and tertiary N; and the halogens [115] generally appeared in relevant nano-QSAR studies.

7.3.2.4.2 *Biological effects induced by ENMs*

It is well known that ENMs are capable of eliciting adverse biological effects by directly or indirectly triggering a series of physical-chemical reactions and ultimately causing cell damage. Reportedly, toxicity of ENMs could occur via a single mechanism or via combinations of the following mechanisms: (i) direct interactions with subcellular structures or biomolecules (e.g., membranes, mitochondria, proteins, and DNA), which could lead to, for instance, mitochondrial damage, denaturation of proteins, and formation of corona; (ii) release of chemical constituents from ENMs such as metal ions; (iii) surface property–based chemical reactivity of ENMs, for example, photochemical, catalytic, and redox properties; and (iv) Trojan horse mechanisms, so-called intruders in which ENMs act as vectors for transporting toxic chemicals.

Generally, there is no doubt that metal-ions leaching from ENMs could act as a key factor causing biological effects of ENMs. Once the ENMs release dissolved ions surrounding the cells, it is often difficult to experimentally distinguish the effects caused by conventional metal ion release from the nano-specific effects. In such a context, the toxicity induced by ENMs is always considered to be intimately correlated with ENM dissolution. Comparable results on the toxicity of ZnO ENMs and Zn salts have been observed for the examples of *Pseudokirchneriella subcapitata* [138], *Thamnocephalus platyurus* and *Daphnia magna* [39, 139], and *E. coli* [140]. Results of studies on the toxicity of CuO ENMs to multiple species were also in agreement with this conclusion [39, 141]. It is commonly believed that ion release could occur after the cellular internalization of ENMs, which results in different mechanistic pathways of nanotoxicity. For instance, Stohs and Bagchi [142] proposed a Haber–Weiss–Fenton cycle describing the stimulation of ion leaching to generate ROS, taking Cu^{2+} as an example:

$$O_2^{\bullet -} + Cu^{2+} \rightarrow O_2 + Cu^{+} \tag{7.18}$$

$$Cu^{+} + H_2O_2 \rightarrow Cu^{2+} + OH^{-} + OH^{\bullet}, \tag{7.19}$$

where the ROS such as superoxide anion radicals ($O_2^{\bullet-}$) could be derived from the one-electron reduction of molecular oxygen O_2:

$$O_2 + e \rightarrow O_2^{\bullet-} \tag{7.20}$$

In the Haber–Weiss–Fenton cycle, Cu^{2+} catalyzes the formation of hydroxyl radicals, which in turn enhances the generation of ROS. Meanwhile, it was suggested that the release of ions could be accompanied by ROS formation as well, such as in the Fenton reaction [28]:

$$Fe + O_2 + 2H^+ \rightarrow Fe^{2+} + H_2O_2 \tag{7.21}$$

$$Fe^{2+} + H_2O_2 \rightarrow Fe^{3+} + OH^{\bullet} + OH^- \tag{7.22}$$

Evidence from nano-QSAR studies also demonstrated the contribution of ion release to nanotoxicity. Influence of metal solubility on nanotoxicity was indicted by the developed models [30, 116]. Puzyn et al. [60] developed a linear model based on the sole descriptor ΔH_{Me+} predicting the toxicity of metal oxide ENMs to *E. coli*. It was explained that ΔH_{Me+} is an efficient descriptor characterizing the stability of metal oxides, which is associated with both the lattice energy of oxides and the sum of the IPs of a given metal. The release of cations with a smaller charge is seen as more energetically favorable than that with a larger charge [118]. This explains the observations of previous studies giving oxide toxicity in the following order: as $Me^{2+} > Me^{3+} > Me^{4+}$ [60]. According to Kar et al. [93], the charge of the metal cation corresponding to a given oxide (χ_{ox}) was also used for the parameterization of nanotoxicity data. In the study by Liu et al. [88], the descriptor ionic index of metal cation Z^2/r was involved in building classification nano-SARs. Z is the ionic charge, and r represents the Pauling ionic radius of the released ions [119]. Z^2/r is a measure of the involvement of a metal ion in electrostatic interactions and is able to provide information on the affinity of a metal ion for water molecules. Likewise, such form of index was also used in a random forest model [91, 121], coupled with a parameter (S_1) describing the van der Waals interactions between surface molecules or cations. Other descriptors related to ionic charge and/or radius are polarization force [118], covalent index, triatomic descriptor of atomic charges, and tetra-atomic descriptor of atomic charges [121].

Accordingly, Gajewicz et al. [28] employed two descriptors, that is, enthalpy of formation of metal oxide nanocluster (ΔH_f^c) and Mulliken's electronegativity of the cluster (χ^c), to linearly explain the cytotoxicity of metal oxide ENMs to HaCaT. The ΔH_f^c is associated with the energy of a single metal-oxygen bond in oxides ($E_{\Delta H^o}$), which can be expressed as [143]

$$E_{\Delta H^o} = \frac{2\Delta H_f^o \cdot 2.612\times10^{19}}{N_A n_e}, \tag{7.23}$$

where N_A is the Avogadro number and n_e is the number of electrons involved in the formation reaction. A high ΔH_f^c value indicates a strongly bound cation of a large formal charge in the oxides, which affects the detachment of the metal cation from the surface of the ENMs. As for χ^c, according to Burello and Worth [114] the electronegativity value of a metal oxide (χ_{oxide}) can be calculated from that of the corresponding cation (χ_{cation}) using the equation [144]

$$\chi_{oxide} \approx 0.45\chi_{cation} + 3.36. \tag{7.24}$$

Therefore, a higher value of χ_{cation} indicates a stronger ability of a cation to attract electrons in the Haber–Weiss–Fenton cycle, which in turn results in higher reactivity of the metal oxide ENMs [28]. The two descriptors ΔH_f^c and χ^c meanwhile also refer to ENMs' surface redox activity. Burello and Worth [114] reported that the energy of a bandgap (E_g) can be obtained on the basis of $E_{\Delta H^o}$:

$$E_g = Ae^{0.34E_{\Delta H^o}} \tag{7.25}$$

and thus the conduction and valence band energies of oxides become

$$E_c = -\chi_{oxide} + 0.5E_g + E_{shift} \tag{7.26}$$

and

$$E_v = -\chi_{oxide} - 0.5E_g + E_{shift}, \tag{7.27}$$

where E_{shift} represents the value changes of band edges in respect to the solution's pH. As hypothesized, the redox potentials of relevant biological reactions could be unbalanced if they lie closer to the E_c or E_v, thereby causing cellular oxidative stress [30]. This was confirmed by Liu et al. [88], who identified E_c and χ_{oxide} for the development of nano-QSAR models, and by Kar et al. [93], Sizochenko et al. [121], Luan et al. [89], and Kleandrova et al. [3], who used metal electronegativity as one of the modeling parameters. Pathakoti et al. [94] also obtained two descriptors (absolute electronegativity of the metal and metal oxide) for describing the toxicity of metal oxide ENMs to *E. coli* under darkness. Other descriptors considered to be associated with the surface redox properties of ENMs and causing oxidative stress are ΔH_{IE}, E_{Amz} [29, 88], CI, S_3 [91], C_p, ALZLUMO [94], and polarizability [3, 89, 117] in relevant nano-QSAR studies.

On the other hand, other than the general consensus taking ion release and ROS generation as driving factors in nanotoxicity, it is evident that other mechanisms of toxicity also play a vital role in certain cases. Xiao et al. [62] reported that for both Cu and ZnO ENMs, the particles per se, rather than the dissolved ions, provided the major contribution to the toxicity to *Daphnia magna* (26% and 31%, respectively). Similarly, Hua et al. [108] also revealed a dominant contribution of ZnO ENMs over the Zn ion tested for zebrafish embryos, for which the dissolution-driven mechanism of ENM toxicity apparently does not apply. More precisely, it was shown that the shape of ENMs significantly affect ENMs' toxicity as needle-shaped ZnO ENMs were proven to be the most toxic to *Phaeodactylum tricornutum* as compared to morphologically different ENMs with equal solubility and ion release [145]. Observations of nanotoxicity affected by the shape of ENMs were also reported for ZnO nanospheres, nanosticks, and cuboidal submicron particles [108]. Computational studies proved the involvement of surface property-related descriptors in nano-QSAR modeling, such as the surface area and coating [83, 90, 116, 121], hydrophobicity and polar surface area of surface molecules [3, 89], surface-area-to-volume ratio [121], zeta potential [120], and the Wigner–Seitz radius of an oxide's molecule (which describes the available fraction of molecules on the surface of ENMs) [91, 121]. The Wigner–Seitz radius also relates to MW and density and, therefore, also to the molecular volume. All of them have been included in the models [3, 89, 119, 121]. Descriptors related to size [3, 29, 88–90, 117, 119, 120], material composition [29, 83, 90, 119, 121], and aggregation behavior [91, 119–121] were also found to affect nanotoxicity.

In summary, the characteristics of ENMs may pose effects on the toxicity of ENMs as the consequence of either a single mechanism or a combination of mechanisms. Analysis of the descriptors employed in existing nano-QSAR studies assists in offering a mechanistic interpretation on the basis of the main driving factors. As discussed above, ENMs' surface properties, such as lipophilicity, hydrogen bonding capacity, electronegativity, shape, size, and flexibility of the surface modifiers, are vital for their uptake by cells. As for ENM-triggered toxicity, properties correlating with the ability of ion release and ROS generation could be important indicators, along with the information about the size, surface redox properties, and composition of ENMs.

7.4 Conclusions and Outlook

Development of reliable nano-QSARs constitutes a great opportunity in reducing the ethical considerations as well as the time needed for and costs of conventional experimental evaluations of ENMs. Regulatory hazard assessment and risk evaluation strategies for ENMs as well as assessment of novel ENMs that are safer by design benefit from the availability of properly validated nano-QSARs. Even though the promising potency of extending QSAR methodologies into nanotoxicity has been addressed, the nano-QSAR approach is still in its infancy. A general scarcity of properly documented and reliable experimental data is to be regarded as one of the major drawbacks in building nano-QSAR models. The information provided in Table 7.3 clearly evidences that only a limited number of data sets is available for the purpose of deriving nano-QSARs. Two key issues underlie this general observation and can be seen as the primary challenges in the development of nano-QSARs:

- Most of the studies reported so far do not meet the basic quality criteria that are commonly applied to evaluate experimental data. This is, among others, due to a lack of basic characterization data on the pristine particles, a lack of proper description of the experimental methods used, or even a lack of reporting of a consistent endpoint.
- Because of the lack of properly reported data, it is virtually impossible to combine and integrate existing experimental data from various data sources. Consequently, the various data sets and the nano-QSAR results cover only a small fraction of the chemical space relevant for ENMs.

Therefore, in light of advancing computational nanotoxicology, a summary and further organization of potentially useful data on nanotoxicity are essential. Besides the data quantity, especially the quality of the experimental data collected should be taken care of for establishing a data-driven nano-QSAR approach. A key challenge therefore is to properly evaluate the quality of experimental information assembled from various sources.

A second key requirement for deriving QSARs for nanotoxicology is the availability of proper descriptions of the key physicochemical substance properties of ENMs that drive their fate and adverse effects. In turn, evaluation of the predictive properties of nano-specific descriptors will assist in underpinning the mechanistic background to the models derived, which subsequently will allow for increasing confidence in the models developed. Whereas the current generation of nano-QSARs is virtually restricted to descriptions of the core elements of the particles tested, a key challenge is to focus on the surface properties of ENMs as these are vital modulators of the interactions between biological entities and ENMs. Rather than being characterized as a defined entity, ENMs can generally only be defined in a somewhat arbitrary way before being described in terms of descriptors suitable for modeling. Often, ENMs are best described as being complex assemblies that are commonly coated or functionalized with molecules of a diverse nature, the composition of which may change over time. This creates the challenge of defining ENMs as strict entities that interact with biological species and that are responsible as a strict entity for causing toxicity to these biological species. Thereupon, the high extent of complexity of the 3D structure of ENMs challenges the calculation of descriptors on the basis of common computational approaches. Uncertainty surrounding the 3D structure of an ENM persists even when the ENM composition appears to have been properly reported [124]. Concern has been expressed in this respect with regard to the question whether it is possible to build nano-QSARs without considering the speciation of ENMs in the exposure medium. It is well known that once having entered a medium, ENMs are likely to strongly react with the components of the test medium and undergo dramatic changes of surface properties. These surface transformations would in turn affect the reactivity and subsequent biological behavior (e.g., cellular uptake and interaction with subcellular structures) of ENMs. Although it is recognized that speciation of ENMs is important with regard to the bioavailability of ENMs, there is not yet scientific consensus on how to deduce bioavailable exposure estimates. In such a context, modeling solely on the basis of the information of ENMs' pristine structures could be biased and could result in poor predictability and reliability of the models generated. The possibility actually exists that the characteristics of pristine ENMs influence the biological effects of ENMs by affecting the ENMs' dynamic transformations in media, and it can be hypothesized that even though ENM properties could change in the exposure media, the characteristics of the pristine ENMs may still be linked to the observed adverse biological effects. In this circumstance, constructing nano-QSARs on the basis of only the physicochemical characteristics of pristine ENMs could offer the opportunity to develop high-throughput protocols for nontesting nanotoxicity evaluation.

Two ENM-specific challenges further stand out, the first of these being the proper inclusion of size into computational parameters. An attempt to overcome this challenge is the study in Ref. [146], in which a set of novel, theoretical size–dependent nanodescriptors for ENMs was developed. However, the key problem is that the size of ENMs in reality is never a fixed value but rather a distribution of sizes. Whereas particles have been shown, as discussed above, to significantly modulate ENM toxicity, the preparation of fully homogeneously sized batches of ENMs is simply impossible at this moment. An opportunity to deal with the size issue is the modification of the current generation of nanodescriptors by combining them with fuzzy set theory. The fuzzy set theory permits the gradual assessment of the membership of elements in a set instead of assigning an element into either one set or another [147]. Thus, if a descriptor (D_n) for a cluster of an ENM of size (s) can be expressed as

$$D_{\mathrm{n}} = f(s), \tag{7.28}$$

then the calculation of descriptors combined with fuzzy set theory (D'_{n}) can be described as

$$D'_{\mathrm{n}} = \sum f_{\mathrm{m}}(s) f(s), \tag{7.29}$$

where s is a discrete variable in $f(s)$, or

$$D_n' = \int_a^b f_m(s) f(s) d_s, a \leq s \leq b, \qquad (7.30)$$

where s is a continuous variable in $f(s)$ and f_m is the membership function extracted from the information on the ENM size distribution.

Secondly, it is to be acknowledged that dosimetry of ENMs is an issue that is commonly overlooked in predictive nanotoxicology. As first stated by Oberdörster et al. [148], mass should not be considered as the sole option for expressing the effective dose that is actually inducing adverse effects and that should be taken into account in modeling. Alternative dose metrics, including, for instance, particle size, aspect ratio, surface area/volume ratio, or total volume, might provide alternative opportunities to properly describe the effective dose that is to be taken into account for predictive modeling [149].

Acknowledgments

The research described in this chapter was supported by the European Union's Horizon 2020 research and innovation programme under grant agreement no. 760813 "PATROLS." Martina Vijver was funded by NWO VIDI 864.13.010. The Chinese Scholarship Council (CSC) is gratefully acknowledged for its financial support to Guangchao Chen.

References

1. Maynard, A. D., Aitken, R. J., Butz, T., Colvin, V., Donaldson, K., Oberdörster, G., Philbert, M. A., Ryan, J., Seaton, A., Stone, V., Tinkle, S. S., Tran, L., Walker, N. J. and Warheit, D. B. (2006). Safe handling of nanotechnology, *Nature*, **444**, 267–269.
2. European Commission (EC). Commission recommendation of 18 October 2011 on the definition of nanomaterial (2011/696/EU, 2011), Brussels, Belgium.
3. Kleandrova V. V., Luan, F., González-Díaz, H., Ruso, J. M., Melo, A., Speck-Planche, A. and Cordeiro, M. N. (2014). Computational ecotoxicology: simultaneous prediction of ecotoxic effects of nanoparticles under different experimental conditions, *Environ. Int.*, **73**, 288–294.
4. Puzyn, T., Leszczynska, D. and Leszczynski, J. (2009). Toward the development of "nano-QSARs": advances and challenges, *Small*, **5**, 2494–2509.

5. Project on Emerging Nanotechnologies. Consumer Products Inventory. Available at http://www.nanotechproject.org/cpi (Last accessed on 23.11.2016).
6. Vance, M.E., Kuiken, T., Vejerano, E. P., McGinnis, S. P., Hochella, M. F. Jr, Rejeski, D. and Hull, M. S. (2015). Nanotechnology in the real world: redeveloping the nanomaterial consumer products inventory, *Beilstein J. Nanotechnol.*, **6**, 1769–1780.
7. Lynch, I. (2016). Water governance challenges presented by nanotechnologies: tracking, identifying and quantifying nanomaterials (the ultimate disparate source) in our waterways, *Hydrol. Res.*, **47**, 552–568.
8. Borm, P. A., Robbins, D., Haubold, S., Kuhlbusch, T., Fissan, H., Donaldson, K., Schins, R., Stone, V., Kreyling, W., Lademann, J., Krutmann, J., Warheit, D. and Oberdörster, E. (2006). The potential risks of nanomaterials: a review carried out for ECETOC, *Part. Fibre Toxicol.*, **3**, 11–46.
9. Dowling, A., Clift, R., Grobert, N., Hutton, D., Oliver, R., O'Neill, O., Pethica, J., Pidgeon, N., Porritt, J., Ryan, J., Seaton, A., Tendler, S., Welland, M. and Whatmore, R. (2004). Nanoscience and nanotechnologies: opportunities and uncertainties, Royal Society and Royal Academy of Engineering, London, UK.
10. Mulvaney, P. and Weiss, P. S. (2016). Have nanoscience and nanotechnology delivered? *ACS Nano*, **10**, 7225–7226.
11. Nel, A., Xia, T., Mädler, L. and Li, N. (2006). Toxic potential of materials at the nanolevel, *Science*, **311**, 622–627.
12. Valsami-Jones, E. and Lynch, I. (205). NANOSAFETY. How safe are nanomaterials? *Science*, **350**, 388–389.
13. Gottschalk, F., Nowack, B. and Gawlik, B. (2010). Report on exposure scenarios and release of nanomaterials to the environment. NANEX Work Package 5, European Union, FP7 Project Number 247794.
14. Keller, A. A. and Lazareva, A. (2014). Predicted releases of engineered nanomaterials: from global to regional to local, *Environ. Sci. Technol. Lett.*, **1**, 65–70.
15. Keller, A. A., Vosti, W., Wang, H. and Lazareva, A. (2014). Release of engineered nanomaterials from personal care products throughout their life cycle, *J. Nanopart. Res.*, **16**, 2489.
16. Methner, M., Hodson, L., Dames, A. and Geraci, C. (2010). Nanoparticle Emission Assessment Technique (NEAT) for the identification and measurement of potential inhalation exposure to engineered nanomaterials. Part B: Results from 12 field studies, *J. Occup. Environ. Hyg.*, **7**, 163–176.

17. Magnuson, B. A., Jonaitis, T. S. and Card, J. W. (2011). A brief review of the occurrence, use, and safety of food-related nanomaterials, *J. Food Sci.*, **76**, R126–133.

18. McCracken, C., Dutta, P.K. and Waldman, W. J. (2016). Critical assessment of toxicological effects of ingested nanoparticles, *Environ. Sci. Nano*, **3**, 256–282.

19. Griffitt, R. J., Weil, R., Hyndman, K. A., Denslow, N. D., Powers, K., Taylor, D. and Barber, D. S. (2007). Exposure to copper nanoparticles causes gill injury and acute lethality in zebrafish (*Danio rerio*), *Environ. Sci. Technol.* **41**, 8178–8186.

20. Kwok, K. W., Auffan, M., Badireddy, A. R., Nelson, C. M, Wiesner, M. R., Chilkoti, A., Liu, J., Marinakos, S. M. and Hinton, D.E. (2012). Uptake of silver nanoparticles and toxicity to early life stages of Japanese medaka (*Oryzias latipes*): effect of coating materials, *Aquat. Toxicol.*, **120–121**, 59–66.

21. Oberdörster, G., Oberdörster, E. and Oberdörster, J. (2005). Nanotoxicology: an emerging discipline evolving from studies of ultrafine particles, *Environ. Health Perspect.*, **13**, 823–839.

22. Eichert, T., Kurtz, A., Steiner, U. and Goldbach, H. E. (2008). Size exclusion limits and lateral heterogeneity of the stomatal foliar uptake pathway for aqueous solutes and water-suspended nanoparticles, *Physiol. Plant.*, **134**, 151–160.

23. Miralles, P., Church, T. L. and Harris, A. T. (2012). Toxicity, uptake, and translocation of engineered nanomaterials in vascular plants, *Environ. Sci. Technol.*, **46**, 9224–9239.

24. Klaine, S. J., Alvarez, P. J., Batley, G. E., Fernandes, T. F., Handy, R. D., Lyon, D. Y., Mahendra, S., McLaughlin, M. J. and Lead, J. R. (2008). Nanomaterials in the environment: behavior, fate, bioavailability, and effects, *Environ. Toxicol. Chem.*, **27**, 1825–1851.

25. Kumar, A., Pandey, A. K, Singh, S. S., Shanker, R. and Dhawan, A. (2011). A flow cytometric method to assess nanoparticle uptake in bacteria, *Cytometry A*, **79**, 707–712.

26. Bondarenko, O., Juganson, K., Ivask, A., Kasemets, K., Mortimer, M. and Kahru, A. (2013). Toxicity of Ag, CuO and ZnO nanoparticles to selected environmentally relevant test organisms and mammalian cells in vitro: a critical review, *Arch. Toxicol.*, **87**, 1181–1200.

27. Shaw, S. Y., Westly, E. C., Pittet, M. J., Subramanian, A., Schreiber, S. L. and Weissleder, R. (2008). Perturbational profiling of nanomaterial biologic activity, *Proc. Natl. Acad. Sci. U. S. A.*, **105**, 7387–7392.

28. Gajewicz, A., Schaeublin, N., Rasulev, B., Hussain, S., Leszczynska, D., Puzyn, T. and Leszczynski, J. (2015). Towards understanding mechanisms governing cytotoxicity of metal oxides nanoparticles: hints from nano-QSAR studies, *Nanotoxicology*, **9**, 313–325.

29. Liu, R., Rallo, R., George, S., Ji, Z., Nair, S., Nel, A. E. and Cohen, Y. (2011). Classification NanoSAR development for cytotoxicity of metal oxide nanoparticles, *Small*, **7**, 1118–1126.

30. Zhang, H., Ji, Z., Xia, T., Meng, H., Low-Kam, C., Liu, R., Pokhrel, S., Lin, S., Wang, X., Liao, Y. P., Wang, M., Li, L., Rallo, R., Damoiseaux, R., Telesca, D., Mädler, L., Cohen, Y., Zink, J. I. and Nel, A. E. (2012). Use of metal oxide nanoparticle band gap to develop a predictive paradigm for oxidative stress and acute pulmonary inflammation, *ACS Nano*, **6**, 4349–4368.

31. Zhou, H., Mu, Q., Gao, N., Liu, A., Xing, Y., Gao, S., Zhang, Q., Qu, G., Chen, Y., Liu, G., Zhang, B. and Yan, B. (2008). A nano-combinatorial library strategy for the discovery of nanotubes with reduced protein-binding, cytotoxicity, and immune response, *Nano Lett.*, **8**, 859–865.

32. Donaldson, K., Stone, V., Clouter, A., Renwick, L. and MacNee, W. (2001). Ultrafine particles, *Occup. Environ. Med.*, **58**, 211–216.

33. Juganson, K., Ivask, A., Blinova, I., Mortimer, M. and Kahru, A. (2015). NanoE-Tox: new and in-depth database concerning ecotoxicity of nanomaterials, *Beilstein J. Nanotechnol.*, **6**, 1788–1804.

34. Oberdörster, G. (2000). Toxicology of ultrafine particles: *in vivo* studies, *Philos. Trans. R. Soc. Lond. Ser. A*, **358**, 2719–2740.

35. Angel, B. M., Batley, G. E., Jarolimek, C. V. and Rogers, N. J. (2013). The impact of size on the fate and toxicity of nanoparticulate silver in aquatic systems, *Chemosphere*, **93**, 359–365.

36. Dalai, S., Pakrashi, S., Nirmala, M. J., Chaudhri, A., Chandrasekaran, N., Mandal, A. B. and Mukherjee, A. (2013). Cytotoxicity of TiO_2 nanoparticles and their detoxification in a freshwater system, *Aquat. Toxicol.*, **138**, 1–11.

37. Navarro, E., Piccapietra, F., Wagner, B., Marconi, F., Kaegi, R., Odzak, N., Sigg, L. and Behra, R. (2008). Toxicity of silver nanoparticles to *Chlamydomonas reinhardtii*, *Environ. Sci. Technol.*, **42**, 8959–8964.

38. Dams, R. I., Biswas, A., Olesiejuk, A., Fernandes, T. and Christofi, N. (2011). Silver nanotoxicity using a light-emitting biosensor Pseudomonas putida isolated from a wastewater treatment plant, *J. Hazard. Mater.*, **195**, 68–72.

39. Heinlaan, M., Ivask, A., Blinova, I., Dubourguier, H. C. and Kahru, A. (2008). Toxicity of nanosized and bulk ZnO, CuO and TiO_2 to bacteria

Vibrio fischeri and crustaceans *Daphnia magna* and *Thamnocephalus platyurus*, *Chemosphere*, **71**, 1308–1316.

40. Hu, X., Cook, S., Wang, P. and Hwang, H. M. (2009). *In vitro* evaluation of cytotoxicity of engineered metal oxide nanoparticles, *Sci. Total Environ.*, **407**, 3070–3072.
41. Ivask, A., Bondarenko, O., Jepihhina, N. and Kahru, A. (2010). Profiling of the reactive oxygen species-related ecotoxicity of CuO, ZnO, TiO_2, silver and fullerene nanoparticles using a set of recombinant luminescent *Escherichia coli* strains: differentiating the impact of particles and solubilised metals, *Anal. Bioanal. Chem.*, **398**, 701–716.
42. Samberg, M. E., Orndorff, P. E. and Monteiro-Riviere, N. A. (2011). Antibacterial efficacy of silver nanoparticles of different sizes, surface conditions and synthesis methods, *Nanotoxicology*, **5**, 244–253.
43. Gao, J., Youn, S., Hovsepyan, A., Llaneza, V. L., Wang, Y., Bitton, G. and Bonzongo, J. C. (2009). Dispersion and toxicity of selected manufactured nanomaterials in natural river water samples: effects of water chemical composition, *Environ. Sci. Technol.*, **43**, 3322–3328.
44. Jo, H. J., Choi, J. W., Lee, S. H. and Hong, S.W. (2012). Acute toxicity of Ag and CuO nanoparticle suspensions against Daphnia magna: The importance of their dissolved fraction varying with preparation methods, *J. Hazard. Mater.*, **227–228**, 301–308.
45. Li, M., Czymmek, K. J. and Huang, C. P. (2011). Responses of *Ceriodaphnia dubia* to TiO_2 and Al_2O_3 nanoparticles: a dynamic nano-toxicity assessment of energy budget distribution, *J. Hazard. Mater.*, **187**, 502–508.
46. Lopes, S., Ribeiro, F., Wojnarowicz, J., Łojkowski, W., Jurkschat, K., Crossley, A., Soares, A. M. and Loureiro, S. (2014). Zinc oxide nanoparticles toxicity to Daphnia magna: size-dependent effects and dissolution, *Environ. Toxicol. Chem.*, **33**, 190–198.
47. Hall, S., Bradley, T., Moore, J. T., Kuykindall, T. and Minella, L. (2009). Acute and chronic toxicity of nano-scale TiO_2 particles to freshwater fish, cladocerans, and green algae, and effects of organic and inorganic substrate on TiO_2 toxicity, *Nanotoxicology*, **3**, 91–97.
48. Massarsky, A., Dupuis, L., Taylor, J., Eisa-Beygi, S., Strek, L., Trudeau, V. L. and Moon, T. W. (2013). Assessment of nanosilver toxicity during zebrafish (*Danio rerio*) development, *Chemosphere*, **92**, 59–66.
49. Wang, Z., Chen, J., Li, X., Shao, J. and Peijnenburg, W. (2012). Aquatic toxicity of nanosilver colloids to different trophic organisms: contributions of particles and free silver ions, *Environ. Toxicol. Chem.*, **31**, 2408–2413.

50. Zhu, X., Tian, S. and Cai, Z. (2012). Toxicity assessment of iron oxide nanoparticles in zebrafish (*Danio rerio*) early life stages, *PLoS One*, **7**, e46286.

51. Galindo, T. P., Pereira, R., Freitas, A. C., Santos-Rocha, T. A., Rasteiro, M. G., Antunes, F., Rodrigues, D., Soares, A. M., Gonçalves, F., Duarte, A. C. and Lopes, I. (2013). Toxicity of organic and inorganic nanoparticles to four species of white-rot fungi, *Sci. Total. Environ.*, **458–460**, 290–297.

52. Kasemets, K., Suppi, S., Künnis-Beres, K. and Kahru, A. (2013). Toxicity of CuO Nanoparticles to yeast *Saccharomyces cerevisiae* BY4741 wild-yype and its nine isogenic single-gene deletion mutants, *Chem. Res. Toxicol.*, **26**, 356–367.

53. Kvitek, L., Vanickova, M., Panácek, A., Soukupova, J., Dittrich, A. P. M., Valentova, E., Prucek, R., Bancirova, M., Milde, D. and Zboril, R. (2009). Initial study on the toxicity of silver nanoparticles (NPs) against *Paramecium caudatum*, *J. Phys. Chem. C*, **113**, 4296–4300.

54. Ma, H., Bertsch, P. M., Glenn, T. C., Kabengi, N. J. and Williams, P. L. (2009). Toxicity of manufactured zinc oxide nanoparticles in the nematode *Caenorhabditis elegans*, *Environ. Toxicol. Chem.*, **28**, 1324–1330.

55. Mortimer, M., Kasemets, K. and Kahru, A. (2010). Toxicity of ZnO and CuO nanoparticles to ciliated protozoa *Tetrahymena thermophile*, *Toxicology*, **269**, 182–189.

56. Shi, J. P., Ma, C. Y., Xu, B., Zhang, H. W. and Yu, C. P. (2012). Effect of light on toxicity of nanosilver to *Tetrahymena pyriformis*, *Environ. Toxicol. Chem.*, **31**, 1630–1638.

57. Tyne, W., Lofts, S., Spurgeon, D. J., Jurkschat, K. and Svendsen, C. (2013). A new medium for *Caenorhabditis elegans* toxicology and nanotoxicology studies designed to better reflect natural soil solution conditions, *Environ. Toxicol. Chem.*, **32**, 1711–1717.

58. Yang, X., Gondikas, A. P., Marinakos, S. M., Auffan, M., Liu, J., Hsu-Kim, H. and Meyer, J. N. (2012). Mechanism of silver nanoparticle toxicity is dependent on dissolved silver and surface coating in *Caenorhabditis elegans*, *Environ. Sci. Technol.*, **46**, 1119–1127.

59. Lei, R., Wu, C., Yang, B., Ma, H., Shi, C., Wang, Q., Wang, Q., Yuan, Y. and Liao, M. (2008). Integrated metabolomic analysis of the nano-sized copper particle-induced hepatotoxicity and nephrotoxicity in rats: a rapid in vivo screening method for nanotoxicity, *Toxicol. Appl. Pharmacol.*, **232**, 292–301.

60. Puzyn, T., Rasulev, B., Gajewicz, A., Hu, X., Dasari, T. P., Michalkova, A., Hwang, H. M., Toropov, A., Leszczynska, D. and Leszczynski, J. (2011). Using nano-QSAR to predict the cytotoxicity of metal oxide nanoparticles, *Nat. Nanotechnol.*, **6**, 175–178.

61. Park, E. J., Yi, J., Kim, Y., Choi, K. and Park, K. (2010). Silver nanoparticles induce cytotoxicity by a Trojan-horse type mechanism, *Toxicol. in Vitro*, **24**, 872–878.

62. Xiao, Y., Vijver, M. G., Chen, G. and Peijnenburg, W. (2015). Toxicity and accumulation of Cu and ZnO nanoparticles in *Daphnia magna*, *Environ. Sci. Technol.*, **49**, 4657–4664.

63. Fu, P. P., Xia, Q., Hwang, H. M., Ray, P. C. and Yu, H. (2014). Mechanisms of nanotoxicity: generation of reactive oxygen species, *J. Food Drug Anal.*, **2**, 64–75.

64. Nel, A. E., Mädler, L., Velegol, D., Xia, T., Hoek, E. M., Somasundaran, P., Klaessig, F., Castranova, V. and Thompson, M. (2009). Understanding biophysicochemical interactions at the nano-bio interface, *Nat. Mater.*, **8**, 543–557.

65. Von Moos, N. and Slaveykova, V. I. (2014). Oxidative stress induced by inorganic nanoparticles in bacteria and aquatic microalgae-state of the art and knowledge gaps, *Nanotoxicology*, **8**, 605–630.

66. Commission of the European Communities (CEC) (1996). Technical guidance document in support of commission directive 93/67/EEC on risk assessment for new notified substances. Part II: Environmental risk assessment, Office for Official Publications of the European Communities, Luxembourg.

67. European Chemicals Agency (ECHA) (2011). Guidance on information requirements and chemical safety assessment. Part B: Hazard assessment. Version 2.1, European Chemicals Agency, Helsinki, Finland.

68. Sealy, C. (2011). 'Safe-by-design' nanoparticles show reduced toxicity, *Nano Today*, **6**, 113–114.

69. Mustad, A. P., Smeets, B., Jeliazkova, N., Jeliazkov, V. and Willighagen, E. L. (2014). Summary of the Spring 2014 NSC database survey. Available at: https://figshare.com/articles/ Summary_of_the_Spring_2014_NSC_Database_Survey/1195888. Accessed on 07.02.2017.

70. StatNano (2010). Available at http://statnano.com/. Accessed on 09.02.2017.

71. Nanotechnology Products Database (2016). Available at: http://product.statnano.com/. Accessed on 05.12.2016.

72. Russell, W. M. S. and Burch, R. L. (1959). *The Principles of Humane Experimental Technique* (Methuen, London).

73. Reisfeld, B. and Mayeno, A. N. (2012). What is computational toxicology? *Methods Mol. Biol.*, **929**, 3–7.

74. Peijnenburg, W. (2009). Structure-activity relationships for biodegradation: a critical review, *Pure Appl. Chem.*, **66**, 1931–1941.

75. Chen, G., Li, X., Chen, J., Zhang, Y. N. and Peijnenburg, W. (2014). Comparative study of biodegradability prediction of chemicals using decision trees, functional trees, and logistic regression, *Environ. Toxicol. Chem.*, **3**, 2688–2693.

76. Modarresi, H., Modarress, H. and Dearden, J. C. (2007). QSPR model of Henry's law constant for a diverse set of organic chemicals based on genetic algorithm-radial basis function network approach, *Chemosphere*, **66**, 2067–2076.

77. Singh, K. P., Gupta, S., Kumar, A. and Mohan, D. (2014). Multispecies QSAR modeling for predicting the aquatic toxicity of diverse organic chemicals for regulatory toxicology, *Chem. Res. Toxicol.*, **27**, 741–753.

78. European Chemicals Agency (ECHA) (2008). Guidance on information requirements and chemical safety assessment. Chapter R.6: QSARs and grouping of chemicals. European Chemicals Agency, Helsinki, Finland.

79. Raies, A. B. and Bajic, V. B. (2016). *In silico* toxicology: computational methods for the prediction of chemical toxicity, *Wiley Interdiscip. Rev. Comput. Mol. Sci.*, **6**, 147–172.

80. Sizochenko, N. and Leszczynski, J. (2016). Review of current and emerging approaches for quantitative nanostructure–activity relationship modeling: the case of inorganic nanoparticles, *J. Nanotoxicol. Nanomed.*, **1**, 1–16.

81. Tantra, R., Oksel, C., Puzyn, T., Wang, J., Robinson, K. N., Wang, X. Z., Ma, C. Y. and Wilkins, T. (2015). Nano(Q)SAR: challenges, pitfalls and perspectives, *Nanotoxicol.*, **9**, 636–642.

82. Winkler, D. A. (2016). Recent advances, and unresolved issues, in the application of computational modelling to the prediction of the biological effects of nanomaterials, *Toxicol. Appl. Pharmacol.*, **299**, 96–100.

83. Epa, V. C., Burden, F. R., Tassa, C., Weissleder, R., Shaw, S. and Winkler, D. A. (2012). Modeling biological activities of nanoparticles, *Nano Lett.*, **12**, 5808–5812.

84. Ehret, J., Vijver, M. and Peijnenburg, W. (2014). The application of QSAR approaches to nanoparticles, *ATLA*, **42**, 43–50.

85. Fourches, D., Pu, D., Tassa, C., Weissleder, R., Shaw, S. Y., Mumper, R. J. and Tropsha, A. (2010). Quantitative nanostructure-activity relationship modeling. *ACS Nano*, **4**, 5703–5712.

86. Ghorbanzadeh, M., Fatemi, M. H. and Karimpour, M. (2012). Modeling the cellular uptake of magnetofluorescent nanoparticles in pancreatic cancer cells: a quantitative structure activity relationship study, *Ind. Eng. Chem. Res.*, **51**, 10712–10718.

87. Kar, S., Gajewicz, A., Puzyn, T. and Roy, K. (2014). Nano-quantitative structure-activity relationship modeling using easily computable and interpretable descriptors for uptake of magnetofluorescent engineered nanoparticles in pancreatic cancer cells, *Toxicol. in Vitro*, **28**, 600–606.

88. Liu, R., Zhang, H. Y., Ji, Z. X., Rallo, R., Xia, T., Chang, C. H., Nel, A. and Cohen, Y. (2013). Development of structure-activity relationship for metal oxide nanoparticles, *Nanoscale*, **5**, 5644–5653.

89. Luan, F., Kleandrova, V. V., González-Díaz, H., Ruso, J. M., Melo, A., Speck-Planche, A. and Cordeiro, M. N. (2014). Computer-aided nanotoxicology: assessing cytotoxicity of nanoparticles under diverse experimental conditions by using a novel QSTR-perturbation approach, *Nanoscale*, **6**, 10623–10630.

90. Singh, K. P. and Gupta, S. (2014). Nano-QSAR modeling for predicting biological activity of diverse nanomaterials, *RSC Adv.*, **4**, 13215–13230.

91. Sizochenko, N., Rasulev, B., Gajewicz, A., Kuz'min, V., Puzyn, T. and Leszczynski, J. (2014). From basic physics to mechanisms of toxicity: the "liquid drop" approach applied to develop predictive classification models for toxicity of metal oxide nanoparticles, *Nanoscale*, **6**, 13986–13993.

92. Toropov, A. A., Toropova, A. P., Puzyn, T., Benfenati, E., Gini, G., Leszczynska, D. and Leszczynski, J. (2013). QSAR as a random event: modeling of nanoparticles uptake in $PaCa_2$ cancer cells, *Chemosphere*, **92**, 31–37.

93. Kar, S., Gajewicz, A., Puzyn, T., Roy, K. and Leszczynski, J. (2014). Periodic table-based descriptors to encode cytotoxicity profile of metal oxide nanoparticles: a mechanistic QSTR approach, *Ecotoxicol. Environ. Saf.*, **107**, 162–169.

94. Pathakoti, K., Huang, M. J., Watts, J. D., He, X. and Hwang, H. M. (2014). Using experimental data of *Escherichia coli* to develop a QSAR model for predicting the photo-induced cytotoxicity of metal oxide nanoparticles, *J. Photochem. Photobiol. B*, **130**, 234–240.

95. Toropov, A. A., Toropova, A. P., Benfenati, E., Gini, G., Puzyn, T., Leszczynska, D. and Leszczynski, J. (2012). Novel application of the CORAL software to model cytotoxicity of metal oxide nanoparticles to bacteria *Escherichia coli*, *Chemosphere*, **89**, 1098–1102.

96. Ivask, A., Juganson, K., Bondarenko, O., Mortimer, M., Aruoja, V., Kasemets, K., Blinova, I., Heinlaan, M., Slaveykova, V. and Kahru, A. (2014). Mechanisms of toxic action of Ag, ZnO and CuO nanoparticles to selected ecotoxicological test organisms and mammalian cells in vitro: a comparative review, *Nanotoxicology*, **8**(Suppl 1), 57–71.

97. Blaise, C., Gagne, F., Ferard, J. F. and Eullaffroy, P. (2008). Ecotoxicity of selected nano-materials to aquatic organisms, *Environ. Toxicol.*, **23**, 591–598.

98. Sanderson, H., Johnson, D. J., Wilson, C. J., Brain, R. A. and Solomon, K. R. (2003). Probabilistic hazard assessment of environmentally occurring pharmaceuticals toxicity to fish, daphnids and algae by ECOSAR screening, *Toxicol. Lett.*, **144**, 383–395.

99. Kahru, A. and Dubourguier, H. C. (2010). From ecotoxicology to nanoecotoxicology, *Toxicology*, **269**, 105–119.

100. Gladisa, F., Eggertb, A., Karstena, U. and Schumanna, R. (2010). Prevention of biofilm growth on man-made surfaces: evaluation of antialgal activity of two biocides and photocatalytic nanoparticles, *Biofouling*, **26**, 89–101.

101. Winkler, D. A., Mombelli, E., Pietroiusti, A., Tran, L., Worth, A., Fadeel, B. and McCall, M. J. (2013). Applying quantitative structure-activity relationship approaches to nanotoxicology: current status and future potential, *Toxicology*, **313**, 15–23.

102. Tiede, K., Hassellöv, M., Breitbarth, E., Chaudhry, Q. and Boxall, A. B. (2009). Considerations for environmental fate and ecotoxicity testing to support environmental risk assessments for engineered nanoparticles, *J. Chromatogr. A*, **1216**, 503–509.

103. El Badawy, A. M., Luxton, T. P., Silva, R. G., Scheckel, K. G., Suidan, M. T. and Tolaymat, T. M. (2010). Impact of environmental conditions (pH, ionic strength, and electrolyte type) on the surface charge and aggregation of silver nanoparticles suspensions, *Environ. Sci. Technol.*, **44**, 1260–1266.

104. Baalousha, M., Manciulea, A., Cumberland, S., Kendall, K. and Lead, J. R. (2008). Aggregation and surface properties of iron oxide nanoparticles: influence of pH and natural organic matter. *Environ. Toxicol. Chem.*, **27**, 1875–1882.

105. Tso, C. P., Zhung, C. M., Shih, Y. H., Tseng, Y. M., Wu, S. C. and Doong, R. A. (2010). Stability of metal oxide nanoparticles in aqueous solutions, *Water Sci. Technol.*, **61**, 127–133.

106. El Badawy, A. M., Silva, R. G., Morris, B., Scheckel, K. G., Suidan, M. T. and Tolaymat, T. M. (2011). Surface charge-dependent toxicity of silver nanoparticles, *Environ. Sci. Technol.*, **45**, 283–287.

107. Handy, R. D., von der Kammer, F., Lead, J. R., Hassellöv, M., Owen, R. and Crane, M. (2008). The ecotoxicology and chemistry of manufactured nanoparticles, *Ecotoxicology*, **17**, 287–314.

108. Hua, J., Vijver, M. G., Richardson, M. K., Ahmad, F. and Peijnenburg, W. (2014). Particle-specific toxic effects of differently shaped zinc oxide nanoparticles to zebrafish embryos (*Danio rerio*), *Environ. Toxicol. Chem.*, **33**, 2859–2868.

109. Suresh, A. K., Pelletier, D. A. and Doktycz, M. J. (2013). Relating nanomaterial properties and microbial toxicity, *Nanoscale*, **5**, 463–474.

110. Organization for Economic Co-operation and Development (OECD) (2007). Guidance document on the validation of (quantitative) structure-activity relationships [(Q)SAR] models, ENV/JM/MONO(2007)2, Paris, France.

111. Chaudhry, Q., Castle, L. and Watkins, R. (2010). *Nanotechnologies in Food* (Royal Society of Chemistry Publishers, Cambridge, UK).

112. Park, M., Lankveld, D., Van Loveren, H. and de Jong, W. (2009). The status of in vitro toxicity studies in the risk assessment of nanomaterials, *Nanomedicine*, **4**, 669–685.

113. Oksel, C., Ma, C. Y., Liu, J. J., Wilkins, T. and Wang, X. Z. (2015). (Q)SAR modelling of nanomaterial toxicity: a critical review, *Particuology*, **21**, 1–19.

114. Burello, E. and Worth, A. P. (2011). A theoretical framework for predicting the oxidative stress potential of oxide nanoparticles, *Nanotoxicology*, **5**, 228–235.

115. Chau, Y. T. and Yap, C. W. (2012). Quantitative nanostructure–activity relationship modelling of nanoparticles, *RSC Adv.*, **2**, 8489–8496.

116. Chen, G., Peijnenburg, W., Kovalishyn, V. and Vijver, M. G. (2016). Development of nanostructure–activity relationships assisting the nanomaterial hazard categorization for risk assessment and regulatory decision-making, *RSC Adv.*, **6**, 52227–52235.

117. Liu, X., Tang, K., Harper, S., Harper, B., Steevens, J. A. and Xu, R. (2013). Predictive modeling of nanomaterial exposure effects in biological systems, *Int. J. Nanomed.*, **8**, 31–43.

118. Mu, Y., Wu, F., Zhao, Q., Ji, R., Qie, Y., Zhou, Y., Hu, Y., Pang, C., Hristozov, D., Giesy, J. P. and Xing, B. (2016). Predicting toxic potencies of metal oxide nanoparticles by means of nano-QSARs, *Nanotoxicology*, **10**, 1207–1214.

119. Pan, Y., Li, T., Cheng, J., Telesca, D., Zink, J. I. and Jiang, J. (2016). Nano-QSAR modeling for predicting the cytotoxicity of metal oxide nanoparticles using novel descriptors, *RSC Adv.*, **6**, 25766–25775.

120. Papa, E., Doucet, J. P. and Doucet-Panaye, A. (2015). Linear and non-linear modelling of the cytotoxicity of TiO_2 and ZnO nanoparticles by empirical descriptors, *SAR QSAR Environ. Res.*, **26**, 647–665.

121. Sizochenko, N., Rasulev, B., Gajewicz, A., Mokshyna, E., Kuz'min, V. E., Leszczynski, J. and Puzyn, T. (2015). Causal inference methods to assist in mechanistic interpretation of classification nano-SAR models, *RSC Adv.*, **5**, 77739–77745.

122. Weissleder, R., Kelly, K., Sun, E. Y, Shtatland, T. and Josephson, L. (2005). Cell-specific targeting of nanoparticles by multivalent attachment of small molecules, *Nat. Biotechnol.*, **23**, 1418–1423.

123. Sayes, C. and Ivanov, I. (2010). Comparative study of predictive computational models for nanoparticle-induced cytotoxicity, *Risk Anal.*, **30**, 1723–1734.

124. Fourches, D., Pu, D. and Tropsha A. (2011). Exploring quantitative nanostructure-activity relationships (QNAR) modeling as a tool for predicting biological effects of manufactured nanoparticles, *Comb. Chem. High Throughput Screening*, **14**, 217–225.

125. Gramatica, P. (2007). Principles of QSAR models validation: internal and external, *QSAR Comb. Sci.*, **26**, 694–701.

126. Wold, S., Sjöström, M. and Eriksson, L. (2001). PLS-regression: a basic tool of chemometrics, *Chemom. Intell. Lab. Syst.*, **58**, 109–130.

127. Stewart, J. J. P. (2007). Optimization of parameters for semiempirical methods. V. Modification of NDDO approximations and application to 70 elements, *J. Mol. Model.*, **13**, 1173–1213.

128. Stewart, J. J. P. (2009). MOPAC2009. S.C. Chemistry, editor 2009. Available at: http://openmopac.net/MOPAC2009.html

129. Netzeva, T. I., Worth, A., Aldenberg, T., Benigni, R., Cronin, M. T., Gramatica, P., Jaworska, J. S., Kahn, S., Klopman, G., Marchant, C. A., Myatt, G., Nikolova-Jeliazkova, N., Patlewicz, G. Y., Perkins, R., Roberts, D., Schultz, T., Stanton, D. W., van de Sandt, J. J., Tong, W., Veith, G. and Yang. C. (2005). Current status of methods for defining the applicability domain of (quantitative) structure–activity relationships, *ATLA*, **33**, 155–173.

130. Chen, G., Vijver, M. G., and Peijnenburg, W. (2015). Summary and analysis of the currently existing literature data on metal-based nanoparticles published for selected aquatic organisms: applicability for toxicity prediction by (Q)SARs, *ATLA*, **43**, 221–240.

131. Sushko, I., Novotarskyi, S., Körner, R., Pandey, A. K., Rupp, M., Teetz, W., Brandmaier, S., Abdelaziz, A., Prokopenko, V. V., Tanchuk, V. Y., Todeschini, R., Varnek, A., Marcou, G., Ertl, P., Potemkin, V., Grishina, M., Gasteiger, J., Schwab, C., Baskin, I. I., Palyulin, V. A., Radchenko, E. V., Welsh, W. J., Kholodovych, V., Chekmarev, D., Cherkasov, A., Aires-de-Sousa, J., Zhang, Q. Y., Bender, A., Nigsch, F., Patiny, L., Williams, A., Tkachenko, V. and Tetko, I. V. (2011). Online chemical modeling environment (OCHEM): web platform for data storage, model development and publishing of chemical information, *J. Comput.-Aided Mol. Des.*, **25**, 533–554.

132. Zhao, F., Zhao, Y., Liu, Y., Chang, X., Chen, C. and Zhao, Y. (2011). Cellular uptake, intracellular trafficking, and cytotoxicity of nanomaterials, *Small*, **7**, 1322–1337.

133. Unfried, K., Albrecht, C., Klotz, L., von Mikecz, A., Grether-Beck, S. and Schins R. (2007). Cellular responses to nanoparticles: target structures and mechanisms, *Nanotoxicology*, **1**, 52–71.

134. Yue, T. and Zhang X. (2011). Molecular understanding of receptor-mediated membrane responses to ligand-coated nanoparticles, *Soft Matter*, **7**, 9104–9112.

135. Lin, J., Zhang, H., Chen, Z. and Zheng, Y. (2010). Penetration of lipid membranes by gold nanoparticles: insights into cellular uptake, cytotoxicity, and their relationship, *ACS Nano*, **4**, 5421–5429.

136. Xia, T., Rome, L. and Nel, A. (2008). Nanobiology: particles slip cell security, *Nat. Mater.*, **7**, 519–520.

137. Thevenot, P., Cho, J., Wavhal, D., Timmons, R. B. and Tang, L. (2008). Surface chemistry influences cancer killing effect of TiO_2 nanoparticles, *Nanomedicine*, **4**, 226–236.

138. Franklin, N. M., Rogers, N. J., Apte, S. C., Batley, G. E., Gadd, G. E. and Casey, P. S. (2007). Comparative toxicity of nanoparticulate ZnO, bulk ZnO, and $ZnCl_2$ to a freshwater microalga (*Pseudokirchneriella subcapitata*): the importance of particle solubility, *Environ. Sci. Technol.*, **41**, 8484–8490.

139. Wiench, K., Wohlleben, W., Hisgen, V., Radke, K., Salinas, E., Zok, S. and Landsiedel, R. (2009). Acute and chronic effects of nano- and non-nano-scale TiO_2 and ZnO particles on mobility and reproduction of the freshwater invertebrate *Daphnia magna*, *Chemosphere*, **76**, 1356–1365.

140. Li, M., Zhu, L. and Lin, D. (2011). Toxicity of ZnO nanoparticles to *Escherichia coli*: mechanism and the influence of medium components, *Environ. Sci. Technol.*, **45**, 1977–1983.

141. Bondarenko, O., Ivask, A., Kakinen, A. and Kahru, A. (2012). Sub-toxic effects of CuO nanoparticles on bacteria: kinetics, role of Cu ions and possible mechanisms of action, *Environ. Pollut.*, **169**, 81–89.

142. Stohs, S. J. and Bagchi, D. (1995). Oxidative mechanisms in the toxicity of metal ions, *Free Radical Biol. Med.*, **18**, 321–336.

143. Portier, J., Hilal, H. S., Saadeddin, I., Hwang, S. J., Subramanian, M. A. and Campet, G. (2004). Thermodynamic correlations and band gap calculations in metal oxides, *Prog. Solid State Chem.*, **32**, 207–217.

144. Portier, J., Campet, G., Poquet, A., Marcel, C., Subramanian, M. A. (2001). Degenerate semiconductors in the light of electronegativity and chemical hardness, *Int. J. Inorg. Mater.*, **3**, 1039–1043.

145. Peng, X., Palma, S., Fisher, N. S. and Wong, S. S. (2011). Effect of morphology of ZnO nanostructures on their toxicity to marine algae, *Aquat. Toxicol.*, **102**, 186–196.

146. Tämm, K., Sikk, L., Burk, J., Rallo, R., Pokhrel, S., Mädler, L., Scott-Fordsmand, J. J., Burk, P. and Tämm T. (2016). Parametrization of nanoparticles: development of full-particle nanodescriptors. *Nanoscale*, **8**, 16243–16250.

147. Zimmermann, H.-J. (2010). Fuzzy set theory, *Wiley Interdiscip. Rev. Comput. Stat.*, **2**, 317–332.

148. Oberdörster, G., Oberdörster, E. and Oberdörster, J. (2007). Concepts of nanoparticle dose metric and response metric, *Environ. Health Perspect.*, **115**, A290.

149. Hua, J., Vijver, M., Chen, G., Richardson, M. and Peijnenburg, W. (2016). Dose metrics assessment for differently shaped and sized metal-based nanoparticles, *Environ. Toxicol. Chem.*, **35**, 2466–2473.

150. van de Waterbeemd, H., Smith, D. A. and Jones, B. C. (2001). Lipophilicity in PK design: methyl, ethyl, futile, *J. Comput. Aided. Mol. Des.*, **15**, 273–286.

Chapter 8

Read-Across to Fill Toxicological Data Gaps: Good Practice to Ensure Success with Nanoparticles

Mark T.D. Cronin, Steven J. Enoch, Judith C. Madden, and Andrea-Nicole Richarz
School of Pharmacy and Chemistry, Liverpool John Moores University, Byrom Street, Liverpool, L3 3AF, England
m.t.cronin@ljmu.ac.uk

There is a paucity of suitable toxicological data for nanoparticles. Grouping of similar nanoparticles allowing read-across to fill toxicological data gaps is a potentially rapid and low-cost solution to provide more information about hazard. In addition to savings in time, there are a number of reasons why read-across is increasingly being used to predict toxicological endpoints, including the difficulty of testing these molecules, animal welfare concerns, and compliance with regulatory frameworks. Read-across for nanoparticles can learn much from the experience gained from preparing read-across

Computational Nanotoxicology: Challenges and Perspectives
Edited by Agnieszka Gajewicz and Tomasz Puzyn

ISBN 978-981-4800-64-8 (Hardcover), 978-0-429-34137-3 (eBook)
www.jennystanford.com

arguments for small molecules. For instance, uncertainties relating to read-across should be identified and defined, allowing for an assessment of the confidence that may be placed in a prediction. Much of the uncertainty relates to the assessment of similarity and whether this is appropriate, justifiable, and mechanistically based.

8.1 Introduction

Read-across of toxicological information and data is the process of interpolating the properties of one molecule (with known, preferably high-quality, data) to a similar molecule. The purpose is to provide an estimate of the toxicity of molecules for which data may be lacking and hence to fill "data gaps." Toxicological information is typically required to understand the harmful effects the molecule may have on human health or the environment [1].

Read-across has been applied to small molecules (e.g., drugs, personal care products, and pesticides with a molecular mass less than 1000 Da) for at least two decades, with literature citations going back to 2000, if not before (cf. Hanway and Evans [2]). The process of read-across has moved on from a paradigm of "this molecule looks like this one"—that is, an analogue approach based on homologous series—to something requiring a greater deal of evidence to justify the concept of similarity [3]. At the same time in silico toxicology has undergone a renaissance and is increasingly going back to its roots, requiring mechanistic understanding and embracing the Adverse Outcome Pathway (AOP) philosophy as a means to support modeling [4].

Recent years have seen an inevitable growth in interest in the application of grouping and read-across to fill toxicological data gaps for nanoparticles [5–13]. In many ways, however, this use of read-across has grown independently of the wealth of information and experience that has been built up for small molecules. This chapter attempts to describe some of the current wisdom in applying read-across for small molecules and indicate how this may be applied to filling data gaps regarding nanoparticles.

8.2 Why and When Read-Across Is Used to Predict Toxicity

To understand the current state of the art of the use of read-across to fill data gaps, it is important understand the reasons why read-across has been applied and in what circumstances. These issues are summarized below with reference to both small molecules and nanoparticles.

8.2.1 Ethics

The gathering of toxicological information has traditionally relied on animal testing, with various species of animals being assumed to be surrogates for humans and the whole spectrum of environmental species. With some notable exceptions, the use of animals to provide information on the toxicological properties of molecules has been viewed by many as being adequate, if not wholly accurate. However, for several decades, public opinion has questioned the ethical values behind the use of animals to provide this information, arguing that it is both cruel and unjustifiable scientifically. Public pressure has materialized into chemicals' legislation as well as consumer pressure, encouraging the uptake of alternatives, including in silico models [14]. For some endpoints, there is a real lack of data [5–7] as well as it being appreciated that chronic and development/reproductive endpoints are difficult to assess with traditional quantitative structure–activity relationship (QSAR) models [15]. Thus, read-across is seen as a viable alternative for providing nontest information regarding hazard.

8.2.2 Cost

The complete toxicological assessment of a chemical or a nanoparticle using traditional (animal) tests costs several million US dollars. In addition, long-term bioassays to assess, for example, carcinogenicity may take up to two to three years to complete. With the exception of a relatively small number of new "high-value" products, for example, pharmaceuticals and pesticides, the cost of testing is prohibitive and the time reduces the length of the patent when a substance may be marketed. Thus, especially for existing chemicals (of which up to 100,000 may be in use with few or no data), the cost of testing would be excessive. For some compounds, for certain endpoints, when applied correctly, read-across may provide the same information, to an acceptable level of confidence, as animal testing, thus allowing for savings in both time and money.

8.2.3 Compliance with Regulatory Pressures

The use of chemicals, including nanoparticles, is regulated by national and international agencies. Formal frameworks of regulation are applied in all countries in an attempt to ensure the safety of humans (including accidental, deliberate, and occupational exposure) and the environment. It is common that individual legislation is applied to different uses and industrial sectors, for example, ingredients for pharmaceuticals, pesticides, and personal care products are all regulated independently in the European Union. At times this may cause confusion and contradiction, for example, if an ingredient, preservatives being one such example, is used in various products. Since nanoparticles have numerous applications, they potentially fall under many areas of legislation.

To ensure safety in the use of chemicals, regulation generally requires knowledge of two fundamental pieces of information. Firstly, the concentration, the route of exposure, and the use of an ingredient should be known to give an adequate estimate of exposure. Secondly, the intrinsic hazard of the ingredient should be determined. Risk is the function of exposure and hazard. As noted above, hazard has traditionally been identified by animal tests; however, recent legislation, such as the European Union's Registration, Evaluation, Authorisation and Restriction of Chemicals (REACH) regulation, has encouraged the use of alternatives, including read-across [16]. The European Cosmetics Regulation has gone further by prohibiting animal testing on any ingredient. This has again stimulated much interest in the development and application of read-across approaches for complex and difficult-to-predict endpoints [17, 18]. Thus, globally there are many legislative frameworks for the regulation of chemicals that have promoted the use of read-across for toxicity prediction.

8.2.4 Expectation of Common Properties within a Group

For many groups of compounds, including certain classes of nanoparticles, the toxicity can be assumed to be predictable due to the understanding of the relationship with chemistry. Thus, for endpoints such as genotoxic (DNA reactive) mutagenicity or skin sensitization, both of which require the capability to bind covalently, consistency and predictivity may be assumed within a group (within definable restrictions). There are two obvious scenarios. The first is when a group of compounds has a common functional group that is associated with toxicity—in this case the group members may all be assumed to be toxic. The second scenario is the absence of toxicity, due to the absence of potentially toxic functional groups. The former (assumption of toxicity) scenario is more acceptable in terms of its precautionary nature—the assumption of toxicity or the worst-case scenario. As such, this may, in theory, be possible on the basis of prediction of chemistry alone—although high-quality experimental data would increase the weight of evidence. The latter scenario (absence of toxicity) is more difficult to justify by read-across within a chemical class, group, or category. Such an argument would require biological data on which to anchor the read-across as well as an understanding of how the structural variations within the group may affect biological activity. Thus, a well-justified grouping and read-across approach, with appropriate evidence, should allow for interpolation of toxicity within a well-defined group—ultimately negating the requirement for costly and time-consuming testing of all members of the group.

8.2.5 The Necessity for Alternatives due to the Difficulty of Testing

Not all chemicals can be tested experimentally—and this is especially true for nanoparticles. There are many reasons why a small or large molecule may prove difficult to test. These include the problems associated with physicochemical properties—most notably solubility of nanoparticles and other issues, such as volatility (albeit more of a problem with small molecules!). In addition, (a)biotic transformation or breakdown of a molecule will make testing very difficult. With regard to nanoparticles specifically, issues such as agglomeration of particles in solution are also of great concern. All of these factors, plus inadequate test/experimental design in some cases, mean that some compounds could not, or should not, be tested and caution should be taken in many of the available toxicity data for nanoparticles [19]. If traditional toxicological testing is not feasible, then read-across from suitable surrogate data may prove to be one of the few means of assessing the hazard of some nanoparticles.

8.2.6 Lack of Data for New Nanoparticles and New Toxicological Problems

The application of nanoparticles to many different areas of technology is rapidly increasing. It is much quicker to manufacture new nanoparticles than it is to determine their hazard experimentally. With the acknowledged cost, time, animal use, and experimental difficulty associated with testing, there is hardly any option at this time but to consider alternatives. In silico approaches are among the most viable options for the (initial) evaluation of the hazard of nanoparticles. Read-across, in particular, will optimize limited resources and provide information regarding hazard related to classes of nanoparticles that are currently being expanded or created.

Allied to the issues of lack of data, in silico methods may allow for the probing of new toxicological problems that may be elicited by nanoparticles. For instance, an understanding of similarity as required by read-across may provide some retrospective interpretation of toxicity data. In addition, read-across may help make hypotheses for testing. Lastly, should information for nanoparticle-related hazards become available, then read-across may assist in making this available to a broader number of compounds.

8.2.7 Opportunities to Utilize New Methods and Techniques

Twenty-first-century toxicology embraces a new philosophy in toxicological assessment. This is centered, in part at least, around the consideration of pathways of toxicity and their perturbation. New data streams are becoming available, including high-throughput and high-content information, as well as those from molecular biology [20, 21]. While difficulties remain in obtaining these data for nanoparticles, such nonstandard data will need to find areas of application—mechanistic basis of read-across being one of these. Closely related to this approach is the application of AOPs to nanoparticles [22], which will mutually support read-across and vice versa.

8.3 Good Practice in Read-Across: Ensuring Success

There is increasing knowledge regarding how to perform read-across successfully, specifically with regard to improving the possibilities for acceptance at the regulatory level [23–28], some of which has been recently published as guidance for the read-across of nanoparticles [29]. This body of knowledge has predominantly been developed from read-across examples for small organic molecules and can be separated into identification of common errors [3, 28] and means of identifying uncertainty and improving confidence in read-across [27]. In addition to the authors' acquired experience, this section interprets this knowledge and places it in the context of performing read-across for nanoparticles.

The following describes particular aspects of performing a read-across prediction to fill a data gap that should be useful to the practitioner. It is not intended to be a definitive list but should provide some guidance on how to ensure (and, at times, improve) the quality of read-across assessments for any purpose.

8.3.1 Proper Definition of Structure

For all in silico analysis of toxicological data, it is vital to have the correct structure. The situation is no different for nanoparticles, and for a number of reasons, the definition of the structure is likely to be more complex for larger molecules than for small molecules. For instance, the structure of a nanoparticle must be defined accurately in terms of its size (or distribution of sizes), crystallinity, the chemical structure of the core/shell, shape, ligands, solubility, and so on (cf. Ref. [30]). It is, however, essential that the structure of the nanoparticle be fully, correctly, and unambiguously described and defined, with all relevant dimensions (and their distribution of dimensions as appropriate) noted. There must be consistency between the stated structures, names, and other identifiers of the nanoparticles. This is true not only for the nanoparticle for which the read-across is to be made but also for the nanoparticles from which toxicity is read across.

The definition of the structure includes, inevitably, ensuring that issues with isomers are resolved. If isomers cannot be fully defined, then this limitation should be stated so that the reader/ assessor of the read-across is aware. This leads to the consideration of mixtures, whether this be mixtures of isomers or differently sized nanoparticles. Where possible, the composition of the mixture should be defined. The effect of a mixture on a read-across is another important issue that must be considered as part of the whole read-across justification and description (see Sections 8.3.6 and 8.3.8). Likewise, the core and any external groups of any nanoparticle must be defined as fully as possible [31].

The structure of the nanoparticles in the grouping and read-across must also be stated in any documentation (see Section 8.3.8). This is also important for the calculation of any descriptors or properties that may be obtained and be required for QSARs or quantitative read-across.

8.3.2 Understanding How Structure Affects Toxicology and Mechanism of Action: Appropriate Grouping

Key to identifying analogues for read-across is the appreciation of how chemical structure brings about any potential toxicity or how the absence of any specific functional groups will result in no/low toxicity. This is a complex issue and is at the heart of a read-across through the definition of similarity. For small molecules much is known about structure–activity relationships for many toxicological endpoints [4, 13, 14]. For nanoparticles there is less knowledge, although the fundamental principles are becoming better defined and understood.

An appropriate metric of similarity and understanding of the relationship to the mechanism of action are essential to performing read-across successfully. Once the nanoparticle structure has been defined, the aspect of structure important for grouping must be established. An appropriate and transparent means of establishing similarity is essential for regulatory acceptance of read-across for small molecules. Currently, establishing similarity is relatively unsophisticated for nanoparticles; however, this in itself may not be a restriction. Most grouping approaches have been based around "obvious" groupings of core structure, for example, metal oxides and carbon nanotubes. The unsophisticated nature of this approach is certainly transparent and also pragmatic but does require proof that there is no change in toxicity or that any change is understandable and predictable, as the structure varies in the group. As discussed below, within these types of groups, further evidence may be required to justify the read-across prediction. Various examples of groupings exist [5–13, 32].

The role of AOPs will no doubt be vital to help understand how the chemical structure of nanoparticles impacts the mechanism of action—a framework for establishing AOPs for nanotoxicology is available [22]. Most commonly the key event of the AOP that is most relevant for read-across is the molecular initiating event [4, 33], and knowledge of this may give a mechanistic justification for read-across, where possible, for nanotoxicology. Ensuring the mechanistic justification and defining the AOP naturally go hand in hand, although there is no absolute requirement for mechanistic transparency, provided supporting evidence is available for the chemical basis of the grouping.

8.3.3 High-Quality Experimental Data to Anchor the Read-Across

It is important that the quality and limitations of the toxicological data or information that is read across be known. The read-across assessment can only have the same quality as the data from which it is read. This is especially true for nanotoxicological data, which are prone to variability and may lack experimental precision [13, 32]. There are a number of schemes to evaluate the quality of toxicological information (cf. Refs. [34–36]). However, key criteria for the testing of nanoparticles must be maintaining the integrity of the nanoparticle and its proper and full characterization; determining that the nanoparticle was soluble (nonaggregated) at the concentrations tested; and performing the assay, which is applicable to the type of nanoparticle being tested according to known and recorded guidelines and preferably according to good laboratory practice. Another aspect is looking for consistency across data within a category, that is, reaching the same conclusions from different tests on the same nanoparticle. The need for ensuring the quality of the data will be especially important if there is only one, or a small number, of data points from which to read across.

8.3.4 Use Appropriate Computational Tools for Read-Across

There are a number of computational tools to help with read-across [37, 38]. These must be used prudently and appropriately, with a full record of their use maintained. The main computational tools for read-across are focused on small molecules; however, some are freely available and may be of some use to the modeler [38]. The most frequently used of these tools is the OECD QSAR Toolbox [39]; however, this currently has no real functionality to assist with grouping and read-across for nanoparticles. Despite this, experience with the OECD QSAR Toolbox may prove beneficial to the novice to become familiar with the read-across workflow and types of information required. For instance, the OECD QSAR Toolbox provides access, following verification of the structure, to databases of toxicological, physicochemical, and environmental fate information. The Toolbox also allows for the profiling of a molecule, usually using structural fragments, many of which are related to mechanisms of action. It then allows the user to visualize the resultant category, perform read-across (or trend analysis with quantitative data), and produce a report.

Much of the read-across workflow for small molecules can be replicated in some way for nanoparticles. There are a number of databases that can be searched to gain toxicological information on nanoparticles (cf. Refs. [40–42]). However, these resources do not have the capability to profile molecules (i.e., to place them into toxicologically meaningful groups) or identify potentially useful analogues. Other predictive software is available, for example, the Nano-LAZAR [43], which may, if used cautiously and with expert interpretation, provide some clues as to the grouping of nanoparticles.

Thus, the developer of a read-across for nanoparticles should have knowledge of the available databases and computational tools and be able to use them proficiently, being able to understand the output. However, there are currently few truly usable bespoke tools for the read-across of nanoparticles, so the modeler will have to use intuition and be prepared to do some or all of the work by hand.

8.3.5 Justification of Nanoparticle Similarity and Read-Across Hypothesis

Similarity between nanoparticles in the group where read-across will be performed is fundamental to the success of the prediction. For small molecules it has been shown that definition and a rational explanation of similarity are vital to the regulatory acceptance of a read-across [28]. As mentioned above, for nanoparticles, the similarity is often assumed from the same core structure. In most cases this will be a reasonable starting point for grouping of these molecules, but variations in the structure should also be borne in mind. In addition, it is often helpful to define carefully the limits and boundaries of a group. For instance, talking about grouping "metal oxide" nanoparticles is less helpful than defining specific metal oxides (and valence states, etc.), along with sizes and other groups attached to the core. To prove a read-across is valid, mechanistic justification should also be demonstrable, that is, the basis of the grouping or similarity can be shown to have the same mechanism of action (or absence of relevant mechanisms of action).

8.3.6 Inclusion of Toxicokinetics for Members of a Category

One of the key findings following analysis of read-across case studies for small molecules is the need for better understanding and inclusion of data for toxicokinetics for the members of that category [27]. The consideration of toxicokinetics has often been overlooked in read-across, but it is essential. It is required as it will control potency of toxic action. For instance, if a category is formed on the basis of a common chemical structural element (whether it be the core of the nanoparticle or not), structural modifications in a group have the potential to change its toxicokinetics and hence its uptake and distribution—in other words its bioavailability. Should change in structure within a group increase bioavailability, then the possibility of toxicity may be increased. In a fundamental analogue approach, the consideration of toxicokinetics is effectively overlooked. While a crucial element of read-across, it is often not considered due to the lack of data, not the least for nanoparticles. While toxicokinetics will be difficult to assess either experimentally or in silico, the effect of structural changes should be considered in the toxicokinetics when performing a read-across prediction for toxicity.

8.3.7 Identifying and Defining Uncertainties

To understand the strengths and weaknesses of a read-across the uncertainties at all stages must be identified. Uncertainties can be associated with every aspect, ranging from the definition of nanoparticle structure to the similarity argument and the data on which the read-across is based.

Uncertainties for read-across (in general, although applicable on the whole to nanoparticles) have been defined by a number of workers and were brought together and summarized by Schultz et al. [44]. In addition, Schultz provided templates (with examples) to assist the user in applying and capturing this information. The uncertainties were classified as low, moderate, and high—which in itself is a subjective decision. It is noted that there are other schemes for the classification of uncertainty [45].

The European Chemicals Agency (ECHA) has recently published the Read-Across Assessment Framework (RAAF) [46]. The RAAF defines criteria that are useful in determining the uncertainty or otherwise of a read-across argument; the application of the RAAF was demonstrated through a number of case studies described in Ref. [13].

Thus, defining the uncertainty of a read-across and all its components is a useful means of self-assessment. It is a subjective process, but when performed properly, it should allow a critical appraisal of the weaknesses of a particular read-across and, with reference to the next section, an appreciation of the confidence that may be associated with a particular prediction.

8.3.8 Assessing and Increasing Confidence

Assessing the confidence of any prediction, and especially that based on read-across, is often a subjective process. The assessment can be guided by quantitative analysis, for example, determining the accuracy of prediction given a certain set of circumstances. Traditionally, QSAR analysis has used statistics such as the predictivity of an external test [47] to determine the confidence that may be placed in a prediction—however, even this approach is complex, requiring knowledge of the model's applicability domain, etc. Due to the complexities in the use of read-across to fill toxicity data gaps, assessing the confidence of a prediction, particularly for nanoparticles, is complex and difficult and it is seldom possible to give a prediction more than a low/moderate/high confidence score. The uncertainties defined in Section 8.3.7 provide some guidance on how to establish a level of uncertainty; hence it is important that they by clearly defined. However, the analysis of uncertainties often ends up as a self-assessment of a read-across, which may be subject to the developer's personal bias and needs.

Key to providing increased confidence is to understand the read-across process itself. In the context of building a weight of evidence to support the read-across, mechanistic understanding and evidence to corroborate it are key; this is in the context that the target nanoparticle should have demonstrable similarity in mechanisms to the read-across nanoparticles. Here is an intrinsic problem should mechanism not be known. However, similarity could be defined in terms of the same responses from high throughput/content or in vitro assays and even appropriate in silico predictions and profiling. Thus, our understanding of the uncertainties in grouping and read-across, especially if they could be translated to the context of nanoparticles, means that the confidence level can be assigned in a read-across. Mapping of the uncertainties for nanoparticle read-across is not well refined, but where significant uncertainties are apparent, further evidence could be provided in order to reduce them and hence increase confidence.

8.3.9 Description and Documentation of the Read-Across Prediction

A full description of the grouping and read-across process is essential for the successful regulatory consideration of a prediction [28, 44]. For a full understanding, regulatory submissions should give details and explanation of the rationale for a grouping or read-across. This should be more than simple unannotated printouts from software such as the OECD QSAR Toolbox and may require detailed scientific explanation. There is no formal global reporting format for a read-across prediction, although some elements obviously should be provided. The following list is not exhaustive and is expanded upon in many other publications, reports, and guidance (cf. Refs. [3, 28, 44, 46, 48, 49]). To allow full appreciation and transparency of a read-across prediction it should incorporate some, or all, of the following elements:

- A full description of the structural and relevant physicochemical characteristics of the target and source nanoparticles
- A description of the structural and mechanistic basis of similarity
- Details of the biology (toxicity and other) data on which the read-across is based, along with an assessment of its quality
- Details of computational resources used and how they were applied, for example, databases, calculated properties, and read-across tools
- Determination of the uncertainties of the read-across in accordance with standard protocols, for example, Schultz et al. [44] or the ECHA RAAF [46]
- Details of further evidence, for example, new approach methodologies, that may increase confidence through weight of evidence [27]

The documentation should be sufficiently detailed and critical to allow for the read-across to be repeated from scratch and for the confidence in the prediction to be presented.

8.4 Conclusions

Grouping of similar nanoparticles to allow for read-across to fill data gaps in toxicological information is a useful and increasingly well-established technique. There are a number of reasons why read-across is applied, but mostly it is done to address the data gaps in a cost-efficient, rapid, and ethically acceptable approach to allow for the hazard identification of these materials. At face value, read-across appears simplistic. However, to obtain reliable results care must be taken in the preparation of an argument. Elements of good practice for read-across can be taken from those established for small molecules and applied to nanoparticles. Fundamental among these to perform read-across are the proper and full definition of the nanoparticle structure and relevant characteristics, appropriate (where possible mechanistic based) basis for similarity, and the use of high-quality data to read across from. In the analysis of a read-across prediction the mechanism of action must be demonstrated and evidenced as part of the process of defining uncertainties, allowing for the assessment of confidence. In summary, while read-across for nanoparticles is in its infancy, there are elements of good practice that can be applied to ensure its success.

Acknowledgments

The funding received from the European Union Seventh Framework Programme (FP7/2007-2013) under Grant Agreement No. 309837 (NanoPUZZLES project) is gratefully acknowledged. The authors are also most appreciative of and grateful to Dr. Richard Marchese-Robinson while at Liverpool John Moores University for his excellent contribution and inspiration that supported this work.

References

1. Cronin, M. T. D., Madden, J. C., Enoch, S. J. and Roberts, D. W. (eds.) (2013). *Chemical Toxicity Prediction: Category Formation and Read-Across* (The Royal Society of Chemistry, Cambridge, England).
2. Hanway, R. H. and Evans, P. F. (2000). Read-across of toxicological data in the notification of new chemicals, *Toxicol. Lett.*, **116**, 61.
3. Patlewicz, G., Ball, N., Becker, R. A., Booth. E. D., Cronin, M. T. D., Kroese, D., Steup, D., van Ravenzwaay, B. and Hartung, T. (2014). Read-across approaches – misconceptions, promises and challenges ahead, *ALTEX*, **31**, 387–396.

4. Cronin, M. T. D. and Richarz, A.-N. (2017). Relationship between adverse outcome pathways and chemistry-based *in silico* models to predict toxicity, *Appl. in Vitro Toxicol.*, **3**(4), 286–297.
5. Gajewicz, A. (2017). What if the number of nanotoxicity data is too small for developing predictive nano-QSAR models? An alternative read-across based approach for filling data gaps, *Nanoscale*, **9**, 8435–8448.
6. Gajewicz, A. (2017). Development of valuable predictive read-across models based on "real-life" (sparse) nanotoxicity data, *Environ. Sci. Nano*, **4**, 1389–1403.
7. Gajewicz, A., Jagiello, K., Cronin, M. T. D., Leszczynski, J. and Puzyn, T. (2017). Addressing a bottle neck for regulation of nanomaterials: quantitative read-across (nano-QRA) algorithm for cases when only limited data is available, *Environ. Sci. Nano*, **4**, 346–358.
8. Landsiedel, R. (2016). Concern-driven integrated approaches for the grouping, testing and assessment of nanomaterials, *Environ. Pollut.*, **218**, 1376–1380.
9. Arts, J. H. E., Hadi, M., Irfan, M. A., Keene, A. M., Kreiling, R., Lyon, D., Maier, M., Michel, K., Petry, T., Sauer, U. G., Warheit, D., Wiench, K., Wohlleben, W. and Landsiedel, R. (2015). A decision-making framework for the grouping and testing of nanomaterials (DF4nanoGrouping), *Regul. Toxicol. Pharmacol.*, **71**, S1–S27.
10. Arts, J. H. E., Irfan, M. A., Keene, A. M., Kreiling, R., Lyon, D., Maier, M., Michel, K., Neubauer, N., Petry, T., Sauer, U. G., Warheit, D., Wiench, K., Wohileben, W. and Landsiedel, R. (2016). Case studies putting the decision-making framework for the grouping and testing of nanomaterials (DF4nanoGrouping) into practice, *Regul. Toxicol. Pharmacol.*, **76**, 234–261.
11. Gajewicz, A., Cronin, M. T. D., Rasulev, B., Leszczynski, J. and Puzyn, T. (2015). Novel approach for efficient predictions properties of large pool of nanomaterials based on limited set of species: nano-read-across, *Nanotechnology*, **26**, 015701.
12. Lynch, I., Weiss, C. and Valsami-Jones, E. (2014). A strategy for grouping of nanomaterials based on key physico-chemical descriptors as a basis for safer-by-design NMs, *Nano Today*, **9**, 266–270.
13. Worth, A., Aschberger, K., Asturiol Bofill, D., Bessems, J., Gerloff, K., Graepel, R., Joossens, E., Lamon, L., Palosaari, T. and Richarz, A. (2017). *Evaluation of the Availability and Applicability of Computational Approaches in the Safety Assessment of Nanomaterials*, EUR 28617 EN (Publications Office of the European Union, Luxembourg).

14. Cronin, M. T. D. and Yoon, M. (2019). Computational methods to predict toxicity, in *The History of Alternative Test Methods in Toxicology*, Balls, M., Combes, R. and Worth, A., eds. (Academic Press, London, UK), pp 287–300.

15. Cronin, M. T. D., Enoch, S. J., Mellor, C. L., Przybylak, K. R., Richarz, A.-N. and Madden, J. C. (2017). *In silico* prediction of organ level toxicity: linking chemistry to adverse effects, *Toxicol. Res.*, **33**, 173–182.

16. Spielmann, H., Sauer, U. G. and Mekenyan, O. (2011). A critical evaluation of the 2011 ECHA reports on compliance with the REACH and CLP regulations and on the use of alternatives to testing on animals for compliance with the REACH regulation, *Altern. Lab. Anim.*, **39**, 481–493.

17. Berggren, E., Amcoff, P., Benigni, R., Blackburn, K., Carney, E., Cronin, M., Deluyker, H., Gautier, F., Judson, R. S., Kass, G. E., Keller, D., Knight, D., Lilienblum, W., Mahony, C., Rusyn, I., Schultz, T., Schwarz, M., Schüürmann, G., White, A., Burton, J., Lostia, A. M., Munn, S. and Worth A. (2015). Chemical safety assessment using read-across: assessing the use of novel testing methods to strengthen the evidence base for decision making, *Environ. Health. Perspect.*, **123**, 1232–1240.

18. Lynch, I. and Lee, R. G. (2016). In support of the inclusion of data on nanomaterials transformations and environmental interactions into existing regulatory frameworks, in *Managing Risk in Nanotechnology: Innovation, Technology, and Knowledge Management*, Murphy F., McAlea E. and Mullins M., eds. (Springer, Cham, Switzerland), pp. 145–169.

19. Marchese Robinson, R. L., Lynch, I., Peijnenburg, W., Rumble, J., Klaessig, F., Marquardt, C., Rauscher, H., Puzyn, T., Purian, R., Åberg, C., Karcher, S., Vriens, H., Hoet, P., Hoover, M. D., Ogilvie Hendren, C. and Harper, S. L. (2016). How should the completeness and quality of curated nanomaterial data be evaluated? *Nanoscale*, **8**, 9919–9943.

20. Gaulton, A., Bellis, L. J., Bento, A. P., Chambers, J., Davies, M., Hersey, A., Light, Y., McGlinchey, S., Michalovich, D., Al-Lazikani, B. and Overington, J. P. (2012). ChEMBL: a large-scale bioactivity database for drug discovery, *Nucleic Acids Res.*, **40**, D1100–D1107.

21. Wang, Y., Xiao, J., Suzek, T. O., Zhang, J., Wang, J. and Bryant, S. H. (2009). PubChem: a public information system for analyzing bioactivities of small molecules, *Nucleic Acids Res.*, **37**, W623–W633.

22. Gerloff, K., Landesmann, B., Worth, A., Munn, S., Palosaari, T., Whelan, M., (2017). The adverse outcome pathway approach in nanotoxicology, *Comput. Toxicol.*, **1**, 3–11.

23. Przybylak, K. R., Schultz, T. W., Richarz, A.-N., Mellor, C. L., Escher, S. E. and Cronin, M. T. D. (2017). Read-across of 90-day rat oral repeated-dose toxicity: a case study for selected β-olefinic alcohols, *Comput. Toxicol.*, **1**, 22–32.
24. Mellor, C. L., Schultz, T. W., Przybylak, K. R., Richarz, A. -N., Bradbury, S. P. and Cronin, M. T. D. (2017). Read-across for rat oral gavage repeated-dose toxicity for short-chain mono-alkylphenols: a case study, *Comput. Toxicol.*, **2**, 1–11.
25. Schultz, T. W., Przybylak, K. R., Richarz, A. -N., Mellor, C. L., Escher, S. E., Bradbury, S. P. and Cronin, M. T. D. (2017). Read-across of 90-day rat oral repeated-dose toxicity: a case study for selected n-alkanols, *Comput. Toxicol.*, **2**, 12–19.
26. Schultz, T. W., Przybylak, K. R., Richarz, A.-N., Mellor, C. L., Bradbury, S. P. and Cronin, M. T. D. (2017). Read-across of 90-day rat oral repeated-dose toxicity: a case study for selected 2-alkyl-1-alkanols, *Comput. Toxicol.*, **2**, 28–28.
27. Schultz, T. W. and Cronin, M. T. D. (2017). Lessons learned from read-across case studies for repeated-dose toxicity, *Regul. Toxicol. Pharmacol.*, **88**, 185–191.
28. Ball, N., Cronin, M. T. D., Shen, J., Blackburn, K., Booth, E.D., Bouhifd, M., Donley, E., Egnash, L., Hastings, C., Juberg, D. R., Kleensang, A., Kleinstreuer, N., Kroese, E.D., Lee, A. C., Luechtefeld, T., Maertens, A., Marty, S., Naciff, J. M., Palmer, J., Pamies, D., Penman, M., Richarz, A. N., Russo, D. P., Stuard, S. B., Patlewicz, G., van Ravenzwaay, B., Wu, S. D., Zhu, H. and Hartung, T. (2016). Toward good read-across practice (GRAP) guidance, *ALTEX*, **33**, 149–166
29. European Chemicals Agency (ECHA) (2017). *Guidance on Read-Across. Appendix R6-1: Recommendations for nanomaterials applicable to the Guidance on QSARs and Grouping*, Version 1.0, ECHA, Helsinki.
30. Auffan, M., Rose, J., Bottero, J.-Y., Lowry, G. V., Jolivet, J.-P., and Wiesner, M. R. (2009). Towards a definition of inorganic nanoparticles from an environmental, health and safety perspective, *Nat. Nanotechnol.*, **4**, 634–641
31. Marchese Robinson, R. L., Cronin, M. T. D., Richarz, A. -N. and Rallo, R. (2015). An ISA-TAB-nano based data collection framework to support data-driven modelling of nanotoxicology, *Beilstein J. Nanotechnol.*, **6**, 1978–1999.
32. Golbamaki, N., Rasulev, B., Cassano, A., Marchese Robinson, R. L., Benfenati, W., Leszczynski, J. and Cronin, M. T. D. (2015). Genotoxicity of metal oxide nanomaterials: review of recent data and discussion of possible mechanisms, *Nanoscale*, **7**, 2154–2198.

33. Przybylak, K. R. and Schultz, T. W. (2013). Informing chemical categories through the development of adverse outcome pathways, in *Chemical Toxicity Prediction: Category Formation and Read-Across*, Cronin, M. T. D., Madden, J. C., Enoch, S. J. and Roberts, D. W., eds. (The Royal Society of Chemistry, Cambridge, England), pp. 44–71.

34. Klimisch, H. J., Andreae, M. and Tillmann, U. (1997). A systematic approach for evaluating the quality of experimental toxicological and ecotoxicological data, *Regul. Toxicol. Pharmacol.*, **25**, 1–5.

35. Lubinski, L., Urbaszek, P., Gajewicz, A., Cronin, M. T. D., Enoch, S. J., Madden, J. C., Leszczynska, D., Leszczynski, J. and Puzyn, T. (2013). Evaluation criteria for the quality of published experimental data on nanomaterials and their usefulness for QSAR modeling, *SAR QSAR Environ. Res.*, **24**, 995–1008.

36. Przybylak, K. R., Madden, J. C., Cronin, M. T. D. and Hewitt, M. (2012). Assessing toxicological data quality: basic principles, existing schemes and current limitations, *SAR QSAR Environ. Res.*, **23**, 435–459.

37. Madden, J. C. (2013). Tools for grouping chemicals and forming categories, in *Chemical Toxicity Prediction: Category Formation and Read-Across*, Cronin, M. T. D., Madden, J. C., Enoch, S. J. and Roberts, D. W., eds. (The Royal Society of Chemistry, Cambridge, England), pp. 72–97.

38. Patlewicz, G., Helman, G., Pradeep, P. and Shah, I. (2017). Navigating through the minefield of read-across tools: a review of *in silico* tools for grouping, *Comput. Toxicol.*, **3**, 1–18.

39. Dimitrov, S. D., Diderich, R., Sobanski, T., Pavlov, T. S., Chankov, G. V., Chapkanov, A. S., Karakolev, Y. H., Temelkov, S. G., Vasilev, R. A., Gerova, K. D., Kuseva, C. D., Todorova, N. D., Mehmed, A. M., Rasenberg, M. and Mekenyan, O. G. (2016). QSAR Toolbox – workflow and major functionalities, *SAR QSAR Environ. Res.*, **27**, 203–219.

40. Juganson, K., Ivask, A., Blinova, I., Mortimer, M. and Kahru, A. (2015). NanoE-Tox: new and in-depth database concerning ecotoxicity of nanomaterials, *Beilstein J. Nanotechnol.*, **6**, 1788–1804.

41. Jeliazkova, N., Chomenidis, C., Doganis, P., Fadeel, B., Grafstrom, R., Hardy, B., Hastings, J., Hegi, M., Jeliazkov, V., Kochev, N., Kohonen, P., Munteanu, C. R., Sarimveis, H., Smeets, B., Sopasakis, P., Tsiliki, G., Vorgrimmler, D. and Willighagen, E. (2015). The eNanoMapper database for nanomaterial safety information, *Beilstein J. Nanotechnol.*, **6**, 1609–1634.

42. Kong, L.J., Tuomela, S., Hahne, L., Ahlfors, H., Yli-Harja, O., Fadeel, B., Lahesmaa, R. and Autio, R. (2013). NanoMiner - integrative human

transcriptomics data resource for nanoparticle research, *PLoS One*, **8**, e68414.

43. Helma, C., Rautenberg, M. and Gebele, D. (2017). Nano-Lazar: read across predictions for nanoparticle toxicities with calculated and measured properties, *Front. Pharmacol.*, **8**, 377.

44. Schultz, T. W., Amcoff, P., Berggren, E., Gautier, F., Klaric, M., Knight, D. J., Mahony, C., Schwarz, M., White, A. and Cronin, M. T. D. (2015). A strategy for structuring and reporting a read-across prediction of toxicity, *Regul. Toxicol. Pharmacol.*, **72**, 586–601.

45. Schultz, T. W., Richarz, A. N. and Cronin, M. T. D. (2019). Assessing uncertainty in read-across: questions to evaluate toxicity predictions based on knowledge gained from case studies, *Comput. Toxicol.*, **9**, 1–11

46. European Chemicals Agency (ECHA) (2017). *Read-Across Assessment Framework (RAAF)*, 60pp. ECHA, Helsinki, Finland, ECHA-17-R-01-EN. Available from: https://echa.europa.eu/documents/10162/13628/raaf_en.pdf

47. Aptula, A. O., Jeliazkova, N. G., Schultz, T. W. and Cronin, M. T. D. (2005). The better predictive model: high q2 for the training set or low root mean square error of prediction for the test set? *QSAR Combin. Sci.*, **24**, 385–396.

48. Organisation for Economic Co-operation and Development (OECD) (2014). *Guidance on Grouping of Chemicals*, 2nd ed., OECD Series on Testing and Assessment, No. 194. ENV/JM/MONO(2014)4, OECD, Paris.

49. European Chemicals Agency (ECHA) (2017). *Guidance on Information Requirements and Chemical Safety Assessment*. Chapter R.6: QSARs and grouping of chemicals, ECHA, Helsinki.

Chapter 9

Computational Methods of Interspecies Nanotoxicity Extrapolation: A Step toward the Future

Supratik Kar,[a] Shinjita Ghosh,[b] and Jerzy Leszczynski[a]
[a]*Interdisciplinary Center for Nanotoxicity, Department of Chemistry and Biochemistry, Jackson State University, Jackson, MS 39217, USA*
[b]*School of Public Health, Jackson State University, Jackson, MS 39217, USA*
supratik.kar@icnanotox.org; shinjita89@gmail.com; jerzy@icnanotox.org

The toughest challenge in the shifting nanomaterial (NM) era is how to deal with a hastily expanding system of NMs with respect to health safety and ecosystem risk assessment. To define nanotoxicity, a large number of researchers working in the field of identification and defining regulatory requirements build safe orientated grouping approaches with intelligent testing and nontesting strategies (ITS) using a series of computational approaches. These computational strategies can reduce the number of tests and assist in risk assessment and management of different regulatory bodies. Experimental nanotoxicity evaluation is a time-consuming and

Computational Nanotoxicology: Challenges and Perspectives
Edited by Agnieszka Gajewicz and Tomasz Puzyn

ISBN 978-981-4800-64-8 (Hardcover), 978-0-429-34137-3 (eBook)
www.jennystanford.com

financially expensive practice and employs a large number of animals for studies. Thus, in silico techniques for predictive nanotoxicology are highly anticipated to overcome the loopholes. The quantitative structure–toxicity relationship (QSTR) is competent for evaluating structural features and physicochemical properties responsible for toxicity of NMs. Simple interspecies models and interspecies QSTR (i-QSTR) modeling offer a tool for inference of NM sensitivity with known levels of uncertainty for a diverse pool of species. Interspecies and i-QSTR modeling are able to extrapolate nanotoxicity data for one species to another species for a specific endpoint when the data for the second species are unavailable. This emerging approach can help conquer the cost of manifold toxicity tests along with providing efficient understanding of the mechanism of action (MOA) of NMs for diverse organisms and endpoints. The extrapolation of nanotoxicity is a constructive idea to fill the data gaps where the toxicity value for an NM is absent for a specific endpoint.

9.1 Introduction

The major challenge in the evolution of nanomaterials (NMs) is how to deal with the diversifying system of NMs with respect to their toxicity assessments, along with controlled management. A huge number of commercial applications of nanotechnology and NMs pose anonymous risks to human health and to the environment due to their complex life cycles [1]. Due to their structure complexity and differences in the physicochemical properties, it is very difficult to predict their effects before production and application in daily life [2]. Interestingly, it is fundamentally unfamiliar which properties determine and/or influence the toxicity of particular NMs for a specific endpoint and biological species. Thus, there is an urgent requirement for efficient tools for testing both requisite and potentially unwanted properties of NMs [3]. Not only that, effort should be made to define and develop techniques for exposure, characterization, potential toxicity, fate, and transport, as well as life cycle assessment, of engineered nanoparticles (NPs). In this context, regulatory agencies are looking for possibilities of assessing these dangers as quickly as possible [4].

Due to toxicity testing and risk assessment of diverse NMs, there is an increase in animal testing. According to the report of the European Commission Joint Research Centre, approximately 3.9 million test animals could potentially be employed as an outcome of the introduction of the Registration, Evaluation, Authorisation and Restriction of Chemicals (REACH) program if there are no alternative methods available [5]. On the contrary, alternative methods can reduce animal usage, with potential savings of 1.9 million test animals [6]. Therefore, considering regulatory aspects as well as with regard to the ethical concerns and following the recommendations of the REACH system and the new European Directive 2010/63/EU [7], testing should be executed through tests other than vertebrate animal tests whenever possible. Since experimental estimations of hazardous effects of NMs are expensive and time consuming, the application of efficient in silico approaches is one of the key alternatives before their mass production. In fact, one can reliably predict toxicological effects directly from the physicochemical properties and chemical structures of NM entities [8]. In this context, qualitative or quantitative structure–activity, structure–property, and structure–toxicity relationship (QSAR, QSPR, and QSTR) models or the grouping of structurally related substances, that is, read-across theory has gained enormous popularity [9].

The concept of QSAR in the risk assessment and management of NMs has been extensively discussed by major regulatory bodies worldwide. The major names include the Organization for Economic Co-operation and Development (OECD), the World Health Organization, the European Commission (REACH, NanoSafety Cluster), the US Environmental Protection Agency (US EPA), the European Food Safety Authority, and the Scientific Committee on Consumer Products [5, 8, 9]. The QSAR, QSPR, and QSTR approaches are defined on the basis of the mathematical relationship between the variations in molecular structures (molecular descriptors) and a given physicochemical or biological property (endpoint). Encouraging results of investigations confirm that there is a strong need to develop more efficient "nano-QSAR" models that can accurately assess the toxicity of NPs. These approaches have been already applied successfully to predict the toxicity and physicochemical properties of a number of different types of inorganic and organic NPs and to diverse forms of carbon nanotubes (CNTs) by different research groups worldwide [10–20].

There are relatively large toxicity databases available for higher trophic levels of aquatic species, but fewer toxicity data are available for the primary producers in aquatic systems in terms of ecotoxicity. There are nanotoxicity data available for *Escherichia coli*, zebrafish, and a few mammalian cell lines, but their effects on aquatic or terrestrial organisms as well as on rodents are countable on one hand. Therefore, without any doubt, there is a strong requirement for a competent approach by which one can employ toxicity data of one species to predict the toxicity of multiple species. As the cost of multiple toxicity tests is too high and time consuming, the US EPA's Interspecies Correlation Estimation (ICE) (http://www.epa.gov/ceampubl/fchain/webice/) offers the chance to produce multiple toxicity data from a solo known toxicity value. ICE utilizes the preliminary toxicity estimate for one species to produce correlation toxicity values for multiple species, along with the interpretation of the mechanism of action (MOA) of the chemical to specific species [21]. As in the present chapter we are discussing nanotoxicity, QSTR will be mentioned to define any form of quantitative relationships.

Simple interspecies or interspecies QSTR (i-QSTR) models involve a response in the form of toxicity (termed as "endpoint") for a specific species acting as a predictor variable along with other descriptors to model the same endpoint against another species [22]. The endpoint that acts as a predictor variable can emphasize the MOA of a particular molecule to some degree as it is generated by standard experimental bioassay, while other descriptors as employed in simple QSTR models are obtained purely from chemical structure and/or physicochemical properties. Therefore, compared to the normal QSTR models, interspecies models amplify the dependability for further prediction. Additionally, the specific response of a molecule to a particular biological species is reported but its quantitative response to another species is not known. In such circumstances, interspecies models provide a tool for estimating sensitivity toward chemical exposure with known levels of uncertainty for different species [23].

In this background, in the upcoming days, interspecies nanotoxicity extrapolation will give a major push to economically filling data gaps faster than the traditional methods while maintaining the reduction in animal usage. Therefore, considerable efforts should be made for the generation of high-quality experimental data to generate highly efficient interspecies models. The present chapter discusses the fundamental background of interspecies models along with i-QSTR models, their basic differences, and their advantages in nanotoxicity prediction in the present time. Successful models are discussed to promote the understanding of interspecies extrapolation and their application in nanotoxicity. The chapter will help beginners to understand the basic principles and required components for developing a successful interspecies model for nanotoxicity.

9.2 Nanotoxicity

The fundamental structural configuration of NPs and factors like their surface area, the tendency to aggregate, the size and shape of the particles, and their surface charge play critical roles in their distribution through the environment and living organisms, including the human body. Due to their unique nature and capability to affect normal biological activity, NMs have drawn a lot of interest in the last decade as well as in the present scenario. NPs may enter the human body through different ports of entry, such as the lungs, the skin, the eyes, and the gastrointestinal tract. Notably, NMs may penetrate deep into tissues through fine capillaries, travel throughout the body, and interact with organs, tissues, and cells. NMs are also found in various cells, such as mitochondria, fibroblasts, nuclei, lipid vesicles, and macrophages, employing pharmacokinetic studies [1, 5]. A good number of recent in vivo and in vitro studies illustrated that the direct contact of NMs with the cell surfaces may lead to the following hazards: (i) oxidant stress and inflammation due to the formation of reactive oxygen species (ROS), leading to infection and exacerbation of asthma and chronic obstructive pulmonary disease [24], (ii) lipid peroxidation of cellular membranes followed by cell damage [25], and (iii) the up-/down-regulation of genes encoding an explicit protein concerned with inflammatory processes and abnormalities in cell adhesion [26].

A good number of toxicological studies have suggested that NMs can be cytotoxic [1], neurotoxic [27], genotoxic [28], ecotoxic [29], or bactericidal [30]. Single- and multiwalled CNTs (SWCNTs and

MWCNTs) [31] as well as fullerenes (C_{60}) [32] exhibit cytotoxicity to human cell lines. There is strong evidence of pulmonary inflammation in rodents and humans induced through anatase (TiO_2), nano-sized silicon dioxide (SiO_2), and zinc oxide (ZnO) [33]. There is a huge deficiency of information on NM exposure, dosimetry, risk assessment, and health effect. Additionally, we lack the knowledge regarding interactions of NMs with biological systems, as well as with the environment. Therefore, one needs to put great effort into defining and developing methods for exposure, characterization, engineering controls, probable toxicity, fate and transport, and life cycle evaluation of a diverse set of NMs. The most common and probable mechanisms of NMs toxicity are demonstrated in Fig. 9.1.

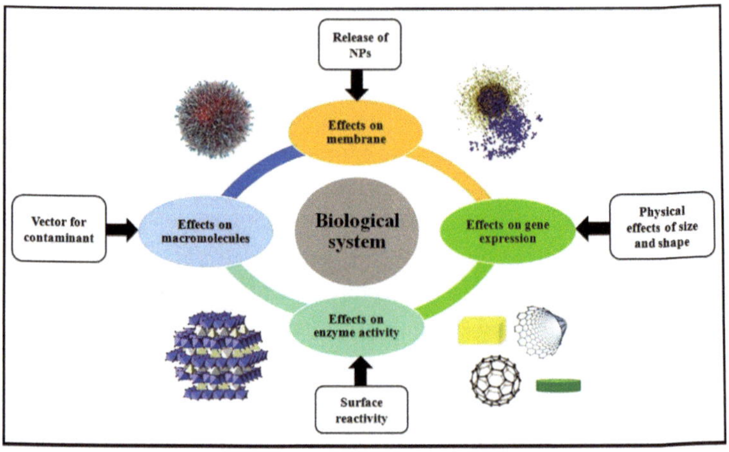

Figure 9.1 Hypothesized mechanisms of NM toxicity.

Insufficient knowledge and potential risks arising from exposure to NMs led the UK government to appeal for advice from the Royal Society and Royal Academy of Engineering for the formation of their Nanoscience and Nanotechnology Working Group. Their report (http://www.nanotec.org.uk) made an extensive range of recommendations about the necessity for supplementary and enhanced information and for a rational approach to these concerns. The United States also engaged in nanotechnology innovations worldwide with the National Nanotechnology Initiative (NNI), launched in 2000 with eight agencies. The NNI consists of the individual and cooperative nanotechnology activities of 25 federal agencies, with a range of research and regulatory roles and responsibilities.

9.2.1 Carbon NMs

Carbon nanostructures such as fullerenes, SWCNTs and MWCNTs, and graphene have been given great importance in the field of nanotechnology, with numerous potential applications in diversified areas due to their small dimensions and unique properties [34]. CNTs are expected to have asbestos-like toxicity (induce lung cancer and mesothelioma) because of their similarity in shape as well as the high ratio between length and diameter [35]. Carbon NMs released into the environment lead to exposure to diverse organisms and have the potential to damage the brain, skin, and lung tissue and accumulate in the body [36]. Not only that, CNTs can rapidly enter and spread in the spleen, heart, bone marrow, kidneys, and liver [36]. Among all the biological systems of the human body, reports suggest that the skin, eyes, and lung have the maximum risk of exposure to NMs during the production, application, and disposal of carbon-based NMs [36]. An experiment illustrated the genotoxicity in human bronchial epithelial cells (BEAS-2B) due to CNTs and graphite NPs. Aggregates of SWCNTs and MWCNTs may induce peribronchiolar lung fibrosis, and DNA damage is also reported with dose-dependent increment [37].

9.2.2 Metal Oxide NMs

Metal oxide NMs can adopt an enormous number of structural geometries, with an electronic structure that can adopt metallic, semiconductor, or insulator characteristics. A huge number of industrial applications are already reported in sensors, circuits, piezoelectric devices, fuel cells, solar cells, and coatings for the passivation of surfaces against corrosion and as catalysts. But, till date, there is restricted information related to systematic data on experimentally measured toxicity for metal oxide NMs. Though a small number of data are available, the majority of them deal with in vitro studies for bacteria cultures [38] and only a few in vivo studies were documented for mammals, predominantly for rats and mice, in recent time [39]. A comprehensive review discussing the cellular responses induced by manufactured metal oxide NPs (MeONPs) has been published by Horie et al. [40]. Karlsson et al. reviewed the cytotoxicity of a series of MeONPs [41]. Fahmy and Cormier [42] suggested that when human lung carcinoma A549 cells were exposed to CuO, TiO_2, ZnO, $CuZnFe_2O_4$, Fe_3O_4, and Fe_2O_3 NPs, the intracellular ROS levels increased in CuO-exposed cells and oxidative stress–related DNA damage was observed in cells exposed to Fe_3O_4, $CuZnFe_2O_4$, ZnO, and CuO NPs. Above all, CuO NPs caused stronger cytotoxicity than CuO fine particles [42]. The assessment of NP toxicity, by in vitro or in vivo studies, offers significant information, especially in terms of the MOA of toxicity for specific organisms. The major drawbacks of these in vitro and in vivo studies are high cost, time consumption, as well as requirement of animal experimentations. Therefore, the need of the hour is computational studies for nanotoxicity prediction and risk assessment.

9.3 Role of Computational Methods in Nanotoxicity Predictions

Current computational approaches show a great potential in predicting activities, properties, toxicities, reactivity, and MOA for a diverse range of molecular systems, from NPs and small organic molecules to large biomolecules. Quantum chemical calculations and molecular dynamics (MD) simulations proved to be helpful in addressing the probable risks associated with NMs. Outcomes of theoretical calculations might offer helpful suggestions to experimentalists working in this research field. Computer simulations could be important in scrutinizing the structure of surfaces and identifying even the smallest change in the position of the edges, angles, corners, and surface steps of atoms and defects of NMs. The interatomic interaction mechanisms between NPs and biological molecules can be well understood, which will assist in the safe production and exploitation of NMs [43].

Computational studies focus on finding out the low-energy geometry of the system under study in order to model representative physical properties or interactions with other systems [44]. Computational models could be very useful in understanding the details of DNA damage by NMs as well as in mapping innovative strategies for protecting DNA from radiation-induced damage. Sun and Bu performed density functional theory and dispersion-corrected calculations for complexes of nucleic acid bases and C_{60} to identify the probable mechanism of damage [45]. Obata and Honda [46] employed computational tools elucidating the effects of surfactants on the toxicity of CNTs in an aqueous environment, useful for recognizing the mechanism of interactions between DNA and CNTs [47]. Mechanistic interpretations are indispensable not only in designing new materials but also in reducing the risk of probable toxicity of the materials. The driving forces accountable for the stabilizing interactions between DNA and CNTs were identified as van der Waals and hydrophobic interactions and the entropy loss due to the confinement of the DNA backbone. MD simulations exhibited the conformational change of double-stranded DNA on the nanotube surface, and SWCNTs induced a spontaneous conformational change to the single-stranded DNA bases [47].

One of the leading computational tools at present is QSTR modeling [48], essential for high-throughput screening, reducing animal testing, providing details of the mechanistic interpretations, and generating hypotheses for risk assessment and managements [49]. The novelty of QSTR modeling depends not only on new molecules' toxicity prediction but also on identifying the features and physicochemical properties responsible for toxicity. QSTR models should be generated taking all five principles of OECD guidelines into account. A reliable and true validated model is very useful for a dependable prediction of any new/external set of NMs. Chemical category, read-across, and trend analysis techniques are also useful for predicting unknown toxicity of NMs using similar NMs (called chemical analogs) with known toxicity from the same chemical category.

To predict the cellular uptake of NMs in pancreatic cancer (PaCa2) cells, a good number of studies have been performed by different research groups [10, 11, 14, 15]. A rational nanoquantitative structure–property relationship (nano-QSPR) model on the catastrophe theory has been developed by Carbó-Dorca and Besalù [17]. The cytotoxicity of metal oxide NMs to bacteria *Escherichia coli* was modeled by Puzyn et al. [3] employing quantum chemical calculations. The same data were also modeled by Toropov et al. [13] and Sizochenko et al. [50] applying web-based software CORAL and liquid drop methods, respectively. Gajewicz et al. [51] generated a nano-QSAR model describing the toxicity of 18 nanometal oxides to a HaCaT cell line as a model for dermal exposure. Liu et al. [19] developed a nano-QSAR model for the toxicity of metal oxide NMs for BEAS-2B and murine myeloid (RAW 264.7) cells. Shao et al. [20] developed QSAR models using the nanotoxicity endpoint of surface-modified MWCNTs.

The necessity for a wider range of computational tools and techniques to replace animal experimentation is significant for the implementation of REACH legislations. In this direction, in silico approaches and expert systems provide reasonable insights into the relationships between particular chemical attributes and specific toxicological effects, especially for nanotoxicity evaluations. A good number of predictive toxicology expert systems are available at present. For example, ASTER, CAESAR, DEREK, ECOSAR, and TOPKAT have a tremendous capability to make fast and accurate predictions [52, 53]. There is no doubt that, at present, predictive toxicology systems are high-quality data sources that support the model development process and safety criteria globally.

9.4 Interspecies Model: Extrapolation through Toxicity–Toxicity Correlation

Due to the presence of a huge number of chemicals, it is impractical to check the toxicity of individual molecules for diverse endpoints as well as for multiple species. An attractive hypothesis is the classification of the mechanistic basis of toxicity posed by chemicals, which is anticipated to direct the pathway of assessing chemical hazard. Simply, toxicity–toxicity correlation defines the identification of correlation of toxicity between two species with respect to a definite endpoint. The result is the correlation of toxicity between two or more genera of organisms to a specific endpoint, the aim being to create a link between the organisms employed. Therefore, if a chemical imparts toxicity in different genera of living organisms by following the same mechanistic path, a correlation might exist when the toxicity between the considered organisms is compared. Such studies can enlighten the mechanistic basis implied in the estimation of the hazardous potential of chemicals. Not only that, the toxicological features can be extrapolated and aid in filling data gaps when making an absolute assessment of chemical hazards.

ICE models are one of the best among the present models and approaches used to extrapolate toxicity data from one species to another [54]. ICE models characterize regression models that describe the relationship between the acute toxicity (EC_{50}; mg/kg bodyweight) of a set of chemicals tested in two species. If toxicity data are accessible for the surrogate species, the toxicity to the predicted taxon can be predicted employing the specific ICE model for that species-taxon pair. Not only that, the modeling of reactive chemicals is complicated as there are diverse structural classes that are difficult to illustrate with common descriptors in a single equation [55]. Therefore, generation of interspecies correlation may overcome the problem if a chemical has similar MOAs in different species.

A straightforward interspecies correlation model for two species can be computed according to the following mathematical expression:

$$\log_{10}\left(1/EC_{50}[\text{Predicted species}] \right) = a + x \times \log_{10}\left(1/EC_{50}[\text{Surrogate species}] \right) \quad (9.1)$$

In Eq. 9.1, EC_{50} (half maximal effective concentration of a chemical that induces a response halfway between the baseline and the maximum after a specified exposure time) is the toxic effect observed and a and x are the intercept and slope of the line, respectively. A model bearing a noteworthy linear relationship (p-value < 0.05) will be employed for further analyses. Note that EC_{50} is given as an example only; one can employ any form of response to the model, for example, IC_{50} (half maximal inhibitory concentration measuring the effectiveness of a chemical in inhibiting a specific biochemical response) and LD_{50} (median lethal dose representing the individual dose required to kill 50% of a population of test animals). The distinct difference between simple interspecies correlation (described in the present section) and i-QSTR for toxicity–toxicity prediction is discussed in the next section.

The exploitation of organisms of lower taxonomical classes as the indicator species in toxicity evaluation utilizes comparatively easy experimental techniques, cutting time and cost incurred, and permits recognition of early toxicological features. Further, the outcome can be extrapolated to the related organisms of higher taxonomical classes, permitting filling of data gaps. The same approach is applicable in extrapolating data from vertebrate organisms to human beings. Therefore, without any doubt, extrapolation of toxicological data is important to bridge the data gaps between diverse species and reducing the usage of experimental animals.

9.5 Fundamental of i-QSTR Correlation

Extrapolation of a specific toxicity endpoint data from one species to another is a complicated task with the involvement of reliability issue of the prediction data. As the MOA of toxicity can be employed as physicochemical properties along with the toxicity of surrogate species, the i-QSTR can provide valuable information regarding nanotoxicity. An i-QSTR model can be employed to conclude the species-specific toxicity of NMs and make a considerably more dependable comparison with a single-endpoint QSTR model as well as with simple interspecies model for toxicity prediction and identification of the MOA for a specific species [56]. Cronin et al. [56] made an important suggestion, that if the toxicity patterns are similar for two species for a defined chemical (in our case NMs) and specific endpoint and they correlate well, then the chemicals can be expected to have comparable and/or similar MOAs in both species. Again, if the toxicity patterns are different and the toxicity response is not well correlated between two species for a specific endpoint, the MOAs of the chemical are expected to be different for both species. Therefore, chemicals may show species-specific toxicities and the toxicity may vary with the type of endpoints also.

A general i-QSTR model can be expressed as follows:

$$\log_{10}\left(\frac{1}{EC_{50}[\text{Predicted species}]} \right) = a + x_1 \times \log_{10}\left(\frac{1}{EC_{50}[\text{Surrogate species}]} \right) + x_2 \times D \tag{9.2}$$

Here, the coefficients of surrogate species toxicity and descriptor (D) are denoted as x_1 and x_2, respectively. To model a more complex MOA, one needs more physicochemical properties as descriptors (may vary from 1 to n) related to toxicity. Other terms are as mentioned in Eq. 9.1.

Zhang et al. [57] mentioned four conditions to recognize the significance of the physicochemical descriptors that can be considered in interspecies correlations:

- **Comparable MOAs and bio-uptake of NMs:** The studied NMs have comparable sites of action, respond through the same receptors, and are followed by similar biochemical responses along with similar toxic effects for two different species. This can happen in closely related species.
- **Comparable MOAs but different bio-uptake of NMs:** The NMs have diverse pathways to reach the sites of action, though they react with similar receptors after bio-uptake. The response can be predicted if a bio-uptake restriction is employed to accurately identify the physiological interspecies difference.
- **Dissimilar MOAs but comparable bio-uptake of NMs:** The NMs have similar pathways to reach the sites of action, but they respond by reacting with different receptors. With the experimental response of one species, the response can be predicted for another species with the i-QSTR model if we identify one or more descriptors to explain the variation of reactivity with receptors.
- **Dissimilar MOAs and bio-uptake of NMs:** The NMs have different capabilities of reaching the sites of action and react with different receptors. In this condition, correlation can be checked between the data points for the toxicity response, but the prediction quality and reliability will be poor.

Therefore, Zhang et al. [57] modified Eq. 9.2 introducing crucial parameters denoting variation in bio-uptake and difference in toxic MOAs between two species for a specific endpoint:

$$\log_{10}\left(1/\mathrm{EC}_{50}[\text{Predicted species}]\right) = a + x \times \log_{10}\left(1/\mathrm{EC}_{50}[\text{Surrogate species}]\right) + P_{\mathrm{B}} + P_{\mathrm{M}} \tag{9.3}$$

Here, P_{B} is a physicochemical parameter to nullify the variation in bio-uptake, P_{M} is a physicochemical parameter to correct the difference in toxic MOAs between species, and the remaining terms are as mentioned in Eq. 9.1.

It is imperative to point out that the disparity between interspecies models and i-QSTR models is only related to bio-uptake factors along with physicochemical parameters to identify the best possible MOA of nanotoxicity in a specific species in the case of the latter approach. There are many articles where both terms are discussed interchangeably, but the i-QSTR model is the sophisticated form of interspecies models that eliminates the drawbacks of classical interspecies models. The fundamental idea and steps for i-QSTR modeling are reported in Figs. 9.2 and 9.3, respectively.

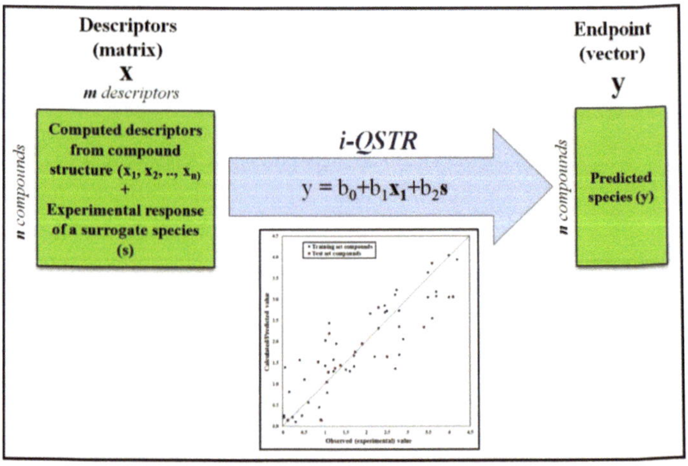

Figure 9.2 Elementary principle of i-QSTR modeling.

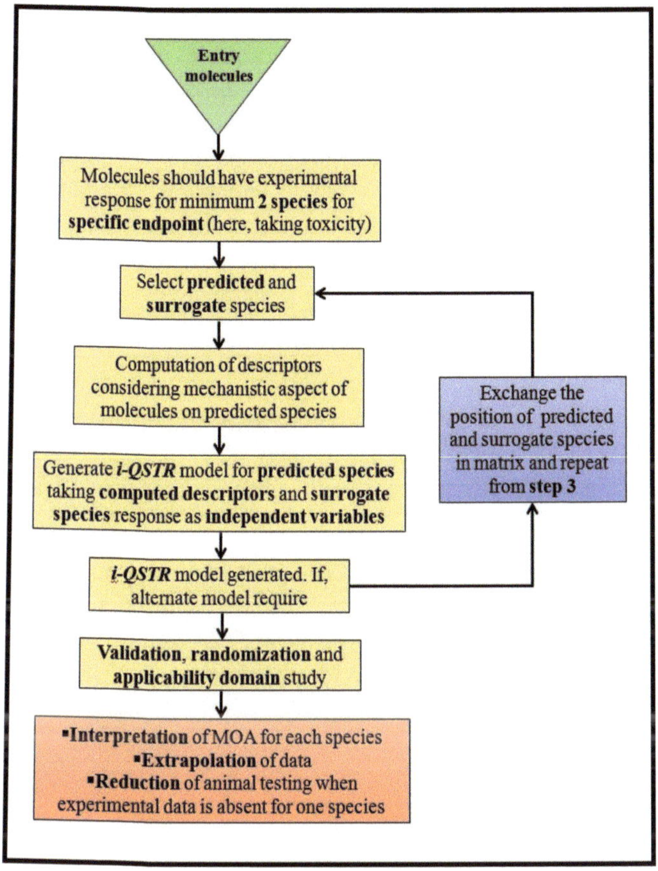

Figure 9.3 Schematic diagram of individual steps of i-QSTR modeling.

9.6 Necessity of Interspecies Extrapolation

Since the inception of interspecies correlation modeling, some attention has given by computational scientists to this particular issue [58]. A long time ago, LeBlanc [58] showed significant interspecies relationships between closely related species (fish vs. fish), more distantly related species (fish vs. invertebrates), and most distantly related species (fish vs. algae), indicating that species have closer correlations between them as they get closer in taxonomy. The chances of getting a good correlation between two different taxonomic species are less as the MOAs of a specific NP are different for two species. On the other hand, the chances of obtaining a better model are higher for species from the same taxonomy. Therefore, if two species have a weak response correlation, it may be picked up by employing physicochemical properties and structural features of studied NPs in the form of i-QSTR modeling. This indicates the importance of i-QSTR model compared to normal interspecies modeling, as illustrated in Fig. 9.4.

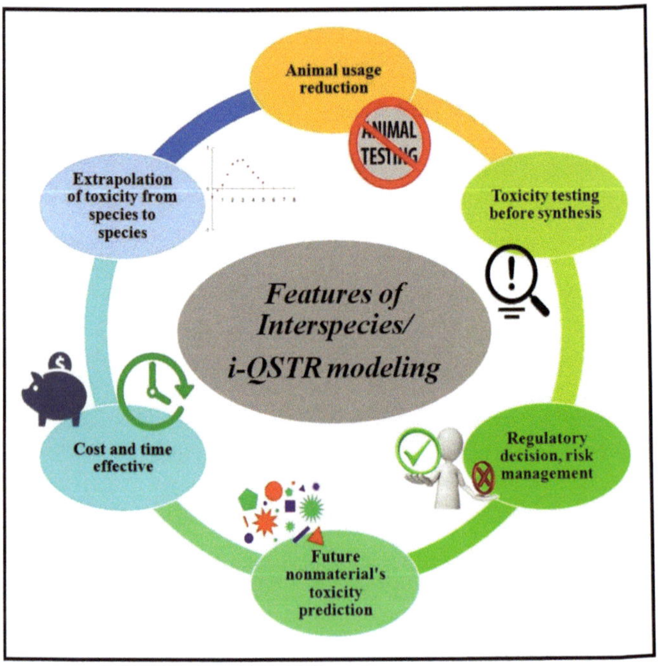

Figure 9.4 Significance of interspecies and i-QSTR modeling in nanotoxicity.

i-QSTR should be employed frequently for the following reasons:

- **Recognition of the MOA behind nanotoxicity:** Evaluation of the mechanistic interpretation behind the toxicity of NMs to a species is the most important criterion of an i-QSTR model, which employs the toxicity of one species as a predictor variable to model the surrogate species and can highlight the MOA to some degree as it is derived by a standard experimental bioassay [59].

- **Extrapolation of data:** The i-QSTR models have the capability to predict the toxicity of NMs in one species on the basis of that in another species through extrapolation of data. The i-QSTR model can be utilized to predict the toxicity of a species employing the toxicity data of the surrogate species when a response is absent for the first species.
- **Species-specific nanotoxicity:** A statistically significant interspecies model suggests that both species share similar MOAs of toxicity. A poor correlation indicates different MOAs of toxicity. Interestingly, Nendza and Wenzel [60] reported that a molecule can show similar toxicity in some species and different in other species due to completely different MOAs of toxicity or completely different target sites. Therefore, the MOA behind the toxicity of NMs can be checked through the phenomena that take place at target sites, which is specific for a given species. This hypothesis proposes species-specific toxicity modeling for a better understanding of the MOA at the molecular level for each species.
- **Data gaps filling:** The use of surrogate assays to predict toxicity in higher species by interspecies relationships methods has been endorsed as a primary source to fill up the data gaps in nanotoxicity databases.
- **Reduction of animal experimentation:** Like normal QSTR models, the i-QSTR model assists in the reduction of animal testing. Not only can one extrapolate data from a lower species to a higher species, but also there is no need to use higher species for toxicity evaluation. This is an added advantage of i-QSTR models over the normal QSTR model.

9.7 Computational Interspecies Nanotoxicity Models

Several groups of authors have presented appropriate insight into the MOAs of a diverse range of chemicals through the development of interspecies correlation models. A huge number of electronic, quantum chemical, structural, as well as topological parameters, along with experimental outcomes, have also been successfully employed in generating i-QSTR models. Not only that, species-specific models were developed employing MOA-based classification schemes. Many researchers also use the web-based ICE (Web-ICE) tool for the prediction of toxicity of pesticides registered in the federal list for threatened and endangered species (https://www3.epa.gov/webice/).

The idea of interspecies modeling has been employed and validated by a large number of researchers worldwide in the last two decades for activity and toxicity modeling between two or multiple species. Dimitrov et al. [61] developed two successful i-QSTR models for acute toxicity of aldehydes to fathead minnow (LC_{50}) correlated with fish and *Tetrahymena pyriformis* (IC_{50}) separately, employing octanol/water partition coefficient ($\log K_{ow}$) and donor delocalizability of the aldehyde oxygen atom–based reactivity ($D_{O\text{-atom}}$) as descriptors. Dyer et al. [21] performed interspecies correlation-based species sensitivity distribution analysis using three surrogate species—rainbow trout, fathead minnow, and water flea—employing experimental toxicity data versus predicted values obtained from the US EPA's Web-ICE module. In another study, toxicity of organothiophosphate pesticides was modeled with two species, *Daphnia magna* and *Cyprinus carpio*, using octanol-water partition coefficient ($\log K_{ow}$) and quantum-chemical parameters by Zvinavashe et al. [62]. Zhang et al. [57] explored the bio-uptake and toxicity profile with interspecies correlation models employing toxicity data of eight diverse species (*Vibrio*, river bacteria, algae, *Daphnia*, carp, *Tetrahymena*, fathead minnow, and guppy) for 141 chemicals. Highly predictive i-QSTR models were reported between *Vibrio Fischeri* with river bacteria and fish species like guppy and fathead minnow. On the contrary, the authors concluded that there was very poor correlation between algae toxicity with *Tetrahymena* or *Daphnia*, suggesting mechanistic changes. The first interspecies modeling for the pharmaceuticals ecotoxicity was reported by Kar and Roy [63] employing 77 molecules for *Daphnia magna* and fish. Structural fragments like carbonyl group and aromatic carbon fragment were identified as important features for toxicity to *Daphnia*, while fish toxicity was encoded by fragments like X=C=R, R–C(=X)–X, and keto group. In another work, Kar and Roy [64] explored the carcinogenicity response to organic chemicals through interspecies formalism with two rodent species, rat and mouse. The authors identified the polynuclear aromatic hydrocarbons, and compounds with bromine atoms were identified as the structural alerts for carcinogenicity and subsequently utilized for the interspecies models for predicting the toxicity of the compounds without experimental data concerning one species. The ICE tool and the acute-to-chronic ratio parameter for the categorization of no-effect concentration of perfluorooctane sulfonic acid derivatives were studied by Qi et al. [65]. Raevsky et al. [66] examined the interspecies correlation of the toxicity response toward organic chemicals of daphnids, *Tetrahymena*, fish, and mice using multiple modeling algorithms. A series of triazoles and benzo-triazoles were modeled through interspecies modeling by Cassani et al. [67] for assessing toxicity toward *Daphnia magna* and Oncorhynchus *mykiss*. Roy et al. [68] attempted i-QSTR models for the first time for ionic liquids between two species *Scenedes musvacuolatus* and *Daphnia magna* employing the partial least squares (PLS) tool. Furuhama et al. [69] developed i-QSTR models using the toxicity response of *Pseudokirchneriella subcapitata*, *Oryziaslatipes*, and *Daphnia magna* toward a series of phenols and aromatic amine derivatives taking the structural features and SMARTS[1] notation as descriptors.

Interestingly, till today, there is only one successful i-QSTR model that was developed by our group employing a nanomaterial's toxicity as a response to the model. In this study, Kar et al. [70] proposed equations to extrapolate the cytotoxicity data of MeONPs for *Escherichia coli* to that for human cell lines and vice versa. Features like standard enthalpy of formation of metal oxide nanocluster (ΔH_f^c) and electronic chemical potential (μ) were identified to predict HaCaT cytotoxicity along with the experimental cytotoxicity of *Escherichia coli*. The first feature defines the detachment of metal cations from the surface of metal oxides, and the second explains the transfer of electrons from the valence band to the conduction band by the intracellular redox processes. To predict *Escherichia coli* cytotoxicity, the charge of the metal cation subsequent to a given oxide (χ_{ox}) and enthalpy of formation of a gaseous cation (ΔH_{Me+}) evolved as noteworthy attributes along with the experimental cytotoxicity to HaCaT cell lines. Kar et al. [70] concluded that the employment of species-specific cytotoxicity of NPs as predictor variables along with calculated descriptors offers superior models

[1]Smiles ARbitrary Target Specification

for prediction compared with the models constructed using only the computed descriptors. The authors summarized the extrapolating equations as follows:

$$f(E.\ coli\_\text{cytotoxicity}) = f(\text{Human\_cytotoxicity}) + f(\Delta H_{\text{Me+}}, \chi_{\text{ox}}) \quad (9.4)$$

$$f(\text{Human\_cytotoxicity}) = f(E.\ coli\_\text{cytotoxicity}) + f((\Delta H_{\text{f}}^{\text{c}}, \mu) \quad (9.5)$$

We strongly believe that the i-QSTR modeling technique is very capable of extrapolating toxicity data between diverse species, especially when the toxicity data are present for one species but absent for the other. Importantly, one has to consider the specificity and sensitivity toward dissimilar indicator species in conjunction with the understanding of the MOA of the chemicals. As the experimental toxicity data for NMs are very limited in the present scenario and restricted to a few specific species (mainly *Escherichia coli*, HaCaT, PaCa2 cells, human BEAS-2B cells, mouse RAW 264.7 cells, zebrafish, and Wistar rat lung cell), there is an urgent need to develop more and more i-QSTR models for NMs.

Additionally, multitarget QSTR (mt-QSTR) can be successfully employed for screening new, untested MeONPs for their safety assessment. The hypothesis behind the i-QSTR and mt-QSTR approaches is different, but the aim is more or less the same—to employ multiple species for making models to extrapolate data for new or untested species to avoid a lot of animal testing and fill up huge data gaps. An mt-QSTR model was successfully developed by Basant and Gupta [71] using multiple toxicity endpoints of MeONPs taking into account four different experimental data sets available on cytotoxicity in *Escherichia coli* and HaCaT cells. The developed mt-QSTR model identified three structural features, namely the oxygen percent (OP), aqueous solubility (LogS), and Mulliken's electronegativity (Xc), which were mechanistically interpretable for the toxicity mechanisms of the MeOPs in *Escherichia coli* and HaCaT cells. The mt-QSTR models reliably predicted multiple toxicity endpoints of diverse MeONPs and yielded high correlations in the test data, revealing high predictivity of the proposed model for all the four toxicity endpoints. The authors also analyzed the performance of the proposed mt-QSTR model with the single-target QSTR models, and they are comparable.

9.8 Species and Endpoints for an Interspecies Model

The selection of a species for a specific endpoint is the most important criterion of interspecies modeling. For instance, the US EPA's web-based ICE application employed aquatic and terrestrial species for ecotoxicity assessments [21, 72]. Again, drug-specific toxicity and/or adverse drug reactions must be checked employing mammals as the species to replicate the basic physiological structure of human systems. In Table 9.1, we have provided the most commonly employed species, organisms, or cell lines that can be used for experimental studies for checking the toxicity of diverse NMs. A comprehensive list of accessible data sets for future modeling of interspecies nanotoxicity [3, 4, 12, 51, 73–81] is provided in Table 9.2. What is interesting is that the modeler can choose any two species for specific endpoints to make an interspecies or i-QSTR model with a condition that the studied compounds must possess experimental responses for both species. For example, as discussed earlier, Kar et al. [70] developed the first i-QSTR model for 16 NMs employing data sets 1 and 3 of Table 9.2.

9.9 Future Direction

Risk assessment and management of nanotoxicity is a multidisciplinary task. To reduce and eventually stop the use of huge numbers of animals in toxicity testing of NMs, in silico methods are emerging as cost-effective and time-saving techniques. A toxicological assessment can be achieved efficiently by implementing a "pathway-based" model rather than "empirical testing" according to the National Research Council's report [82]. The mechanism behind the toxicity of NMs, along with genomics for individual species, should be evaluated for the best-possible extrapolation of data. If this step is possible, interspecies correlation models can bridge the gaps in toxicity assessment, allowing data extrapolation for diverse species. There are many areas where more focus is needed to increase confidence in interspecies modeling. The major areas are as follows:

Table 9.1 Species, organisms, or cell lines used as response as well as surrogate species for the development of interspecies correlation models

Species	Organism
Algae	*Anabaena flos-aquae, Chlamydomonas reinhardtii, Chlorella pyrenoidosa, Chlorella vulgaris, Desmodesmus subspicatus, Dunaliella tertiolecta, Microcystis aeruginosa, Microcystis flos-aquae, Minutocellus polymorphus, Phaeodactylum tricornutum, Pseudokirchneriella subcapitata, Scenedesmus acutus, Scenedesmus quadricauda, Sellaphora seminulum, Skeletonema costatum, Thalassiosira pseudonana*
Birds	American crow (*Corvus brachyrhynchos*), American kestrel (*Falco sparverius*), American kestrel (*Falco sparverius*), American robin (*Turdus migratorius*), black-billed magpie (*Pica hudsonia*), boat-tailed grackle (*Quiscalus major*), brown-headed cowbird (*Molothrus ater*), budgerigar (*Melopsittacus undulatus*), California quail (*Callipepla californica*), Canada goose (*Branta canadensis*), chicken (*Gallus gallus*), chukar (*Alectoris chukar*), common grackle (*Quiscalus quiscula*), eastern screech owl (*Megascops asio*), fulvous whistling duck (*Dendrocygna bicolor*), golden eagle (*Aquila chrysaetos*), golden sparrow (*Passer luteus*), golden-crowned sparrow (*Zonotrichia atricapilla*), gray partridge (*Perdix perdix*), house finch (*Carpodacus mexicanus*), house sparrow (*Passer domesticus*), Japanese quail (*Coturnix japonica*), mallard (*Anas platyrhynchos*), mourning dove (*Zenaida macroura*), northern bobwhite (*Colinus virginianus*), Peking duck (*Anas domestica*), red-billed quelea (*Quelea quelea*), red-legged partridge (*Alectoris rufa*), red-winged blackbird (*Agelaius phoeniceus*), ringed turtledove (*Streptopelia risoria*), ring-necked pheasant (*Phasianus colchicus*), rock dove (*Columba livia*), sharp-tailed grouse (*Tympanuchus phasianellus*), starling (*Sturnus vulgaris*), tricolored blackbird (*Agelaius tricolor*), turkey (*Meleagris gallopavo*), white-crowned sparrow (*Zonotrichia leucophrys*)
Invertebrates	Amphipod (*Allorchestes compressa, Gammarus fasciatus, Gammarus lacustris, Gammarus pseudolimnaeus, Hyalella azteca*), crayfish (*Orconectes nais*), daphnid (*Ceriodaphnia dubia, Daphnia magna, Daphnia pulex, Simocephalus serrulatus*), eastern forktail (*Ischnura verticalis*), midge (*Chironomus plumosus, Chironomus tentans, Paratanytarsus dissimilis, Paratanytarsus parthenogeneticus*), mysid (*Americamysis bahia*), northern river crangonyctid (*Crangonyx pseudogracilis*), oligochaete (*Lumbriculus variegatus*), ostracod (*Cypris subglobosa*), Pink shrimp (*Farfantepenaeus duorarum*), polychaete (*Neanthes virens*), ramshorn snail (*Planorbella trivolvis*) snipefly (*Atherix variegata*), starfish (*Asterias forbesi*), stonefly (*Claassenia sabulosa, Pteronarcella badia*)

Species	Organism
Vertebrates	Apache trout (*Oncorhynchus gilae*), Atlantic salmon (*Salmo salar*), Atlantic silverside (*Menidia menidia*), black bullhead (*Ameiurus melas*), black crappie (*Pomoxis nigromaculatus*), bluegill (*Lepomis macrochirus*), bonytail chub (*Gila elegans*), brook trout (*Salvelinus fontinalis*), brown trout (*Salmo trutta*), bull trout (*Salvelinus confluentus*), cape fear shiner (*Notropis mekistocholas*), channel catfish (*Ictalurus punctatus*), chinook salmon (*Oncorhynchus tshawytscha*), coho salmon (*Oncorhynchus kisutch*), Colorado squawfish (*Ptychocheilus lucius*), common carp (*Cyprinus carpio*), cutthroat trout (*Oncorhynchus clarkii*), eastern oyster (*Crassostrea virginica*), fathead minnow (*Pimephales promelas*), flatworm (*Dugesia tigrina*), fountain darter (*Etheostoma fonticola*), Gila topminnow (*Poeciliopsis occidentalis*), goldfish (*Carassius auratus*), green sunfish (*Lepomis cyanellus*), greenthroat darter (*Etheostoma lepidum*), guppy (*Poecilia reticulata*), inland silverside (*Menidia beryllina)*, isopod (*Asellus aquaticus, Caecidotea brevicauda*), lake trout (*Salvelinus namaycush*), largemouth bass (*Micropterus salmoides*), Leon Springs pupfish (*Cyprinodon bovinus*), medaka (*Oryzias latipes*), mosquitofish (*Gambusia affinis*), Mozambique tilapia (*Oreochromis mossambicus*), northern pike (*Esox lucius*), pinfish (*Lagodon rhomboides*), rainbow trout (*Oncorhynchus mykiss*), razorback sucker (*Xyrauchen texanus*), redear sunfish (*Lepomis microlophus*), sheepshead minnow (*Cyprinodon variegatus*), shortnose sturgeon (*Acipenser brevirostrum*), southern leopard frog (*Rana sphenocephala*), spotfin chub (*Erimonax monachus*), threespine stickleback (*Gasterosteus aculeatus*), walleye (*Sander vitreus*), western toad (*Bufo boreas*), white sucker (*Catostomus commersonii*), yellow perch (*Perca flavescens*)
Mammals	Black-tailed jackrabbit (*Lepus californicus*), deer mouse (*Peromyscus maniculatus*), dog (*Canis familiaris*), guinea pig (*Cavia porcellus*), meadow vole (*Microtus pennsylvanicus*), mouse (*Mus musculus*), mule deer (*Odocoileus hemionus*), Norway rat (*Rattus norvegicus*), rabbit (*Oryctolagus cuniculus*), ricefield rat (*Rattus argentiventer*), roof rat (*Rattus rattus*) *Cell lines:* HaCaT cells, BEAS-2B, RAW 264.7 cells, A 549 cells, Wistar rat lung cells, U937 cells, PaCa2 cells

Table 9.2 Available toxicity data set for NPs for interspecies models

Data set	Organism	Endpoint	No. of NMs	NMs	Reference
1	*Escherichia coli*	Cytotoxicity	17	Al_2O_3, Bi_2O_3, CoO, Cr_2O_3, CuO, Fe_2O_3, In_2O_3, La_2O_3, NiO, Sb_2O_3, SiO_2, SnO_2, TiO_2, V_2O_3, Y_2O_3, ZnO, ZrO_2	3
2	*Escherichia coli*	Cytotoxicity	24	Al_2O_3, CuO, CeO_2, Co_3O_4, CoO, Cr_2O_3, Fe_2O_3, Fe_3O_4, Gd_2O_3, HfO_2, In_2O_3, La_2O3, Mn_2O_3, NiO, Ni_2O_3, Sb_2O_3, SiO_2, SnO_2, TiO_2, Y_2O_3, Yb_2O_3, ZnO, ZrO_2, WO_3	73
3	HaCaT cells	Cytotoxicity	19	Al_2O_3, Bi_2O_3, CoO, Cr_2O_3, CuO, Fe_2O_3, In_2O_3, La_2O_3, MnO, NiO, Sb_2O_3, SiO_2, SnO_2, TiO_2, V_2O_3, WO_3, Y_2O_3, ZnO, ZrO_2	51
4	BEAS-2B cells	Cytotoxicity	9	Al_2O_3, CeO_2, Co_3O_4, TiO_2, ZnO, CuO, SiO_2, Fe_3O_4, WO_3	12
5	264.7 RAW cells	Cytotoxicity	8	Ag, Au, Pt, Al_2O_3, Fe_3O_4, ZnO, SiO_2, QD	74
6	A 549 cells	Cytotoxicity	7	Al_2O_3, Cr_2O_3, Mn_2O_3, Fe_2O_3, NiO, CuO, ZnO	75
7	Rat lung cells	Cytotoxicity	42	different forms of ZnO, TiO_2	76

Data set	Organism	Endpoint	No. of NMs	NMs	Reference
8	BEAS-2B and RAW 264.7 cells	Oxidative stress and pulmonary inflammation	24	Al_2O_3, CuO, CeO_2, Co_3O_4, CoO, Cr_2O_3, Fe_2O_3, Fe_3O_4, Gd_2O_3, HfO_2, In_2O_3, La_2O_3, Mn_2O_3, NiO, Ni_2O_3, Sb_2O_3, SiO_2, SnO_2, TiO_2, Y_2O_3, Yb_2O_3, ZnO, ZrO_2, WO_3	4
9	264.7 RAW cells	Inflammatory and genotoxic responses	7	Ag, Au, Pt, Al_2O_3, Fe_3O_4, SiO_2, ZnO	77
10	Zebrafish	Mortality rate, hatching rate, cardiac rate, and morphological defect	7	Ag, Au, Pt, Al_2O_3, ZnO, SiO_2, QD	78
11	Wistar rat lung cells	Lung inflammation	15	Ag, Al_2O_3, CeO_2 (two sizes), Cr_2O_3, CuO (two sizes), Co_3O_4, MgO, NiO, SiO_2, TiO_2 (rutile), TiO_2 (anatase), ZnO (2 sizes)	79
12	U937 cells	Endocytosis	12	Serum-coated Au	80
13	PaCa2 cancer cells	Uptake of NPs in pancreatic cancer cell	109	NPs with the same metal cores and different linkers	81

- Classification and understanding of the mechanistic basis of the NMs in terms of specific species
- The role of the "chemical domain" extracted from the structure of the NM itself to offer dependable predictions of toxicity of untested or new NMs. Therefore, proper attention is needed to determine the location of a new NM with respect to the chemical domain of the applicability of the model while predicting its toxicity employing the accessible data for surrogate species along with modeled parameters [83].
- Consideration of the "mechanistic domain" of the interspecies model as obtainable information considering the MOA of NPs for help in classifying them into "mechanistic classes."
- During model generation, judging the outlier molecule/ molecules using NMs portraying the same mechanistic behavior.
- Suitable information of bio-uptake of NMs, along with their toxicokinetic and toxicodynamics behavior, if considered for pharmaceutical aspects, can assist the interspecies modeling prototype [84, 85].

In upcoming days interspecies or i-QSTR models can be employed as a rational tool for extrapolation and directing the mechanistic aspects for a diverse range of species. Most importantly, bearing in mind the inadequate chemical space enclosed by single predictive models, manifold models can be employed for the creation of consensus predictions of nanotoxicity covering a wider domain.

9.10 Conclusion

A huge number of nano-QSTR models have been developed in the last decade, and the count is on. Interestingly, the majority of the models were based on single species/cell line assay values for a specific endpoint correlating with structural properties along with experimental parameters in some cases lacking the mechanism and/or reason behind the toxicity of particular NPs inside the species. There lies the importance of interspecies/i-QSTR model construction as the MOA-specific interspecies models may assist in the prediction of species sensitivity by dropping improbability in species extrapolation. The toxicity processes are multifaceted and involve both toxicant transport to the target sites of interaction and interactions between the toxicant and receptors at the sites. Not only that the toxicity mechanisms vary from one species to another and the impediment is increased from higher-class to lower-class species. In interspecies models a number of factors may alter the interspecies models, for example, the bio-uptake process of the toxicants as well as the size and shape of NPs as they are not structurally homogeneous. Therefore, a general mechanism of toxicity cannot be anticipated for all, which is an elementary obligation for QSAR modeling.

There is no doubt that to fill the huge number of nanotoxicity data gaps, the i-QSTR model will be employed by cheminformaticians for the extrapolation of toxicity from a surrogate species (experimental assay is present) to a predicted species (without assay data). Thus, it is urgent to construct more i-QSTR models with a large number of species, especially rodents and mammals, for the extrapolation of nanotoxicity data in the field of ecotoxicity and drug discovery supporting the "three Rs" concept. Along with i-QSTR models, further work is required to illustrate mechanistic interspecies read-across models between diverse species. The US EPA Web-ICE module is a promising tool for estimating toxicity values for molecules with missing species-specific data by employing required interspecies correlation models with the highest confidence. Future directions should include the creation of more NM databases with experimental data for major species and classification of them into diverse classes according to their MOAs as well as development of online tools for the generation of reliable and predictive i-QSTR models.

Acknowledgments

SK and JL are thankful to the National Science Foundation (NSF/CREST HRD-1547754 and NSF/RISE HRD-1547836) for financial support. SG wants to thank NSF-DMR-PREM, grant no. 1826886, for financial help.

References

1. Gajewicz, A., Rasulev, B., Dinadayalane, T., Urbaszek, P., Puzyn, T., Leszczynska, D. and Leszczynski, J. (2012). Advancing risk assessment of engineered nanomaterials: application of computational approaches, *Adv. Drug Delivery Rev.*, **64**, 1663–1693.
2. Winkler, D. A., Mombelli, E., Pietroiusti, A., Tran, L., Worth, A., Fadeel, B. and McCall, M. J. (2013). Applying quantitative structure-activity relationship approaches to nanotoxicology: current status and future potential, *Toxicology*, **313**, 15–23
3. Puzyn, T., Rasulev, B., Gajewicz, A., Hu, X., Dasari, T. P., Michalkova, A., Hwang, H. M., Toropov, A., Leszczynska, D. and Leszczynski, J. (2011). Using nano-QSAR to predict the cytotoxicity of metal oxide nanoparticles, *Nat. Nanotechnol.*, **6**, 175–178.
4. Zhang, H., Ji, Z., Xia, T., Meng, H., Low-Kam, C., Liu, R., Pokhrel, S., Lin, S., Wang, X., Liao, Y. P., Wang, M., Li, L., Rallo, R., Damoiseaux, R., Telesca, D., Mädler, L., Cohen, Y., Zink, J. I. and Nel, A. E. (2012). Use of metal oxide nanoparticle band gap to develop a predictive paradigm for oxidative stress and acute pulmonary inflammation, *ACS Nano*, **6**, 4349–4368.
5. Gajewicz, A., Cronin, M. T. D., Rasulev, B., Leszczynski, J. and Puzyn, T. (2015). Novel approach for efficient predictions properties of large pool of nanomaterials based on limited set of species: nano-readacross, *Nanotechnology*, **26**, 015701.
6. van der Jagt, K., Munn, S., Tørsløv, J. and de Bruijn, J. (2004). Alternative approaches can reduce the use of test animals under REACH JRC report EUR 21405.
7. Directive 2010/63/EU of the European Parliament and of the Council on the protection of animals used for scientific purposes, *J. Euro. Union*, L276/33.
8. Roy, K., Kar, S. and Das, R. N. (2015). *Understanding the Basics of QSAR for Applications in Pharmaceutical Sciences and Risk Assessment* (Academic Press).
9. Roy, K., Kar, S. and Das, R. N. (2015). *A Primer on QSAR/QSPR Modeling: Fundamental Concepts*, SpringerBriefs in Molecular Science (Springer).
10. Fourches, D., Pu, D., Tassa, C., Weissleder, R., Shaw, S. Y., Mumper, R. J. and Tropsha, A. (2010). Quantitative nanostructure-activity relationship modeling, *ACS Nano*, **4**, 5703–5712.
11. Fourches, D., Pu, D. and Tropsha, A. (2011). Exploring quantitative nanostructure–activity relationships (QNAR) modelling as a tool for

predicting biological effects of manufactured nanoparticles, *Comb. Chem. High Throughput Screening*, **14**, 217–225.

12. Liu, R., Rallo, R., George, S., Ji, Z., Nair, S., Nel, A. E. and Cohen, Y. (2011). Classification nanoSAR development for cytotoxicity of metal oxide nanoparticles, *Small*, **7**, 1118–1126.
13. Toropov, A. A., Toropova, A. P., Benfenati, E., Gini, G., Puzyn, T., Leszczynska, D. and Leszczynski, J. (2012). Novel application of the CORAL software to model cytotoxicity of metal oxide nanoparticles to bacteria Escherichia coli, *Chemosphere*, **89**, 1098–1102.
14. Epa, V. C., Burden, F. R., Tassa, C., Weissleder, R., Shaw, S. and Winkler, D. A. (2012). Modeling biological activities of nanoparticles, *Nano Lett.*, **12**, 5808–5812.
15. Kar, S., Gajewicz, A., Puzyn, T. and Roy, K. (2014). Nano-quantitative structure–activity relationship modeling using easily computable and interpretable descriptors for uptake of magnetofluorescent engineered nanoparticles in pancreatic cancer cells, *Toxicol. in Vitro.*, **28**, 600–606.
16. Kar, S., Gajewicz, A., Puzyn, T., Roy, K. and Leszczynski, J. (2014). Periodic table-based descriptors to encode cytotoxicity profile of metal oxide nanoparticles: a mechanistic QSTR approach, *Ecotoxicol. Environ. Saf.*, **107**, 162–169.
17. Carbó-Dorca, R. and Besalù, E. (2011). Construction of coherent nano quantitative structure-properties relationships (nano-QSPR) models and catastrophe theory, *SAR QSAR Environ. Res.*, **22**, 661–665.
18. Xia, X. R., Monteiro-Riviere, N. A., Mathur, S., Song, X., Xiao, L., Oldenberg, S. J., Fadeel, B. and Riviere, J. E. (2011). Mapping the surface adsorption forces of nanomaterials in biological systems, *ACS Nano*, **5**, 9074–9081.
19. Liu, R., Zhang, H. Y., Ji, Z. X., Rallo, R., Xia, T., Chang, C. H., Nel, A. and Cohen, Y. (2013). Development of structure-activity relationship for metal oxide nanoparticles, *Nanoscale*, **5**, 5644–5653.
20. Shao, C. Y., Chen, S. Z., Su, B. H., Tseng, Y. J., Esposito, E. X. and Hopfinger, A. J. (2013). Dependence of QSAR models on the selection of trial descriptor sets: a demonstration using nanotoxicity endpoints of decorated nanotubes, *J. Chem. Inf. Model.*, **53**, 142–158.
21. Dyer, S. D., Versteeg, D. J., Belanger, S. E., Chaney, J. G. and Mayer, F. L. (2006). Interspecies correlation estimates predict protective environmental concentrations, *Environ. Sci. Technol.*, **40**, 3102–3111.
22. Kar, S. and Roy, K. (2012). First report on development of quantitative interspecies structure-carcinogenicity relationship models and

exploring discriminatory features for rodent carcinogenicity of diverse organic chemicals using OECD guidelines, *Chemosphere*, **87**, 339–355.

23. Raimondo, S., Jackson, C. R. and Barron, M. G. (2010). Influence of taxonomic relatedness and chemical mode of action in acute interspecies estimation models for aquatic species, *Environ. Sci. Technol.*, **44**, 7711–7716.

24. Nel, A. (2005). Air pollution-related illness: effects of particles, *Science*, **308**, 804–806.

25. Panessa-Warren, B. J., Warren, J. B., Maye, M. M., van der Lelie, D., Gang, O., Wong, S. S., Ghebrehiwet, B., Tortora, G. T. and Misewich, J. A. (2008). Human epithelial cell processing of carbon and gold nanoparticles, *Int. J. Nanotechnol.*, **5**, 55–91.

26. Cui, D., Tian, F., Ozkan, C. S., Wang, M. and Gao, H. (2005). Effect of single wall carbon nanotubes on human HEK293 cells, *Toxicol. Lett.*, **155**, 73–85.

27. Wiesner, M. R., Lowry, G. V., Alvarez, P., Dionysiou, D. and Biswas, P. (2006). Assessing the risks of manufactured nanomaterials, *Environ. Sci. Technol.*, **40**, 4336–4345.

28. Sharma, V., Singh, S. K., Anderson, D., Tobin, D. J. and Dhawan, A. (2011). Zinc oxide nanoparticle induced genotoxicity in primary human epidermal keratinocytes, *J. Nanosci. Nanotechnol.*, **11**, 3782–3788.

29. Colvin, V. L. (2003). The potential environmental impact of engineered nanomaterials, *Nat. Biotechnol.*, **21**, 1166–1170.

30. Yamakoshi, Y., Umezawa, N., Ryu, A., Arakane, K., Miyata, N., Goda, Y., Masumizu, T. and Nagano, T. (2003). Active oxygen species generated from photoexcited fullerene (C60) as potential medicines: O2-* versus 1O2, *J. Am. Chem. Soc.*, **125**, 12803–12809.

31. Magrez, A., Kasas, S., Salicio, V., Pasquier, N., Seo, J. W., Celio, M., Catsicas, S., Schwaller, B. and Forro, L. (2006). Cellular toxicity of carbon-based nanomaterials, *Nano Lett.*, **6**, 1121–1125.

32. Sayes, C. M., Fortner, J. D., Guo, W., Lyon, D., Boyd, A. M., Ausman, K. D., Tao, Y. J., Sitharaman, B., Wilson, L. J., Hughes, J. B., West, J. L. and Colvin, V. L. (2004). The differential cytotoxicity of water-soluble fullerenes, *Nano Lett.*, **4**, 1881–1887.

33. Chen, Y., Chen, J., Dong, J. and Jin, Y. (2004). Comparing study of the effect of nanosized silicon dioxide and microsized silicon dioxide on fibrogenesis in rats, *Toxicol. Ind. Health*, **20**, 21–27.

34. Sanchez, V. C., Jachak, A., Hurt, R. H. and Kane, A. B. (2012). Biological interactions of graphene-family nanomaterials: an interdisciplinary review, *Chem. Res. Toxicol.*, **25**, 15–34.

35. Muller, J., Huaux, F. and Lison, D. (2006). Respiratory toxicity of carbon nanotubes: how worried should we be? *Carbon*, **44**, 1048–1056.

36. Kayat, J., Gajbhiye, V., Tekade, R. K. and Jain, N. K. (2011). Pulmonary toxicity of carbon nanotubes: a systematic report, *Nanomed. Nanotechnol. Biol. Med.*, **7**, 40–49.

37. Mutlu, G. M., Budinger, G. R., Green, A. A., Urich, D., Soberanes, S., Chiarella, S. E., Alheid, G. F., McCrimmon, D. R., Szleifer, I. and Hersam, M. C. (2010). Biocompatible nanoscale dispersion of single-walled carbon nanotubes minimizes in vivo pulmonary toxicity, *Nano Lett.*, **10**, 1664–1670.

38. Torrens, F. (2005). Calculations on solvents and co-solvents of single-wall carbon nanotubes: cyclopyranoses, *Nanotechnology*, **16**, S181–S189.

39. Cedervall, T., Lynch, I., Lindman, S., Berggård, T., Thulin, E., Nilsson, H., Dawson, K. A. and Linse, S. (2007). Understanding the nanoparticle-protein corona using methods to quantify exchange rates and affinities of proteins for nanoparticles, *PNAS*, **104**, 2050–2055

40. Horie, M., Kato, H., Fujita, K., Endoh, S. and Iwahashi, H. (2012). In vitro evaluation of cellular response induced by manufactured nanoparticles, *Chem. Res. Toxicol.*, **25**, 605–619.

41. Karlsson, K. L., Cronholm, P., Gustafsson, J. and Möller, L. (2008). Copper oxide nanoparticles are highly toxic: a comparison between metal oxide nanoparticles and carbon nanotubes, *Chem. Res. Toxicol.*, **21**, 1726–1732.

42. Fahmy, B. and Cormier, S. A. (2009). Copper oxide nanoparticles induce oxidative stress and cytotoxicity in airway epithelial cells, *Toxicol. in Vitro*, **23**, 1365–1371.

43. Leszczynski, J. (2010). BIONANOSCIENCE nano meets bio at the interface, *Nat. Nanotechnol.*, **5**, 633–634.

44. Lacerda, L., Bianco, A., Prato, M. and Kostarelos, K. (2006). Carbon nanotubes as nanomedicines: from toxicology to pharmacology, *Adv. Drug Delivery Rev.*, **58**, 1460.

45. Sun, W. and Bu, Y. (2011). On the binding strength sequence for nucleic acid bases and C60 with density functional and dispersion-corrected density functional theories: whether C60 could protect nucleic acid

bases from radiation-induced damage, *J. Phys. Chem. C*, **115**, 3220–3228.

46. Obata, S. and Honda, K. (2011). Dynamic behavior of carbon nanotube and bio-/artificial surfactants complexes in an aqueous environment, *J. Phys. Chem. C*, **115**, 19659–19667.

47. Johnson, R. R. Kohlmeyer, A., Johnson, A. T. C. and Klein, M. L. (2009). Free energy landscape of a DNA–carbon nanotube hybrid using replica exchange molecular dynamics, *Nano Lett.*, **9**, 537–541.

48. Puzyn, T., Leszczynska, D. and Leszczynski, J. (2009). Toward the development of "nano-QSARs": advances and challenges, *Small* **5**, 2494–2509.

49. Moore, M. N. (2006). Do nanoparticles present ecotoxicological risks for the health of the aquatic environment?, *Environ. Int.*, **32**, 967–976.

50. Sizochenko, N., Rasulev, B., Gajewicz, A., Kuz'min, V., Puzyn, T. and Leszczynski, J. (2014). From basic physics to mechanisms of toxicity: the "liquid drop" approach applied to develop predictive classification models for toxicity of metal oxide nanoparticles, *Nanoscale*, **6**, 13986–13993.

51. Gajewicz, A., Schaeublin, N., Rasulev, B., Hussain, S., Leszczynska, D., Puzyn, T. and Leszczynski, J. (2015). Towards understanding mechanisms governing cytotoxicity of metal oxides nanoparticles: hints from nano-QSAR studies, *Nanotoxicology*, **9**, 313–325.

52. Roy, K. and Kar, S. (2016). In silico models for ecotoxicity of pharmaceuticals, in *In Silico Methods for Predicting Drug Toxicity*, Benfenati, E., ed., Methods in Molecular Biology, Vol. 1425 (Humana Press, New York), pp. 237–304.

53. Kar, S. and Roy, K. (2012). Risk assessment for ecotoxicity of pharmaceuticals - an emerging issue, *Expert Opin. Drug Saf.*, **11**, 235–274.

54. Asfaw, A., Ellersieck, M. R. and Mayer, F. L. (2003). Interspecies Correlation Estimations (ICE) for Acute Toxicity to Aquatic Organisms and Wildlife. II. User Manual and Software, EPA/600/R-03/106; United States Environmental Protection Agency, Washington, DC.

55. Netzeva, T. H., Pavan, M. and Worth, A. P. (2008). Review of (quantitative) structure-activity relationships for acute aquatic toxicity, *QSAR Comb. Sci.*, **27**, 77–90.

56. Cronin M. T. D., Netzeva T. I., Dearden J. C., Edwards, R. and Worgan A. D. P. (2004). Assessment and modeling of the toxicity of organic

chemicals to Chlorella vulgaris: development of a novel database, *Chem. Res. Toxicol.*, **17**, 545–554.

57. Zhang, X. J., Qin, H. W., Su, L. M., Qin, W. C., Zou, M. Y., Sheng, L. X., Zhao, Y. H. and Abraham, M. H. (2010). Interspecies correlations of toxicity to eight aquatic organisms: theoretical considerations, *Sci. Total Environ.*, **408**, 4549–4555.

58. LeBlanc, G. A. (1984). Interspecies relationships in acute toxicity of chemicals to aquatic organisms, *Environ. Toxicol. Chem.*, **3**, 47–60.

59. Kahn, I., Maran, U., Benfenati, E., Netzeva, T. I., Schultz, T. W. and Cronins, M. T. D. (2007). Comparative quantitative structure-activity-activity relationships for toxicity to *Tetrahymena pyriformis* and *Pimephalespromelas*, *Altern. Lab. Anim.*, **35**, 15–24.

60. Nendza, M. and Wenzel, A. (2006). Discriminating toxicant classes by mode of action: 1. (Eco)toxicity profiles, *Environ. Sci. Pollut. Res.*, **13**, 192–203.

61. Dimitrov, S., Koleva, Y., Schultz, T. W., Walker, J. D. and Mekenyan, O. (2004). Interspecies quantitative structure-activity relationship model for aldehydes: aquatic toxicity, *Environ. Toxicol. Chem.*, **23**, 463–470.

62. Zvinavashe, E., Du, T., Griff, T., van den Berg, H. H. J., Soffers, A. E. M. F., Vervoort, J., Murk, A. J. and Rietjens, I. M. C. M. (2009). Quantitative structure-activity relationship modeling of the toxicity of organothiophosphate pesticides to *Daphnia magna* and *Cyprinuscarpio*, *Chemosphere*, **75**, 1531–1538.

63. Kar, S. and Roy, K. (2010). First report on interspecies quantitative correlation of ecotoxicity of pharmaceuticals, *Chemosphere*, **81**, 738–747.

64. Kar, S. and Roy, K. (2012). First report on development of quantitative interspecies structure carcinogenicity relationship models and exploring discriminatory features for rodent carcinogenicity of diverse organic chemicals using OECD guidelines, *Chemosphere*, **87**, 339–355.

65. Qi, P., Wang, Y., Mu, J. and Wang, J. (2011). Aquatic predicted no-effect-concentration derivation for perfluorooctane sulfonic acid, *Environ. Toxicol. Chem.*, **30**, 836–842.

66. Raevsky, O. A., Razdolskii, A. N., Liplavskii, Ya. V., Raevskaya, O. E. and Yarkov, A. V. (2012). Acute toxicity evaluation upon intravenous injection into mice: interspecies correlations, lipophilicity parameters, and physicochemical descriptors, *Pharm. Chem. J.*, **46**, 3–8.

67. Cassani, S., Kovarich, S., Papa, E., Roy, P. P., van der Wal, L. and Gramatica, P. (2013). Daphnia and fish toxicity of (benzo)triazoles:

validated QSAR models, and interspecies quantitative activity–activity modeling, *J. Hazard. Mater.*, **258–259**, 50–60.

68. Roy, K., Das, R. N. and Popelier, P. L. A. (2015). Predictive QSAR modeling of algal toxicity of ionic liquids and its inter-species correlation with Daphnia toxicity, *Environ. Sci. Pollut. Res.*, **22**, 6634–6641.
69. Furuhama, A., Hasunuma, K. and Aoki, Y. (2015). Interspecies quantitative structure-activity-activity relationships (QSAARs) for prediction of acute aquatic toxicity of aromatic amines and phenols, *SAR QSAR Environ. Res.*, **26**, 301–323.
70. Kar, S., Gajewicz, A., Roy, K., Leszczynski, J. and Puzyn, T. (2016) Extrapolating between toxicity endpoints of metal oxide nanoparticles: predicting toxicity to *Escherichia coli* and human keratinocyte cell line (HaCaT) with Nano-QTTR, *Ecotox. Environ. Saf.*, **126**, 238–244.
71. Basant, N. and Gupta, S. (2017). Multi-target QSTR modeling for simultaneous prediction of multiple toxicity endpoints of nano-metal oxides, *Nanotoxicology*, **11**, 339–350.
72. US EPA (2009). Integrated risk information system. US Environmental Protection Agency, National Center for Environmental Assessment, Washington, DC. Available from <http://www.epa.gov/iris/>
73. Kaweeteerawat, C., Ivask, A., Liu, R., Zhang, H., Chang, C. H., Low-Kam, C., Fischer, H., Ji, Z., Pokhrel, S., Cohen, Y., Telesca, D., Zink, J., Mädler, L., Holden, P.A., Nel, A. and Godwin, H. (2015). Toxicity of metal oxide nanoparticles in Escherichia coli correlates with conduction band and hydration energies, *Environ. Sci. Technol.*, **49**, 1105–1112.
74. Patel, T., Telesca, D., George, S. and Nel, A. E. (2012). Toxicity profiling of engineered nanomaterials via multivariate dose-response surface modeling, *Ann. Appl. Stat.*, **6**, 1707–1729.
75. Chusuei, C. C., Wu, C. H., Mallavarapu, S., Hou, F. Y. S., Hsu, C.-M., Winiarz, J. G., Aronstam, R. S. and Huang, Y. W. (2013). Cytotoxicity in the age of nano: the role of fourth period transition metal oxide nanoparticle physicochemical properties, *Chem. Biol. Interact.*, **206**, 319–326.
76. Papa, E., Doucet, J. P. and Doucet-Panaye, A. (2015). Linear and non-linear modelling of the cytotoxicity of TiO2 and ZnO nanoparticles by empirical descriptors, *SAR QSAR Environ. Res.*, **26**, 647–665.
77. Rallo, R., France, B., Liu, R., Nair, S., George, S., Damoiseaux, R., Giralt, F., Nel, A., Bradley, K. and Cohen, Y. (2011). Self-organizing map analysis of toxicity-related cell signaling pathways for metal and metal oxide nanoparticles, *Environ. Sci. Technol.*, **45**, 1695–1702.

78. George, S., Xia, T., Rallo, R., Zhao, Y., Ji, Z., Lin, S., Wang, X., Zhang, H., France, B., Schoenfeld, D., Damoiseaux, R., Liu, R., Lin, S., Bradley, K. A., Cohen, Y. and Nel, A. (2011). Use of a high-throughput screening approach coupled with in vivo zebrafish embryo screening to develop hazard ranking for engineered nanomaterials, *ACS Nano*, **5**, 1805–1817.

79. Cho, W. S., Duffin, R., Thielbeer, F., Bradley, M., Megson, I. L., MacNee, W. Poland, C. A., Tran, C. L., and Donaldson, K. (2012). Zeta potential and solubility to toxic ions as mechanisms of lung inflammation caused by metal/metal oxide nanoparticles, *Toxicol. Sci.*, **126**, 469–477.

80. Oh, N. and Park, J. H. (2014). Surface chemistry of gold nanoparticles mediates their exocytosis in macrophages, *ACS Nano*, **8**, 6232–6241.

81. Weissleder, R., Kelly, K., Sun, E. Y., Shtatland, T. and Josephson, L. (2005). Cell-specific targeting of nanoparticles by multivalent attachment of small molecules, *Nat. Biotechnol.*, **23**, 1418–1423.

82. National Research Council (2007). Toxicity testing in the 21st century: a vision and a strategy. The National Academies, Washington, DC.

83. Roy, K. and Kar, S. (2015). Importance of applicability domain of QSAR models, in *Quantitative Structure-Activity Relationships in Drug Design, Predictive Toxicology, and Risk Assessment*, Roy, K., ed. (IGI Global), pp. 180–211.

84. Burgess-Herbert, S. L. and Euling, S. Y. (2013). Use of comparative genomics approaches to characterize interspecies differences in response to environmental chemicals: challenges, opportunities, and research needs, *Toxicol. Appl. Pharmacol.*, **271**, 372–385.

85. Kar, S., Das, R. N., Roy, K. and Leszczynski, J. (2016). Can toxicity for different species be correlated?: The concept and emerging applications of interspecies quantitative structure-toxicity relationship (i-QSTR) modeling, *Int. J. Quant. Struct.-Prop. Relat.*, **1**, 23–51.

Chapter 10

On Error Measures for Validation and Uncertainty Estimation of Predictive QSAR Models

Supratik Kar,[a] Kunal Roy,[b] and Jerzy Leszczynski[a]
[a]*Interdisciplinary Center for Nanotoxicity, Department of Chemistry and Biochemistry, Jackson State University, Jackson, MS 39217, USA*
[b]*Drug Theoretics and Cheminformatics Laboratory, Department of Pharmaceutical Technology, Jadavpur University, Kolkata 700032, India*
supratik.kar@icnanotox.org; kunalroy_in@yahoo.com; jerzy@icnanotox.org

Quantitative structure–activity, structure–property, and structure–toxicity relationship (QSAR, QSPR, QSTR) models being used for the prediction of activity, property, and toxicity endpoints of untested chemicals can be exploited for the prioritization plan of synthesis and experimental testing as well as filling data gaps. These models have manifold applications in drug discovery, environmental fate modeling, risk assessment, and property prediction of chemicals and pharmaceuticals. Validation has been recognized as one of the decisive steps for checking the robustness, predictability, and reliability of any QSAR model in order to judge the acceptability of

Computational Nanotoxicology: Challenges and Perspectives
Edited by Agnieszka Gajewicz and Tomasz Puzyn

ISBN 978-981-4800-64-8 (Hardcover), 978-0-429-34137-3 (eBook)
www.jennystanford.com

each step during model development, such as assessing the quality of data, diversity of the data set, predictability on internal and external sets, domain of applicability, and mechanistic interpretability. Validation strategies are largely dependent on the assessment of model quality through various validation metrics. External validation is often considered as the crucial process in evaluating the excellence of predictions of any QSAR model judged through diverse validation metrics, which can be broadly categorized into two key classes, viz, correlation-based metrics (R^2_{test}, R^2_{pred} or Q^2_{F1}, and Q^2_{F2}) and purely error-based measures, like predicted residual sum of squares (PRESS), root mean square error (RMSE), and mean absolute error (MAE). In view of the importance of QSAR validation approaches and different validation parameters in the development of successful and acceptable QSAR models, this chapter is focused on error- as well as correlation-based metrics employed to judge the quality of the regression as well as classification-based QSAR models. Additionally, parameters for model reliability, uncertainty aspects, and prediction confidence are thoroughly discussed, along with the applicability domain and randomization test studies. Different open-source software tools for diverse validation purposes are also mentioned in this chapter.

10.1 Introduction

Quantitative structure–activity relationships (QSARs) have received tremendous attention in drug discovery, materials designing, property and toxicity prediction, and environmental fate modeling of chemicals and pharmaceuticals [1–4]. The term "QSAR" is interchangeably used with QSPR/QSTR (quantitative structure–property relationship and quantitative structure–toxicity relationship) by many researchers when the response or dependent variable is property and toxicity, respectively. Not only that, on the basis of the type of compounds modeled, new derived terms are generated, like nano-QSAR and QNAR (quantitative nanostructure–activity relationship), for nanomaterials modeling [5–8]. Searching using the terms "QSAR," "QSPR," and "QSTR" in Scopus within the time frame from 2007 to 2016, one can find a total of 11,549 publications, which is no doubt a huge number. A graphical yearwise distribution of all publications is illustrated in Fig. 10.1. A good number of QSAR articles demonstrated the following objectives [4]: (i) design of new chemicals or pharmaceuticals with enhanced property or activity profile, (ii) improved understanding and exploration of the mode of action of chemicals and pharmaceuticals regarding their response and toxicity, (iii) optimization of the lead molecule among congeners with reduced toxicity, (iv) rationalization of synthetic experimentation as an alternative to the medium-throughput in vitro and low-throughput in vivo assays, (v) cutback in cost, time, and manpower constraint through the development of more effective and less exhaustive approaches, and finally (vi) an alternative to animal experimentation in conformity with the REACH[1] guidelines [9] and the 3Rs concept [10] signifying "reduction, replacement, and refinement" of animal experiments. Therefore, without any doubt, the QSAR technique has emerged as an efficient tool for the designing, development, and/or screening of new drug molecules/chemicals.

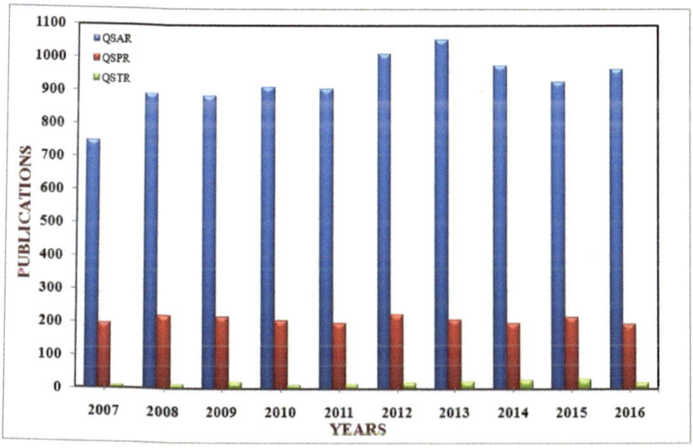

Figure 10.1 Graphical yearwise illustration of the number of publications reporting QSAR, QSPR, and QSTR terms.

Different regulatory agencies worldwide are implementing QSAR models to assess the response, physiological properties, and toxicity of diverse chemicals for decision-making frameworks in chemical safety, environmental risk measures, and management

[1]Registration, Evaluation and Authorisation of Chemicals

[2, 4]. Occasionally, QSAR models are classified into two major categories, regression-based QSAR models for quantitative analysis and classification-based QSAR models for qualitative explanation. To ensure the reliability of a QSAR model for its prediction capability for new or untested molecules, a validation process has an immense role to play, and it is recognized as one of the backbone steps for QSAR model development [11–13]. It is well accepted that validation is a more holistic process for assessing the quality of data and the applicability and mechanistic interpretability of the developed model [14]. Various methodological aspects of validation of QSARs have been the subject of strong debate within the academic and regulatory communities. In many cases, correlation and error-based metrics are largely considered for the identification of the best model in terms of goodness of fit and predictive quality. Interestingly, a number of questions are raised before validation and application of a QSAR model:

- Which validation approaches should be considered?
- What is the primary criterion for setting up the scientific validity of a QSAR model?
- How should QSAR models be used for regulatory purposes?
- Is it possible to use any QSAR model for new molecules?

Here arises the importance of the applicability domain (AD), which allows one to estimate the uncertainty in the prediction of a particular compound on the basis of how similar it is to the training compounds that are used in the model development.

In view of the size and extent of different QSAR validation approaches, we herein focus on error- as well as correlation-based validation metrics in the generation of reliable and acceptable QSAR models. Definitions and thorough explanations of different metrics are provided for classification as well as regression-based QSAR models. Correlation- and error-based measures and their importance are explained to identify and choose the best predictive models. Along with the validation metrics, the important concepts of conformal predictions, uncertainty estimation of predictive QSAR models, and the AD approach to building acceptable QSAR models are discussed meticulously. This chapter will provide QSAR learners as well as experts an aerial perspective on the available validation techniques and their corresponding metrics, along with available open access software tools, for evaluating predictive models.

10.2 Concept and Significance of Validation of QSAR Models

Fast computation time and open access to many cheminformatics tools make it possible to calculate a good number of properties (descriptors). Furthermore, using a variety of optimization measures, it is now possible to find models that can fit well with the experimental data, but there may remain a considerable risk of overfitting. Fitting of data does not confirm the prediction capability of a model; it only provides goodness of fit (GoF). This necessitates validation of the developed models in terms of predictivity and robustness. A QSAR model is judged for its inherent statistical quality as well as its competence to predict endpoints of compounds that are not considered under model generation. Rigorously validated QSAR models have been considered reliable by researchers and regulatory bodies for practical application in synthesis prioritization, risk assessment, and risk management [15, 16]. Statistical model quality (MQ) can be defined as a function of GoF, validation (V), and predictivity (P):

$$MQ = f(\text{GoF}, V, P) \qquad (10.1)$$

Interestingly, the definition of this function is not so straightforward. The major problem is to identify the set of parameters for measuring GoF, validation, and predictivity. Significantly, there is no single parameter that can be used for explaining all the mentioned functions.

As a consequence, the Organization for Economic Cooperation and Development (OECD) [8] has agreed to five principles that should be followed to set up the scientific validity of a QSAR model, thereby facilitating its recognition for regulatory purposes. The OECD has also presented a checklist to offer the required path to the interpretation of the mentioned principles [17, 18]. The OECD principle 4 refers to the demand to establish "appropriate measures of goodness-of-fit, robustness and predictivity" for any QSAR model. It identifies the need to validate a model through internal validation (as represented by GoF and robustness) as well as external validation of the model (predictivity). Validation techniques are expressed in the form of validation metrics to show the quality and predictability of a model. Therefore, the present challenge in the construction of a QSAR model is not only to build a statistically sound and robust model to predict the response within the domain of applicability defined by the training set but also to develop a model with the capacity to precisely predict the activity of untested chemicals.

10.3 Validation Strategies

Validation strategies not only determine the inherent statistical quality of the model but also are recognized as the decisive steps to ensuring the acceptability of the constructed models for their probable use on a new set of data, in order to judge the confidence of predictions. The most frequently employed validation strategies are discussed next [19].

10.3.1 Internal Validation or Cross-Validation

Internal validation is performed with the compounds involved in the QSAR model development process [19, 20]. It involves activity prediction of the studied compounds followed by an estimation of parameters for detecting the precision of predictions. The cross-validation technique mainly involves internal validation, where a sample of n observations is partitioned into calibration (i.e., training) and validation (i.e., test) subsets. The calibration subset is used to build a model, while the validation subset is employed to test how well the model predicts the new data that are not utilized in the

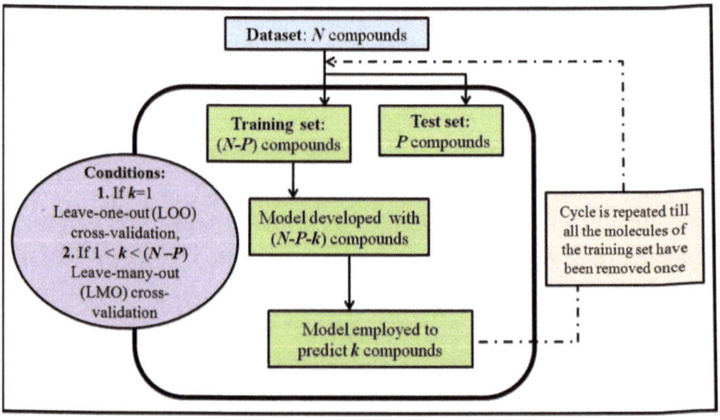

Figure 10.2 Steps for the calculation of LOO and LMO cross-validation.

calibration procedure. To judge the quality and GoF of the model, internal validation is a perfect approach. Most frequently leave-one-out (LOO) and leave-many-out (LMO) techniques are employed in the internal validation approach (Fig. 10.2). But the important shortcoming of this strategy is the lack of predictability of the model when it is applied to an entirely new data set.

10.3.2 External Validation

As internal validation judges the chemicals belonging to the same set of compounds used for the development of the model, one cannot confidently evaluate the predictive capacity of the developed model from internal validation. In many cases, truly external data points being unavailable for prediction reason, the original data set is divided into training and test sets [20, 21]. Consequently, a model is developed with the training set and then the developed model is used to perform the external validation utilizing the test set molecules that are not employed in the model development process. External validation makes sure the predictability and applicability of the developed QSAR model for the prediction of untested and new compounds. From this perspective, the choice of training and test sets should be based on the proximity of the test set members to representative points of the training set in the multidimensional descriptor space [21]. There are three approaches available for the selection of the training and test sets. They are as follows:

- Random selection
- Based on *Y*-response
- Based on *X*-descriptor matrix

Following the third approach mentioned above, most commonly used tools for coherent division of the data set into training and test sets are:

- *k*-Means clustering
- Kohonen's self-organizing map selection
- Principal component analysis (PCA)
- Kennard–Stone selection
- Euclidean distance–based selection
- Sphere exclusion
- Extrapolation-oriented test set selection

10.3.3 True External Validation

This refers to the application of the QSAR model on a data set that is not considered for model development and is obtained from a different source with a similar endpoint, for which the model has been developed. Unfortunately, in the majority of cases, an external data set is absent for the same endpoint to ensure the predictivity of the developed model. In such a case, the original data set is divided into training and test sets for the external validation purpose.

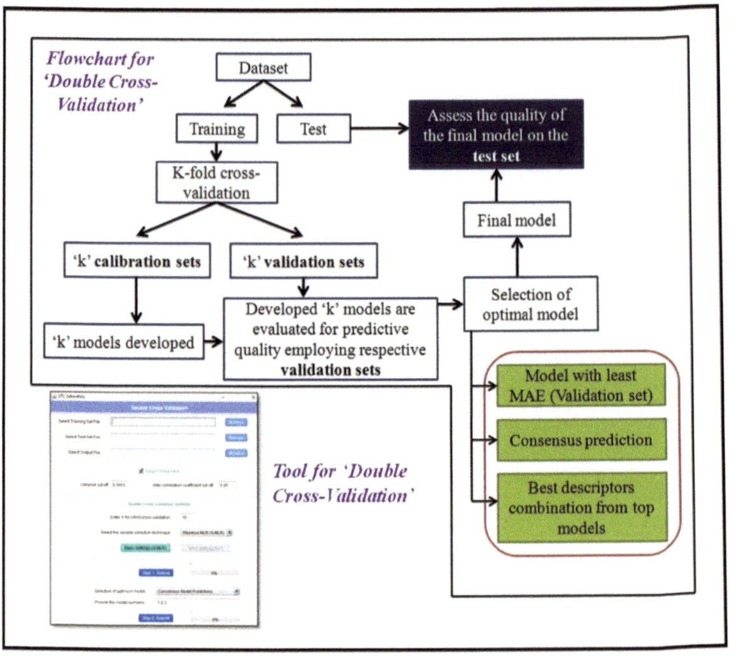

Figure 10.3 Flowchart and open access software tool for double cross-validation.

10.3.4 Double Cross-Validation

As discussed earlier, external validation is generally performed by employing the hold-out method, where the data set is divided into training and test sets; the training set is employed for model building and model selection, while the test set is solely used in model assessment. As the composition of the training set remains fixed in this technique, it is not definite that the resultant model will be optimal as there may be a bias in the descriptor choice. This problem is more obvious for the multiple linear regression (MLR) models than for the more robust and generalized partial least squares and principal component regression models. Thus, employing the double cross-validation technique could be a better choice, in which the training set is further divided into n calibration and validation sets, resulting in diverse compositions [22]. How the best models can be selected is discussed elsewhere [22]. A detailed flowchart of each step and the available open access tool for double cross-validation is presented in Fig. 10.3.

10.3.5 Data Randomization or Y-Scrambling

Randomization or Y-scrambling (often considered under internal validation) executed on the data matrix provides a valuable tool for evaluating the existence of any chance correlation in the QSAR model [23]. A detailed discussion has been provided in Section 10.7.

10.3.6 Bias and Variance in Prediction Errors

It is important to analyze prediction errors of test set compounds in order to search for the presence of any systematic error or bias in model predictions [24]. A QSAR model shows the change of a response (dependent variable Y) with the changes in values of physicochemical properties and/or structural features (independent variables $X_1, X_2, \ldots, X_n$). Therefore, a basic QSAR relationship can be expressed as follows:

$$Y = f(X_1, X_2, \ldots, X_n) \tag{10.2}$$

Considering the relationship a linear one, the approximated relationship obtained from least-squares regression analysis may look like:

$$Y = a_0 + a_1X_1 + a_2X_2 + \ldots + X_n \tag{10.3}$$

In Eq. 10.3, a_0, a_1, a_2, etc., are regression coefficients. It is also to be understood that the true response is unknown, while one only uses a measure of the true value as the observed value. Therefore,

$$Y_{\text{obs(experimental)}} = Y_{\text{true(unknown)}} + \text{measurement error(noise)} \tag{10.4}$$

The measurement error is hidden in the observed values, and we cannot get the true errors from model predictions as the true Y values are unknown to us. The residuals (prediction errors) have two components: (i) error due to bias (systematic error) and (ii) error due to variance (random error). The true (squared) error can be split into bias, variance, and noise in the following manner:

$$\text{Squared error} = \text{Variance} + \text{bias}^2 + \text{noise} \tag{10.5}$$

As the true response is unknown to us, the relationship given above can be approximated as follows:

$$\text{Square residuals} = \text{Variance} + \text{bias}^2 \tag{10.6}$$

The terms "bias^2," "variance," and mean square error (MSE) can be computed using the following expressions:

$$\text{Bias}^2 = \frac{1}{N_\text{c}} \sum_{i=1}^{N_\text{c}} \left(\overline{Y}^\text{B}_{\text{pred}(i)} - Y_{\text{obs}(i)} \right)^2 \tag{10.7}$$

$$\overline{Y}^\text{B}_{\text{pred}(i)} = \frac{\sum_{j=1}^{N_\text{B}} Y^{\text{B}(j)}_{\text{pred}(i)}}{N_\text{B}} \tag{10.8}$$

$$\text{Variance} = \frac{1}{N_\text{c}} \sum_{i=1}^{N_\text{c}} \frac{1}{N_\text{b}} \sum_{j=1}^{N_\text{B}} \left(Y^{\text{B}(j)}_{\text{pred}(i)} - \overline{Y}^\text{B}_{\text{pred}(i)} \right)^2 \tag{10.9}$$

$$\text{MSE} = \sum_{i=1}^{N_\text{C}} \frac{\left(Y_{\text{obs}(i)} - Y_{\text{pred}(i)} \right)^2}{N_\text{C}} \tag{10.10}$$

It is understandable that meager consideration of the ratio of the number of positive errors (NPEs) and the number of negative errors (NNEs) is not adequate. It is also required to compare the absolute (ABS) value of average error (AE) with the average absolute error (AAE), to compare the mean positive error (MPE) with the mean negative error (MNE), and to verify the occurrence of any serial correlation among residuals and correlation of the residuals with the response values. Considering these aspects, Roy et al. [24] have implemented the following criteria for determining any systematic error in quantitative predictions:

- NPE/NNE > 5 or NNE/NPE > 5
- ABS(MPE/MNE) > 2 or ABS(MNE/MPE) > 2
- AAE – ABS(AE) < 0.5 × AAE

- R^2(i^{th} vs. $(i-1)^{th}$ residuals) > 0.5 for residuals sorted on Y_{obs}
- R^2(Y vs. residuals) > 0.5

If any of the above conditions are fulfilled, then an analysis of prediction errors for additional test set compounds is recommended when the number of test set compounds is not more than 30; else the model demands rebuilding to eliminate the bias in prediction errors (and rejected in its current form for quantitative predictions) when the number of test set compounds is above 30. A schematic diagram is illustrated in Fig. 10.4 to show the steps for identifying any systematic error in predictions.

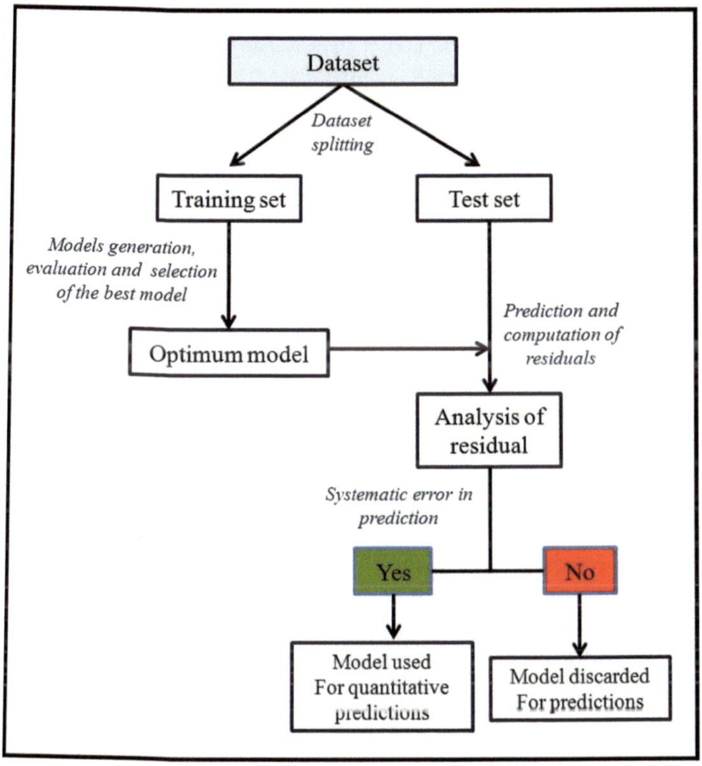

Figure 10.4 Estimation of prediction errors to recognize any systematic error in predictions.

Roy et al. [24] have also developed a software tool named XternalValidationPlus (downloadable from http://teqip.jdvu.ac.in/QSAR_Tools/ and http://dtclab.webs.com/software-tools) for determining the quality of QSAR-derived predictions. This software uses a strategy where there is a serious possibility of the presence of a systematic error in predictions if any one or more of the above conditions are met, provided that a sufficient number of test set compounds (which should be at least 10 or more in order to have an acceptable degree of freedom for statistical reliability) are available.

Additionally, the evaluation of the AD of the QSAR model and selection of outliers are other vital aspects in the course of developing a trustworthy QSAR model with the spirit of OECD principles. It is enormously useful that the user of the QSAR model have sufficient information about the AD of the developed model to identify interpolation (true prediction) or extrapolation (less reliable prediction). Details regarding AD are discussed in Section 10.8.

10.4 Validation Metrics

Validation metrics are used as the decisive criteria to judge the performance of the QSAR models [25–27]. For ease of understanding of readers, we have discussed validation metrics under two different categories, validation metrics for classification-based QSAR models and validation metrics for regression-based QSAR model, which are further classified into two classes: correlation-based metrics and error-based metrics.

10.4.1 Metrics for Classification-Based QSAR Models

Classification-based QSAR models perform a task similar to that performed by MLR by predicting an outcome when the response property has categorical values (active-inactive, positive-negative, etc.) and molecular descriptors are continuous variables. Validation metrics assess the performance of the classification-based models in terms of precise qualitative prediction of the dependent variable [28]. The validation metrics for a classification model typically implemented for two-class problems, where the molecules are classified either as actives/positives/true or as inactives/negatives/false, are defined in detail in Table 10.1.

Table 10.1 Mathematical representation and definition of statistical validation metrics for classification-based QSAR models

Metrics for classification-based QSAR models	
Goodness-of-fit and quality measures	
$\lambda = \det\left(\frac{W_g}{B_g + W_g}\right)$	Wilks λ is a metric for testing the significance of the discriminant model function, calculated from the scalar transformations of the covariance matrices of between- and within-groups variances. It spans 0 to 1, where 0 corresponds to different values of group means (signifying a good level of discrimination achieved by the variables) and 1 refers to similar group mean values (meaning no discrimination achieved by the variables. Suppose B_g and W_g are the random $p \times p$ independent variable matrix with the distribution $W_p(q, \Sigma)$ and $W_p(n, \Sigma)$, respectively, considering $n > p$. The term "det" signifies the determinant of the matrix generated employing descriptors.
$R_c = \sqrt{\frac{\lambda_i}{1+\lambda_i}}$	The measure of the strength of the relationship between the two variates is expressed as a canonical correlation coefficient. Here, λ_i is referred to as the eigen value of the matrix.
$\chi^2 = \sum_{i=1}^{t} \frac{(f_i - F_i)^2}{F_i}$	The chi-square (χ^2) statistics identifies the liberty between two groups, signifying that a higher value of this metric will point out superior separability between groups, that is, good classification analysis. Here, f_i is the observed response, F_i is the predicted response, and t is the number of observations.

$d_{\text{Mahalanobis}}(x_i, x_j) = \sqrt{(x_i - x_j)^T \Sigma^{-1} (x_i - x_j)}$	The square of the Mahalanobis distance is measured under linear discriminant analysis (LDA) to determine the likelihood of a compound being classified in a specific group in the discriminant space. In a referred discrimination space or transformed space, Euclidean distances among data points become equal to Mahalanobis distances. In a multivariate normal distribution with covariance matrix Σ, the Mahalanobis distance between any two data points x_i and x_j can be defined as the mentioned equation. x_i and x_j are two random data points, T is the transpose of a matrix, and Σ^{-1} is the inverse of the covariance matrix.
Internal and external validation metrics	
$\text{Senstivity} = \text{Recall} = \dfrac{\text{TP}}{\text{TP}+\text{FN}}$ $\text{Specificity} = \dfrac{\text{TN}}{\text{TN}+\text{FP}}$ $\text{Accuracy} = \dfrac{\text{TP}+\text{TN}}{\text{TP}+\text{FP}+\text{TN}+\text{FN}}$	The compounds classified under the classification-based model can be divided into four categories on the basis of a comparison between the predicted and observed response: (i) true positives (TP), the active (positive) compounds that have been correctly predicted as actives; (ii) false positives (FP), the inactive compounds that have been erroneously classified as actives; (iii) false negatives (FN), the active (positive) compounds wrongly classified as negatives, that is, inactives; and (iv) true negatives (TN), the inactive compounds that have been accurately predicted as inactives. On the basis of this classification and the number of test set molecules, a two-by-two confusion matrix or contingency table can be constructed that corresponds to the dispositions of the molecules under consideration.

$$F\text{-measure} = \frac{2(\text{recall})(\text{precision})}{\text{recall} + \text{precision}}$$ $$\text{Precision} = \frac{\text{TP}}{\text{TP}+\text{FP}} = \text{FP rate}$$	The *F*-measure refers to the harmonic mean of recall and precision, where recall refers to the accuracy of real prediction and precision defines the accuracy of a predicted class. Higher values of recall and precision associate a higher *F*-measure value, which in turn leads to a better classification.
$$\text{MCC} = \frac{(\text{TP}\times\text{TN}) - (\text{FP}\times\text{FN})}{\sqrt{(\text{TP}+\text{FP})\times(\text{TP}+\text{FN})\times(\text{TN}+\text{FP})\times(\text{TN}+\text{FN})}}$$	The MCC measures the quality of binary classifications. It considers true and false positives and negatives and is regarded as a balanced measure that can be used even if the classes are of diverse sizes. It is simply a correlation coefficient between the observed and predicted binary classifications, and it returns a value between −1 and +1. A coefficient of +1 indicates a perfect prediction, 0 an average random prediction, and −1 an inverse prediction.
$$\text{Cohen's}\ \kappa = \frac{P_r(a) - P_r(e)}{1 - P_r(e)}$$ $$P_r(a) = \frac{(\text{TP}+\text{TN})}{(\text{TP}+\text{FP}+\text{FN}+\text{TN})}$$ $$P_r(e) = \frac{\{(\text{TP}+\text{FP})\times(\text{TP}+\text{FN})\} + \{(\text{TN}+\text{FP})\times(\text{TN}+\text{FN})\}}{(\text{TP}+\text{FN}+\text{FP}+\text{TN})^2}$$	Cohen's kappa (κ) can be utilized to determine the agreement between classification (predicted) models and known classifications. P_r(a) is the relative observed agreement between the predicted classification of the model and the known classification, and $P_r(e)$ is the hypothetical probability of chance agreement. The $P_r(a)$ and $P_r(e)$ values are calculated from the confusion matrix. The resultant Cohen's kappa values between −1.0 and 0.4 indicate that the model is poor, values between 0.4 and 0.6 signify that the model is average, values between 0.6 and 0.8 imply that the model is acceptable, and values between 0.8 and 1.0 indicate that the model is highly predictive.

$G-\text{mean} = \sqrt{\text{sensitivity} \times \text{specificity}}$	Merging sensitivity and specificity into a single parameter via the geometric mean (*G*-mean) allows for a straightforward way to assess the model's ability to perfectly classify active and inactive samples.
Metrics for ROC analysis	
$\text{TP rate} \approx \frac{\text{Positives (active molecules) correctly classified}}{\text{Total positives}}$ $\text{FP rate} \approx \frac{\text{Negatives (inactive compounds) incorrectly classified}}{\text{Total negatives}}$	The ROC curve is a visual illustration of the success and error observed in a classification model. The curve is plotted taking the TP rate on the *y* axis and the FP rate on the *x* axis, and the characteristics of the curve provide easier recognition of the precision of prediction.
$d_i = \sqrt{(1-\text{Se}_r)^2 + (1-\text{Sp}_r)^2}$	If the best model is one whose depiction is positioned as close as possible to the upper-left corner in the ROC graph, a good indicator would be a measure of this distance. The Euclidean distance between the perfect and a real classifier (d_i) is expressed by the equation given on the left. Se_r and Sp_r are the sensitivity and specificity values of the real classifier, respectively.
$\text{ROCED} = (\lvert d_1 - d_2 \rvert + 1) \times (d_1 + d_2) \times (d_2 + 1)$	As two distances (d_i) corresponding to the training and test sets should be as small as possible; the parameter can be defined as the equation given on the left, where d_1 and d_2 depict the distances in an ROC graph for the training and test sets, respectively. ROCED can take values between 0 (which signify ideal classification for both training and test sets) and 4.5 (d_1 = 0.5 random classifier and d_2 = 1). Models with values greater than 2.5 indicate that these models have random responses.

$\text{FIT}(\lambda)=\frac{(1-\lambda)\times(n-p-1)}{(n+p^2)\times\lambda}$ $\text{ROCFIT}=\frac{\text{ROCED}}{\text{FIT}(\lambda)}$	ROC graph Euclidean distance corrected with Wilks λ (ROCFIT) introduced to avoid the probable loss of significance in the variables of the models generated by LDA.
$\text{AUC-ROC}=1-\frac{\sum_{i=1}^{n} r_i}{n\times(N-n)}+\frac{n+1}{2\times(N-n)}$	AU-ROC is equal to a straightforward average of the ranks of the actives. The good performance of "early recognitions" is offset rapidly by "late recognitions." Here, the number of actives and the total number of compounds are denoted by n and N, respectively. AU-ROC defined in the equation is linearly related to the rank sum of actives, which is also called Mann–Whitney U test. r_i is the rank of the i_{th} active.
Metrics for PDD	
$E_{\text{toxicity}}=\frac{\%\text{ of toxic}}{(\%\text{ of nontoxic}+100)}$ $E_{\text{nontoxicity}}=\frac{\%\text{ of nontoxic}}{(\%\text{ of toxic}+100)}$	A pharmacological distribution diagram (PDD) is a frequency distribution plot of a dependent variable where expectancy values of the variable are plotted on the y axis against numeric intervals of the variable on the x axis. Expectancy refers to the probability of categorization of a compound in a specific group for a specific value of the discriminant function (DF). During the LDA, a DF is developed, used for the calculation of discriminant scores of individual compounds. Then the discriminant function values of all compounds are taken in the abscissa in the form of range and the expectancy values are plotted in the ordinate against those ranges.

10.4.2 Metrics for Regression-Based QSAR Models

In consonance with the OECD guidelines regarding the model fitness, robustness, as well as predictivity of regression-based models, a large number of statistical metrics are used by different research groups in this field. The regression-based measure determination coefficient (R^2) between the observed and predicted response values provides a relative trend of changes in the values of the observed response with respect to the predicted ones. Therefore, this metric can deliver satisfactory values for a constant magnitude of errors (as long as the trend of changes follows a pattern) for all the samples even if it is very high. The regression through origin (RTO) approach where the best fitted line is deliberately forced through origin (Y_{obs} = 0, Y_{pred} = 0) penalizes the R^2 values generated from the consequent standard regression analysis where large prediction errors are present [29]. On the basis of the RTO-derived method, researchers have formulated model acceptability criteria such as Golbraikh and Tropsha criteria [30], as well as different r_m^2 metrics [31–34]. However, the RTO approach is able to identify prediction errors of a model as long as the data are devoid of any "systematic error" and/or model bias.

The external predictivity of QSAR models is commonly described by employing various validation metrics, and these can be broadly categorized into two major classes, (i) correlation-based metrics [35] and (ii) error-based metrics [14, 36]. Thus, a model may be considered acceptable when the values of these R^2-based external predictive metrics [R^2_{pred} or $Q^2_{(F1)}$ and $Q^2_{(F2)}$] are more than the stipulated value of 0.5; the closer the values to unity, the greater the confidence in prediction precisions. Even though this formalism seems rational, the outcomes from the comparison with the performance of the training/test set response mean can occasionally be misleading as they are significantly influenced by the range of the training/test set response data and the mean response value. Another important feature for the over and underestimation of prediction errors of these metrics is the distribution of the response data around the mean. While the correlation-based external metrics may sometimes give misleading results regarding the quality of predictions, prediction error–based metrics like predicted residual sum of squares (PRESS), RMSE calibration/prediction (RMSEc/RMSEv), and mean absolute error (MAE) [14, 36] give more straightforward results. The determination of error-based metrics gives more direct information about the prediction errors since they do not compare the prediction errors with other aspects, like performance of the mean. As most of the statistical tests are typically performed at the probability level of 5%, Roy et al. [36] proposed that MAE -based criteria might be checked after deleting 5% test set compounds with high residual values with the objective to obviate the role of any outlier predictions. The conclusion based on 95% of the data will be extremely useful for test sets containing a small number of badly predicted data points where the MAE-based criteria for 100% data can overpenalize the model predictivity.

Mathematical definitions of frequently employed correlation and error-based validation metrics for the regression-based QSAR models are demonstrated in Tables 10.2 and 10.3. Graphical illustrations of available validation metrics and steps to compute r_m^2 metrics are reported in Figs. 10.5 and 10.6, respectively. We have comprehensively demonstrated elsewhere how one can easily calculate mentioned metrics from a classification- and regression-based QSAR models with simple and practical examples [25, 27].

Table 10.2 Mathematical representation and definition of correlation-based metrics for regression-based QSAR models

Correlation-based metrics for the regression-based QSAR models	
Goodness-of-fit and quality measures	
$R^2 = 1 - \frac{\Sigma(Y_{obs} - Y_{calc})^2}{\Sigma(Y_{obs} - \bar{Y}_{training})^2}$	The determination coefficient R^2 measures the variation of observed data with the predicted ones. The maximum possible value for R^2 is 1, which defines a perfect correlation. $\bar{Y}_{training}$ is the mean observed activity of the training set compounds.
$R_a^2 = \frac{\{(n-1) \times R^2\} - p}{n-p-1}$	To overcome the drawbacks associated with the increasing value of R^2 with an increase in the number of variables. Explanatory terms in a model can be adjusted by this modified R^2. Here, n is the number of compounds and p is the number of descriptors.

Internal metrics for checking robustness	
$Q^2(Q^2_{\text{LOO}}) = 1 - \frac{\Sigma(Y_{\text{obs(training)}} - Y_{\text{pred(training)}})^2}{(Y_{\text{obs(training)}} - \bar{Y}_{\text{training}})^2}$	Cross-validated $R^2(Q^2)$ checked for internal validation. $Y_{\text{obs(train)}}$ is the observed activity and $Y_{\text{pred(train)}}$ is the predicted activity of the training set molecules on the basis of the LOO technique. The accepted threshold value of Q^2 is 0.5.
External predictivity metrics	
$Q^2_{(F1)} = 1 - \frac{\Sigma(Y_{\text{obs(test)}} - Y_{\text{pred(test)}})^2}{\Sigma(Y_{\text{obs(test)}} - \bar{Y}_{\text{training}})^2}$	The metric $Q^2_{\text{(F1)}}$ or R^2_{pred} replicates the measure of correlation between the observed and predicted data of test set. $Y_{\text{obs(test)}}$ and $Y_{\text{pred(test)}}$ signify the observed and predicted response data for the test set molecules, while $\bar{Y}_{\text{training}}$ denotes the mean observed activity of the training set. The stipulated threshold value is 0.5.
$Q^2_{(F2)} = 1 - \frac{\Sigma(Y_{\text{obs(test)}}) - Y_{\text{pred(test)}})^2}{\Sigma(Y_{\text{obs(test)}} - \bar{Y}_{\text{test}}}$	In the case of $Q^2_{\text{(F2)}}$, the mean ($\bar{Y}_{\text{test}}$) used in the denominator refers to the values of external predictive parameters. Almost equal or closer values of $Q^2_{\text{(F2)}}$ and $Q^2_{\text{(F1)}}$ indicates that the training set mean lies in close proximity to that of the test set, which indicates that the test set utilized for modeling covers the whole response domain of the model. The threshold value is 0.5.
$Q^2_{\text{(F3)}} = 1 - \frac{\left[\Sigma(Y_{\text{obs(test)}} - Y_{\text{pred(test)}})^2\right] / n_{\text{test}}}{\left[\Sigma(Y_{\text{obs(train)}} - \bar{Y}_{\text{train}})^2\right] / n_{\text{train}}}$	This metric measures the model's predictability. It is sensitive to the selection of training data set and tends to penalize models fitted to a very homogeneous data set, even if predictions are close to the truth, with a threshold value of 0.5. n_{tr} and n_{ext} denote the number of training set and test set compounds, respectively.

$\mathrm{CCC} = \bar{\rho}_c = \dfrac{2\sum\limits_{i=1}^{n}(x_i-\bar{x})(y_i-\bar{y})}{\sum\limits_{i=1}^{n}(x_i-\bar{x})^2+\sum\limits_{i=1}^{n}(y_i-\bar{y})^2+n(\bar{x}-\bar{y})}$	The concordance correlation coefficient (CCC) measures both precision and accuracy, detecting the distance of the observations from the fitting line and the degree of deviation of the regression line from that passing through the origin, respectively. Any deviation of the regression line from the concordance line gives a value of CCC smaller than 1. $x_{\mathrm{obs(test)}}$ and $y_{\mathrm{pred(test)}}$ refer to the observed and predicted values of the test set, n denotes the number of compounds, and $\overline{x_{\mathrm{obs(test)}}}$ and $\overline{y_{\mathrm{pred(test)}}}$ signify the mean of observed and predicted values, respectively.
Golbraikh and Tropsha criteria: $Q^2_{\mathrm{training}} > 0.5$. $R^2_{\mathrm{test}} > 0.6$. $\dfrac{r^2 - r_0^2}{r^2} < 0.1$ and $0.85 \le k \le 1.15$ or $\dfrac{r^2 - r'^2_0}{r^2} < 0.1$ and $0.85 \le k' \le 1.15$. $\left\| r_0^2 - r'^2_0 \right\| < 0.3$.	In the case of least squares, the experimental versus fitted and the fitted versus experimental plots are not always equivalent. According to Golbraikh and Tropsha, regressions of y_{obs} against y_{pred} or y_{pred} against y_{obs} through the origin should be characterized by either k or k' (slopes of the corresponding regression lines) being close to 1. The parameters r^2, r_0^2, and r'^2_0 and the terms k and k' are explained under the r_m^2 metrics in the next row.

Scaled r_m^2 metrics for internal, external, and overall predictivity

$$\overline{r_m^2} = \left(r_m^2 + r_m'^2\right)/2 \quad \text{and}$$

$$\Delta r_m^2 = \left|r_m^2 - r_m'^2\right|,$$

where $r_m^2 = r^2 \times (1 - \sqrt{r^2 - r_0^2})$

$$r_m'^2 = r^2 \times (1 - \sqrt{(r^2 - r_0'^2)})$$

The parameters r^2 and r_0^2 are defined as follows:

$$r_0^2 = 1 - \frac{\Sigma(Y_{obs} - k \times Y_{pred})^2}{\Sigma(Y_{obs} - \overline{Y}_{obs})^2}$$

and

$$r_0'^2 = 1 - \frac{\Sigma(Y_{pred} - k' \times Y_{obs})^2}{\Sigma(Y_{pred} - \overline{Y}_{pred})^2}$$

The terms k and k' are defined as:

$$k = \frac{\Sigma(Y_{obs} \times Y_{pred})}{\Sigma(Y_{pred})^2} \quad \text{and}$$

$$k' = \frac{\Sigma(Y_{obs} \times Y_{pred})}{\Sigma(Y_{obs})^2}$$

The Y_{obs} and Y_{pred} values have been scaled at the beginning using the following formula:

$$Y_{i(scaled)} = \frac{Y_i - Y_{min(obs)}}{Y_{max(obs)} - Y_{min(obs)}}$$

The mean response value of the training set molecules and the distance of the mean from the response values of each compound play a decisive role in the computation of Q^2. The Q^2 value increases with the rise in the value of the denominator. Thus, even for great disparity in the predicted and observed response values, satisfactory Q^2 values may be obtained if the molecules exhibit a considerably broad range of response data. Thus, the r_m^2 metrics for internal validation [$\overline{r_m^2}_{(LOO)}$ and $\Delta r_m^2{}_{(LOO)}$] prevent this error and reflect model predictability in a better way. Similarly, high values of R^2_{pred} may be obtained when the test set molecules bear a wide range of activity data; but this may not signify that the predicted values are very near the corresponding observed data. With respect to finding out the proximity between the observed and predicted activity data for the test set of compounds, the metrics $r_m^2{}_{(test)}$ [$\overline{r_m^2}_{(test)}$, $\Delta r_m^2{}_{(test)}$] were also introduced by Roy and coworkers. They extended the metric to the entire data set by employing the LOO predicted data for the training set and the predicted data for the test set molecules. The metrics have been defined as $\overline{r_m^2}_{(overall)}$ and $\Delta r_m^2{}_{(overall)}$, which reflect the predictability of the model for the entire data set. For the acceptable prediction, the value of all Δr_m^2 metrics should preferably be lower than 0.2, provided that the value of all $\overline{r_m^2}$ metrics is more than 0.5.

r_m^2 rank metrics	
$r_{m\ (rank)}^{2}=r_{(rank)}^{2}\times\left(1-\sqrt{r_{(rank)}^{2}-r_{0\ (rank)}^{2}}\right)$	It is computed utilizing the correlation of the ranks generated from the observed and corresponding predicted data to measure the proximity. In an ideal ranking where the observed and predicted response data perfectly match with each other for all molecules, the obtained $r_m^2{}_{(rank)}$ metric value is 1. The acceptable threshold value is 0.5.

Table 10.3 Mathematical representation and definition of error-based metrics for regression-based QSAR models

Error-based metrics for the regression-based QSAR models	
Goodness-of-fit and quality measures	
$\text{SEE}=s=\sqrt{\dfrac{\sum(Y_{obs}-Y_{calc})^2}{n-p-1}}$	The standard error of estimate (SEE) is calculated from the square root of sum of squares of the residuals divided by the degree of freedom. Lower values of SEE correspond to improved model fitting. Y_{obs} and Y_{calc} are the observed (experimental) and estimated scores, respectively, while n is the number of compounds and p is the number of descriptors.
$\text{PRESS}=\sum(Y_{obs}-Y_{pred})^2$	The model predictivity is evaluated using the predicted residual sum of squares (PRESS). Y_{obs} and Y_{pred} correspond to the observed and LOO-predicted values
$\text{RMSE}_C=\sqrt{\dfrac{\Sigma(Y_{obs(training)}-Y_{calc(training)})^2}{n_{training}}}$	Assessment of the observed activity and the calculated response of the training set molecules through calculation of a metric referred to as the root mean square error in prediction.
$\text{RMSE}_p=\sqrt{\dfrac{\sum(Y_{obs(test)}-Y_{pred(test)})^2}{n_{test}}}$	Assessment of the observed activity and the predictions of the test set molecules through calculation of a metric referred to as root mean square error in prediction

$$\text{MAE} = \frac{\sum \lvert Y_{\text{obs}} - Y_{\text{pred}} \rvert}{n}$$	This is also known as the average absolute error (AAE). This can be considered to be a better index of errors in the context of predictive modeling studies. Because of the involvement of the squared term of the prediction errors in the expression of RMSE, the variance of errors may be influenced for a set of data. That is, squaring the higher prediction error values will have more weight than the lower errors in the formalism of RMSE, while MAE provides an equal weight to all the errors. Thus, MAE is considered to be a simpler and more straightforward determinant of prediction errors.
Good prediction MAE ≤ 0.1 × training set range; and MAE + 3 × σ ≤ 0.2 × training set range	An error of 10% of the training set range should be acceptable, while an error value more than 20% of the training set range should be a very high error. The σ value denotes the standard deviation of the absolute error values for the test set. Considering a normal distribution pattern, mean ±3σ covers 99.7% of the data.
Bad prediction MAE > 0.15 × training set range; or MAE + 3 × σ > 0.25 × training set range	A value of MAE more than 15% of the training set range should be high, while an error more than 25% of the training set range is considered very high. The predictions that do not fall under either of the two conditions may be considered to be of moderate quality. The above criteria should be applied for judging the quality of test set predictions when the number of data points is at least 10 and there is no systematic error in model predictions.

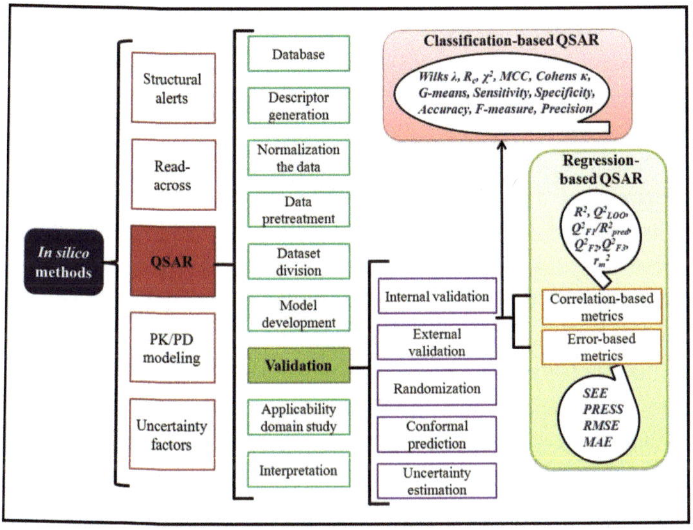

Figure 10.5 Major validation metrics for classification and regression-based QSAR models.

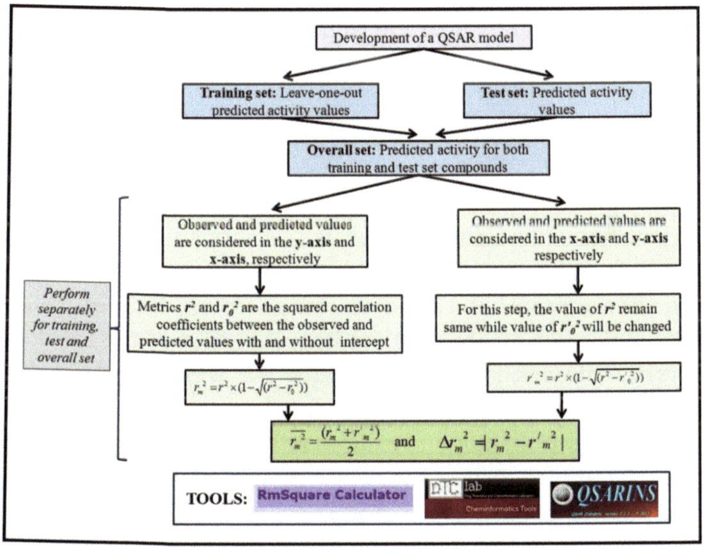

Figure 10.6 Flowchart for the calculation of r_m^2 metrics.

10.5 Model Uncertainty Aspects

Four types of model uncertainty aspects are identified in QSAR modeling [37]. The first type is associated with the quality of the experimental data as no model can be better in quality than the experimental data. The second one originates from the capacity of the statistical model to distinguish the relationship between structure and activity. The third one is related to class assignment that is sensitive to a given cutoff value adopted to discriminate between active and inactive compounds, and for low cutoff values, it is probable that uncertainty will be higher. The fourth and final one is that chemicals can be modeled by a diverse set of descriptors but only a specific subset will be able to maximize predictivity.

10.5.1 Derivation of Uncertainty in QSAR Predictions: Mathematical Formalization

For a set of compounds positioned in a chemical hyperspace defined by a set of orthogonal axes, the certainty attached to the prediction for molecule *i* is defined as the sum of certainty contributions from all the other compounds *j*, which depends on the distance between *i* and *j*:

$$C_\mathrm{i} = \sum \exp\left(-\frac{d(i,j)^2}{\sigma^2}\right) \tag{10.11}$$

Here, individual contribution is a Gaussian function. Pery et al. [38] chose this type of function by analogy with kernel density estimation approaches. The parameter σ accounts for the application of the molecular database to predict properties of the molecule of interest. This methodology estimates the width of the potential functions and the original definition of the relationship between certainty and probability. The difference between 1 and C_i accounts for uncertainty. If C_i is greater than 1, there is no uncertainty on the estimated probability for the chemical under consideration to be active or inactive. Estimation of σ for a given database can be performed as follows:

First calculate the certainty for a molecule to be active (C_{i+}) and the certainty for the molecule to be inactive (C_{i-}) as a function of σ.

$$C_{i+} = \sum \exp\left(-\frac{d(i,j)^2}{\sigma^2}\right) \times 1\ (j = \text{active}) \tag{10.12}$$

$$C_{i-} = \sum \exp\left(-\frac{d(i,j)^2}{\sigma^2}\right) \times (1 - 1\ (j = \text{active})) \tag{10.13}$$

Here, 1 (j = active) equals 1 if the compound j is active and 0 if it is inactive.

Again, if $(C_{i+} + C_{i-}) \geq 1$, then the predicted probability of molecule i to be active is defined as:

$$P_i = \frac{C_{i+}}{C_{i+} + C_{i-}} \text{ with certaint } y\ 1 \tag{10.14}$$

If $(C_{i+} + C_{i-}) < 1$, then the predicted probability for the compound i to be active is the certainty to be active C_{i+} plus the percentage of active molecules in the training data set multiplied by the uncertainty. Indeed, in the strictive case where predictive certainty for a given molecule is 0, the a priori probability for such a molecule to be active equals the proportion of active molecules within the training set.

$$P_i = C_{i+} + \frac{N_+}{N_+ + N_-}(1 - C_{i+} - C_{i-}) \tag{10.15}$$

With uncertainty equal at 1 – $(C_{i+} + C_{i-})$, N_+ and N_- are the number of active and inactive molecules within the data set. From the latter equation, it is clear that if certainty is low, the predicted probability to be active equals the percentage of active molecules in the data set. Finally, for a given database, σ is estimated by maximizing the following likelihood:

$$\text{Likelihood}(\sigma) = \prod P_i\ 1(j = \text{active}) \times (1 - P_i)[1 - 1(j = \text{active})] \tag{10.16}$$

For any new molecule to be predicted, uncertainty calculations are performed as described previously to derive P_i and C_i values. It is worth noting that predictions can also be binarized by the introduction of probability thresholds chosen to discriminate between active and inactive chemicals. These discriminative properties of the σ parameter are demonstrated, in turn, to be important in estimating prediction uncertainty on the basis of Euclidean distances among molecules.

According to this approach, a query molecule can be considered to belong to the AD of a QSAR model if the uncertainty in prediction is lower than a user-defined threshold. The analysis shows that predictions for chemicals belonging to well-designed test sets are portrayed by a predictive uncertainty of 0 (i.e., certainty = 1). Adopting a threshold that equals 0 would therefore ensure that the query chemical is adequately described by the structures of the molecules that compose the training set of the model. Assessing uncertainty in prediction even for chemicals that are far away from the training set is in itself an important piece of information that could be useful in decision making.

10.6 Prediction Confidence and Conformal Predictions

Conformal prediction [39] is introduced as an alternative approach to AD inference. The advantages of using conformal prediction are as follows:

- The approach is based on a dependable and well-defined mathematical framework.
- The concept of the confidence level in conformal predictions is clear-cut, for example, a confidence level of 0.85 means that the conformal predictor will commit, at most, 15% errors.
- The confidence level can be changed depending on the condition where the model is to be applied, and the consequences of such modification are readily understandable. That means if prediction ranges are increased or decreased, then the changes can immediately be inspected [40–44].

A confidence predictor is a prediction algorithm that offers a prediction region, contrary to the single-valued (regression) or single-label (classification) predictions output by typical prediction algorithms. A prediction region is a set $\Gamma\varepsilon$ that contains the true value of an example with the probability of at least $1-\varepsilon$. In the case of regression, the prediction region is a real valued interval (or set of intervals). For a binary classifier with the two classes represented by A and B, the prediction region could be any of the following sets: {A}, {B}, {A,B} (both) or {null} (the empty set). To estimate the performance of confidence predictors, one needs to understand the concepts of validity and efficiency. A confidence predictor is said to be applicable if the frequency of errors (i.e., the fraction of true values outside the prediction region) it commits does not exceed ε at a selected confidence level 1 – ε and efficient if the prediction regions it outputs are as small as possible. In classification models, a measure of efficiency of confidence predictors could be the number of prediction sets containing two or more labels (at a given confidence level). In regression models, a natural measure of efficiency is the size of the prediction interval.

A conformal predictor is a type of confidence predictor. The prediction regions generated by a conformal predictor have the property of always being valid, and one needs to worry about their efficiency only. To construct a conformal predictor's prediction regions, at first the nonconformity measure needs to be defined. Intuitively, this is a way of measuring how different a new example is from old examples. Therefore, a nonconformity measure serves a role similar to that of an AD measure approach. Interestingly, the majority of AD approaches can be employed as nonconformity measures. Conformal prediction is an algorithm-independent technique that works with any predictive method, such as a support vector machine and a random forest (RF), that outputs confidence regions (CRs) for individual predictions in the case of regression and *p* values for categories in a classification setting [39, 45].

10.6.1 Conformal Prediction Errors in Regression

In the conformal prediction [39, 45], the data points in the training set are employed to describe how doubtful a new element is with respect to the data presented to the model in the training set. The conformity for a given data point, *x*, with respect to the training set is quantified with a nonconformity score, α, computed with a nonconformity measure [39], which we define here as:

$$\alpha = \frac{|y - \tilde{y}|}{\tilde{\rho}}, \tag{10.17}$$

where α is the nonconformity score, y and $\tilde{y}$ are the observed and the predicted values, respectively, calculated with an e point prediction model, and $\tilde{\rho}$ is the predicted error for *x* calculated with an error model.

To calculate the confidence intervals (CIs), a point prediction model is required to predict the response variable (y) and an error model to predict errors in prediction ($\tilde{\rho}$). The point prediction and error models can be created with any machine learning algorithm. Both point prediction and error models need to be trained with cross-validation in order to calculate the vector of nonconformity scores for the training set, $D_i = \{x_i\}_i^{N_{tr}}$.

The cross-validation predictions are generated when the training point prediction model serves to calculate the errors in prediction for the data points in the training set, $y_i - \widetilde{y_i}$. This is followed by the development of the error model by training a machine learning model on the training set employing these errors as the dependent variable. The cross-validated predictions from the point prediction model and the cross-validated errors in prediction from the error model are used to make the vector of nonconformity scores for the training set. This vector, after being sorted in increasing order, can be defined as:

$$\alpha_{\text{tr}} = \{\alpha_{\text{tr}\,i}\}_i^{N_{\text{tr}}}, \tag{10.18}$$

where, N_{tr} is the number of data points in the training set.

To generate the CIs for an external set, $D_{\text{EXT}} = \{x_{\text{ext}}\}_j^{N_{\text{ext}}}$, a confidence level ε needs to be defined. Then, the α value related to the user-defined confidence level, α_ε, is calculated as:

$$\alpha_\in = \alpha_{\text{tr}\,i} \text{ if } i \equiv \left|N_{\text{tr}} * \in\right|, \tag{10.19}$$

where ε indicates equality.

Next, the errors in prediction, $\tilde{\rho}_{\text{ext}}$, and the value for the response variable, $\tilde{y}_{\text{ext}}$, for the data points in the external data set are predicted with the error and the point prediction models, respectively. Individual CIs for each data point in the external set are derived from

$$\text{CI}_{\text{ext}\,j} = \left|y_{\text{ext}\,j} - \tilde{y}_{\text{ext}\,j}\right| = \alpha_\epsilon * \tilde{\rho}_{\text{ext}\,j}, \tag{10.20}$$

where y_{ext} corresponds to the true value (unknown for the external data) of y (i.e., the value of the dependent variable for those datapoints in the external data set). The CR is finally defined as:

$$\text{CR} = \tilde{y}_{\text{ext}\,j} + / - \text{CI}_{\text{ext}\,j} \tag{10.21}$$

The method of calculation of conformal prediction errors in regression is schematically presented in Fig. 10.7.

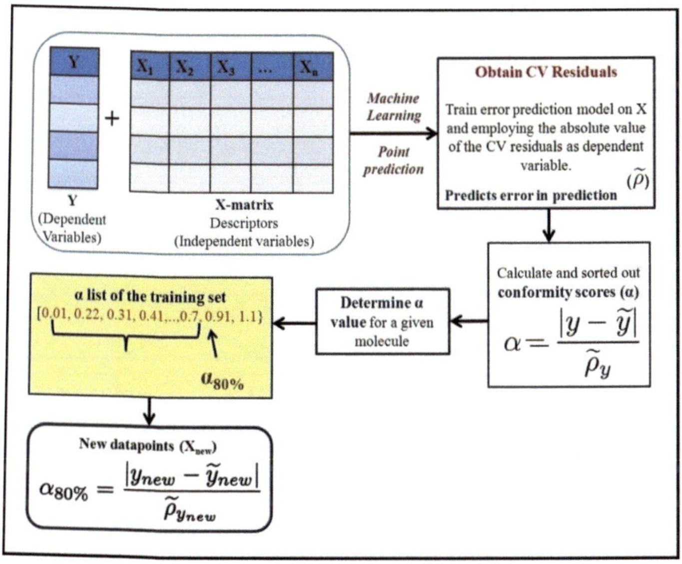

Figure 10.7 Steps for the calculation of conformal prediction errors in regression.

10.6.2 Conformal Prediction Errors in Classification

In the case of a classification model, the nonconformity scores are calculated on a per-class basis. These are calculated as the ratio between the number of trees in the forest voting for a given class divided by the total number of trees (label-wise Mondrian off-line inductive conformal prediction) [46]. For example, in the case of binary classification, if 85 trees from an RF model comprising 100 trees classify a data point to belong to class A, the nonconformity score for this class would be 0.85 (80%) whereas its value for class B would be 0.15. This process generates a matrix (nonconformity scores matrix) where the rows correspond to the data points in the training set and the columns to the number of distinct classes (see Fig. 10.8A). Next, each column of the matrix should be sorted in increasing order. These columns are called Mondrian class lists (Fig. 10.8A). Like a regression model, a confidence level, ε, needs to be specified for the classification models. Therefore, the significance should be defined as 1 – ε. The model trained on the whole data set is employed to classify the data points comprised in the external data

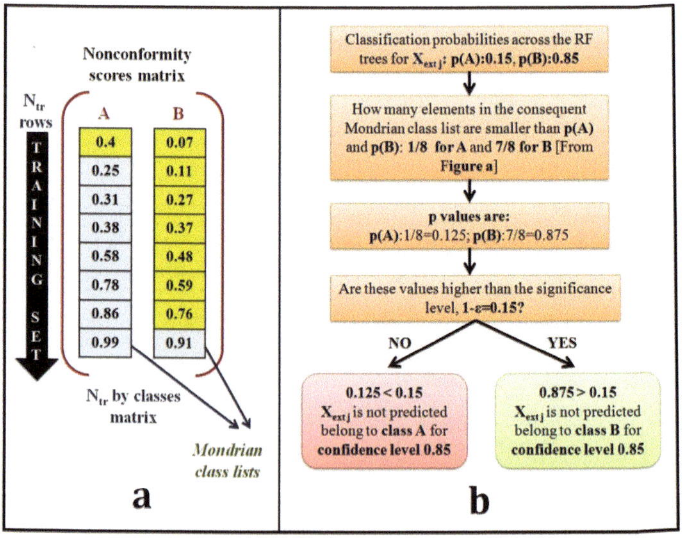

Figure 10.8 Steps for the calculation of conformal prediction errors in a binary classification considering a confidence level of 0.85.

set (see Fig. 10.8B). For instance, for one data point from the external set, namely $x_{\text{ext j}}$, the number of trees in the RF voting for each class enables the calculation of the nonconformity scores or probabilities (p) for that point, $x_{\text{ext j}}$. In the binary case, this is defined as:

$$p(x_{\text{ext j}};A) = \frac{N_{\text{trees}} \text{voting } A}{N_{\text{trees}}} \tag{10.22}$$

$$p(x_{\text{ext j}};B) = \frac{N_{\text{trees}} \text{voting } B}{N_{\text{trees}}} \tag{10.23}$$

To estimate the p values for each class, the number of elements in the consequent Mondrian class list lower than the probability values, that is, $p(x_{\text{ext j}};A)$ and $p(x_{\text{ext j}};B)$, is divided by the number of datapoints in the training set, N_{tr}:

$$p\text{value}(x_{\text{ext j}};A) = \frac{\left|\left\{\text{MCL}(A) < P(x_{\text{ext j}};A)\right\}\right|}{N_{\text{tr}}} \tag{10.24}$$

$$p\text{value}(x_{\text{ext j}};B) = \frac{\left|\left\{\text{MCL}(B) < P(x_{\text{ext j}};B)\right\}\right|}{N_{\text{tr}}} \tag{10.25}$$

Finally, these p values are compared to the significance level defined by the user $(1 - \varepsilon)$. For a datapoint to be predicted to belong to a given class, the p value needs to be superior to the significance level. For example, if p value $(x_{\text{ext j}};A) = 0{:}45$ and p value $(x_{\text{ext }j};B) = 0{:}17$, with a significance level of 0.2, x_{extj} would be predicted to belong to class A but not to B. If both p value $(x_{\text{ext j}};A)$ and p value $(x_{\text{ext j}};B)$ were higher than the significance level, $x_{\text{ext j}}$ would be predicted to belong to both classes. Correspondingly, if both p values were lower than the significance level, $x_{\text{ext j}}$ would be predicted to belong to neither class A nor class B. The method of calculation of conformal prediction errors in classification is schematically presented in Fig. 10.8.

10.7 Randomization: Assessment of Chance Correlation

Randomization is a validation tool used for QSAR models whereby the performance of the original model in data description (R^2) is compared to that of models built for a permuted or randomly shuffled response based on the original descriptor pool and the original model building procedure [47]. Randomization is a practice to ensure the robustness of the QSAR model. On the basis of the matrix permutation, randomization can be two types:

- ***Y*-randomization:** This test is performed to check the robustness of the QSAR model by building several random models by *shuffling the dependent/response variable (Y)* while keeping the independent variables (descriptors or X matrix) unchanged. The resultant random models are expected to have significantly low R^2 *and* Q^2 *values for several trials to pass the test.*
- ***X*-randomization:** This process randomly shuffles the entire descriptor matrix (X) while keeping the values of response (Y) the same followed by model generation.

Again, randomization can be performed two ways on the basis of the selection of descriptors for the QSAR model:

- **Process randomization:** The values of the dependent variable (Y) are randomly scrambled, and the selection of descriptors is done freshly from the total pool of descriptor matrix.
- **Model randomization:** The Y column data points are scrambled, and new QSAR models are developed employing the already selected set of variables as present in the nonrandom model. Generally, it is performed at varying confidence levels.

10.7.1 Randomization Metrics $^{c}R_{p}^{2}$

The difference in the values of the squared mean correlation coefficient of the randomized model (R_r^2) and squared correlation coefficient of the nonrandom model (R^2) is reflected in the value of $^{c}R_{p}^{2}$ [23] parameter. This metric penalizes the model R^2 for a small difference between the values of the squared correlation coefficients of the nonrandom (R^2) and the randomized (R_r^2) models as per the following equation:

$$^{c}R_{p}^{2} = R \times \sqrt{R^2 - R_r^2} \qquad (10.26)$$

The threshold value of $^{c}R_{p}^{2}$ is 0.5, and a QSAR model beyond this predetermined value might be correctly considered to have been obtained by chance only.

10.7.2 Q^2_{yrand} and R^2_{yrand}

The assessment of chance correlation deals with a set of decision inequalities utilizing the values of Q^2_{yrand} and R^2_{yrand} and their relationship as that $R^2_{\text{yrand}} > Q^2_{\text{yrand}}$. This approach has been proposed by Wold and Eriksson [19] with the following rules to determine any chance correlation in the QSAR model.

- $Q^2_{\text{yrand}} < 0.2$ and $R^2_{\text{yrand}} < 0.2 \rightarrow$ no chance correlation
- Any Q^2_{yrand} and $0.2 < R^2_{\text{yrand}} < 0.3 \rightarrow$ negligible chance correlation
- Any Q^2_{yrand} and $0.3 < R^2_{\text{yrand}} < 0.4 \rightarrow$ tolerable chance correlation
- Any Q^2_{yrand} and $R^2_{\text{yrand}} > 0.4 \rightarrow$ recognized chance correlation

Thus, the number of randomization trials resulting in false positive correlations is counted up as the correlation frequency from the Q^2_{yrand} vs. the R^2_{yrand} plot.

10.8 Applicability Domain and Reliability of Predictions

The AD [48–52] is a theoretical chemical space encompassing the model descriptors and modeled response. Reliability and acceptability of any QSAR model largely depend on the domain of applicability of molecules for computing the uncertainty in the prediction of a new chemical entity (NCE) on the basis of similarity to the molecules utilized to construct the QSAR model. Thus, the prediction of a modeled response using QSAR is valid only when the compound being predicted falls within the AD of the model, as it is impracticable to predict an entire universe of chemicals using a single QSAR model. Again, AD can be described as the physicochemical, structural, or biological space, knowledge, or information on the basis of which the training set of the model is developed, and the model is applicable to predictions for NCE within the specific domain. To make confident predictions, a QSAR model should always be employed for those molecules that are within its AD.

The AD is estimated by determining interpolation regions as defined through the training set in the model descriptor space, but the region varies depending on the implemented approach. It is interesting to note that all the implemented AD approaches have their own potential and flaws. A few approaches have multifaceted algorithms behind the detection of interpolation regions, and on the contrary, a few have a strong and easy statistical background behind the AD estimation. It is essential to understand that if a researcher uses miscellaneous AD approaches for the QSAR model obtained from a single data set, the result may fluctuate and therefore none of these recognized methods can be considered as the best or ultimate one. Thus, to evaluate the AD space for a specific QSAR model, one should employ available different AD strategies before assessing an NCE to get the result with full confidence. An outline of hypotheses of data distribution and domain of applicability space of commonly used AD methods is given in Table 10.4 and graphically demonstrated in Fig. 10.9.

Table 10.4 Hypothesis, threshold criteria, and drawbacks of available AD studies

AD approach	Subtype	Hypothesis	Threshold criteria
Ranges in the descriptor space	Bounding box/ Fixed or probabilistic boundaries	The ranges of each descriptor are considered with a uniform distribution defining an *n*-dimensional hyper-rectangle developed on the basis of the highest and lowest values of the individual descriptor employed to construct the model with sides parallel to the coordinate axes. $d(x,y) = x - y$	The boundary is generated on the basis of the highest and lowest values of descriptors to develop the model and response for which the equation is formed of the training set.
	Principal component ranges/PCA bounding box	Principal components (PCs) convert the original data into a new orthogonal coordinate system by the rotation of axes and facilitate correcting for correlations among descriptors. Newly formed axes are defined as PCs presenting the maximum variance of the total compounds.	Combining the bounding box method with PCA can help overcome the setback of correlation between descriptors.
	TOPKAT optimal prediction space	Variation of PCA is implemented by the optimum prediction space (OPS) from TOPKAT OPS 2000. In the PCA approach, instead of the standardized mean value, the data are centered around the mean of the individual parameter range ($[x_{max} - x_{min}]/2$).	The OPS boundary is defined by the minimum and maximum values of the generated data points on each axis of the OPS

AD approach	Subtype	Hypothesis	Threshold criteria
		Thus, it establishes a new orthogonal coordinate system known as the OPS coordinate system. The basic process is the same, extracting eigenvalues and eigenvectors from the covariance matrix of the transformed data.	coordinate system.
Geometrical methods	Convex hull	This estimates the direct coverage of an n-dimensional set employing the convex hull calculation. The convex hull calculation is performed on the basis of complex but efficient algorithms. The approach recognizes the boundary of the data set considering the degree of data distribution.	The interpolation space is defined by the smallest convex area containing the entire training set.
Distance-based or "distance-to-centroid" principle	Leverage approach	The leverage (h) of a compound in the original variable space is calculated on the basis of the HAT matrix as, $H = (X^T(X^TX)^{-1}X)$. Here, H is an $[n \times n]$ matrix that orthogonally projects vectors into the space spanned by the columns of X. The leverage values (h) are calculated (in the x axis) for each compound and plotted versus cross validated standardized residuals (σ) (y axis) referred to as the Williams plot.	AD is defined as a squared area within the ±3 band for standardized residuals and the leverage threshold [$h^* = 3(p + 1)/n$], where p denotes the number of variables and n the number of compounds.

(Continued)

Table 10.4 (*Continued*)

AD approach	Subtype	Hypothesis	Threshold criteria
	Euclidean distance	It is the distance from every other point to a particular point in the data set. A distance score, d_{ij}, for two different compounds X_i and X_j can be measured by the Euclidean distance: $d_{ij}=\sqrt{\sum_{k=1}^{m}(x_{ik}-x_{jk})^2}$ The mean distances of one sample to the residual ones are calculated as follows: $\overline{d_i}=\sum_{j=1}^{n}d_{ij}\Big/ n-1$, where, $i = 1, 2, \ldots, n$. The mean distances are then normalized within the interval of 0 to 1.	Compounds with distance values adequately higher than those of the most active probes are considered to be outside the domain of applicability.
	Mahalanobis distance	The approach considers the distance of an observation from the mean values of the independent variables but not taking the impact on the predicted value.	Observations with values much higher than those of the remaining ones may be considered to be outside the AD.
	City block distance	It is the summed difference across dimensions and is computed employing the following equation: $d(x,y)=\sum_{i=1}^{n}\lvert x_i-y_i\rvert$ It examines the complete	The method is predominantly useful for the distinct type of descriptors. It is used only for training sets that are

AD approach	Subtype	Hypothesis	Threshold criteria
		differences between coordinates of a pair of objects (x_i and y_i), and it assumes a triangular distribution.	uniformly distributed with respect to count-based descriptors.
	Hotelling T^2 test	It is a multivariate student's t-test and proportional to leverage and Mahalanobis distance approach that is used to evaluate the statistical impact of the difference on the means of two or more variables between two groups. Hotelling T^2 corrects for collinear descriptors through the use of the covariance matrix.	On the basis of the t-value, the significant compounds within the domain are determined.
	k-nearest-neighbor approach	The approach is based on similarity search for an NCE with respect to the space shaped by the training set. The similarity is identified by finding the distance of a query chemical from the nearest training compound or its distances from k nearest neighbors in the training set. Thus, similarity to the training set molecules is significant for associating a query chemical with reliable prediction.	If the calculated distance values of test compounds are within the user defined threshold set by the training set, then the prediction of these compounds is considered to be reliable.

(*Continued*)

Table 10.4 (*Continued*)

AD approach	Subtype	Hypothesis	Threshold criteria
	DModX (distance to the model in *X* space)	The basic theory lies in the residuals of *Y* and *X*, which are of diagnostic value for the quality of the model. As there are a number of *X* residuals, one needs a summary for each observation. This is accomplished by the residual SD of the *X* residuals of the corresponding row of the residual matrix *E*. As this SD is proportional to the distance between the data point and the model plane in *X* space, it is also usually called DModX.	A DModX value larger than around 2.5 times (in some cases 3) the overall SD of the *X* residuals (corresponding to an *F*-value of 6.25) signifies that the observation is outside the AD of the model.
	Tanimoto similarity	The Tanimoto index measures similarity between two compounds on the basis of the number of common molecular fragments. To calculate Tanimoto similarity, all single fragments of a particular length in two compounds are computed.	The distance between two compounds and the distance of a compound to the model is the least distance between the query compound and compounds from the training set.
	Standard deviation of the ensemble predictions (STD)	The method depicts that if dissimilar models give appreciably different predictions for a particular molecule, then the prediction is more likely to be erratic. The sample STD can be employed as an estimator of model uncertainty.	Reliable prediction has a low prediction spread, and unreliable prediction has a higher prediction spread.

AD approach	Subtype	Hypothesis	Threshold criteria
Probability density distribution	Parametric method	It assumes a standard distribution. such as Gaussian or Poisson distribution.	Real data distribution is captured without the help of any reference data point. Individual data points are checked for whether they belong to the set.
	Nonparametric method	It does not rely on such assumptions considering the data distribution.	
Miscellaneous approaches	Stepwise based approach	This method reflects mechanistic rationality and transparency of the QSAR model and is practiced in four stages in a sequential manner: general requirements domain, structural domain, mechanistic domain, and "domain of metabolic stimulator."	A query compound is obligatory to gratify all the conditions specified within the stated four steps to be considered within AD.
	Decision trees and decision forests	This approach identifies the AD in terms of prediction confidence and domain extrapolation on the basis of the consensus prediction of decision trees (DTs) and decision forests (DFs). The basic principle is to diminish the overfitting, which can be attained by merging the DTs and keeping the differences within different DTs to the maximum possible.	Predictions from all the combined DTs are averaged in order to find the prediction confidence for a test molecule, while domain extrapolation provides the prediction precision for that compound outside the training space.

(*Continued*)

Table 10.4 (*Continued*)

AD approach	Subtype	Hypothesis	Threshold criteria
	Kernel-based applicability domain	Most machine learning approaches for QSAR greatly depend on a vectorial illustration of the molecules. Thus, the AD is expressed as a subspace of the vector space with one dimension for the individual descriptor used.	It permits the employment of structured similarity measures like the optimal assignment kernel and its extension, instead of a numerical encoding.
	Intelligent *k*-means clustering	This is an intelligent version of the *k*-means clustering algorithm-based measure of distance to domain for QSAR models designed to avert the difficulties of the existing methods in terms of primitiveness and complexity.	A test compound is assigned fuzzy membership of each individual cluster, from which an overall distance may be calculated.

Without any doubt, in recent years, the AD of a prediction model has become an integral concern in QSAR modeling, especially from the perspective of human safety assessment. Hanser et al. [51] described the topic AD from a broader decision perspective rather than the prediction level and addressed three well-defined concerns with specific answers.

- **Applicability:** The first step is to check whether it is valid to apply the model for the compound being predicted.
- **Reliability:** Once one has established that the model is suitable for use for a particular case, one can proceed to the next stage by making a prediction and consider its reliability, which informs the user about the quantity, quality, and relevance of the information available to the model to perform the prediction task.
- **Decidability:** Finally, once the prediction has been judged valid and reliable, one can consider its actual conclusion. The degree of assertiveness of the conclusion that can be derived

from the prediction will depend on the weight of evidence that supports this conclusion.

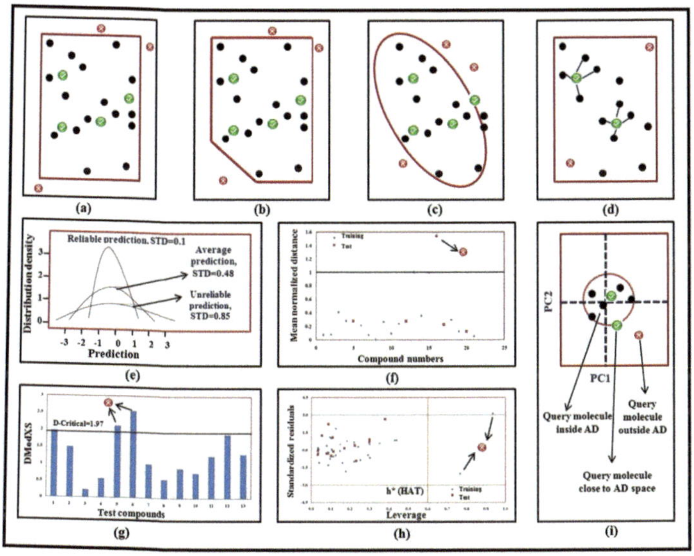

Figure 10.9 Various AD plots. (a) Bounding box, (b) convex hull, (c) distance-to-centroid, (d) *k* nearest neighbors, (e) standard deviation of the ensemble predictions, (f) Euclidean distance, (g) DModX, (h) Williams (leverage), and (i) PCA bounding box.

10.9 Open-Source Software Tools for QSAR Model Development and Validation

The term "cheminformatics" is an applied field of chemistry involving informatics and defines the implication of computational resources for answering unexplained problems arising in chemical, pharmaceutical, food, agrochemicals, and allied industries. Cheminformatics has transformed pharmaceutical, chemical, and material research faster economically and efficiently in the last two decades. Software tools under cheminformatics provide diverse functionalities, like molecular modeling through QSAR and machine learning techniques, virtual screening, read-across techniques, pharmacophore, docking, and molecular dynamics packages, which

Table 10.5 List of open-source software tools for QSAR model development and validation

Software	Tool name	Function	References
DTC lab tools	Normalization Data 1.0	To normalize the data by scaling between 0 to 1.	http://teqip.jdvu.ac.in/QSAR_Tools/, http://dtclab.webs.com/software-tools, http://nanobridges.eu/software
	Data Pre-treatment GUI 1.2	To remove the constant and highly correlated descriptors on the basis of user-specified variance and correlation cut-off values using V-WSP algorithm.	
	Bias-Variance Estimator	To help understand the contribution of two components of the prediction errors, namely, bias (or systematic) error and variance (or random) error, which are obtained from the developed model. This tool employs "bootstrapping" as a resampling technique.	
	Data Curation	Screenings of large chemical libraries against panels of biological targets have led to the rapid expansion of publicly available databases, such as ChEMBL, PubChem, and BindingDB.	

	Modified K-Medoid GUI 1.2	Tends to select *k* most middle objects as initial medoids. The algorithm involves calculation of the distance matrix once and uses it for finding new medoids at every iterative step.
	Data set division GUI 1.2	Includes three different methods, Kennard–Stone based, Euclidean distance based (or diversity based), and activity-/property-based data division (divided into training and test sets).
	Double Cross-Validation Tool v.2.0	Comprises two nested cross-validation loops, referred to as internal and external cross-validation loops. In the outer (external) loop of double cross-validation, all data objects are divided into two subsets, referred to as training and test sets. The training set is used in the inner (internal) loop of double cross-validation for model building and model selection, while the test set is exclusively used for model assessment.
	Genetic Algorithm v4.1	A search heuristic method that mimics the process of natural selection. Where an exhaustive search is impractical, heuristic methods are used to speed up the process of finding a satisfactory solution.

(*Continued*)

Table 10.5 (*Continued*)

	MLR BestSubset Selection 2.1	To select the best descriptor combination out of a set of descriptors by evaluating all possible combinations of descriptors in the input file. Along with the conventional parameters, like R^2, Q^2, Q^2_{F1}, and Q^2_{F2}, the prediction quality of training as well as test set is judged using MAE-based criteria.
	Stepwise MLR v1.2	To perform stepwise MLR using two methods: (i) using alpha value (*t*-test) and (ii) using *F*-value. The user can also select the data pretreatment option to remove constant and intercorrelated descriptors prior to computation.
	Xternal Validation Plus 1.2	To compute all the required external validation parameters, judge the performance of the prediction quality of a QSAR model on the basis of the MAE-based criteria, and check the presence of systematic errors in the model.
	AD (using standardization approach) v1.0	To identify compounds (test set/query compounds) that are outside the applicability domain (AD) of the built QSAR model and also to detect outliers present in the training set compounds by standardization technique.
	AD-MDI[1] GUI 1.2	To define the AD of unknown samples.

	Euclidean AD 1.0	To ensure that the compounds of the test set are representative of the training set. It is based on distance scores calculated by the Euclidean distance norm.	
	MLR Y-Randomization Test 1.2	This test is performed by building several random models by shuffling the dependent variables while keeping the independent variables as they are.	
	NanoProfiler v1.2	To predict the respective property/endpoint of a nanoparticle using nano-QSAR models present in the provided database that are already reported in the literature and then to perform clustering to find analogues based on the respective predicted endpoint.	
QSARINS (QSAR INSubria)	QSARINS, QSARINS-Chem	Software for the development and validation of MLR models by ordinary least squares (OLS) and implemented according to the statistical approach for predictivity. It offers data normalization, splitting of the data sets, OLS model development, GA for variable selection, validation, and AD by leverage.	http://www.qsar.it/

(Continued)

Table 10.5 (*Continued*)

CORAL	CORAL-QSAR/QSPR	To generate special optimal descriptors and build QSAR models. Analysis is performed for substances represented by simplified molecular input line entry system (SMILES) by the Monte Carlo method.	http://www.insilico.eu/coral
PaDEL	PaDEL-ADV	To perform virtual screening using AutoDockVina.	http://www.yapcwsoft.com/dd/
	PaDEL-DDPredictor	To calculate pharmacodynamics, pharmacokinetics, and toxicological properties of compounds.	
	PaDEL-Descriptor	To calculate molecular descriptors and fingerprints. The software currently calculates 1875 descriptors (1444 1D and 2D descriptors and 431 3D descriptors) and 12 types of fingerprints (total 16,092 bits).	
QSAR4U	SiRMS	A simple tool to generate 2D SiRMS descriptors for single compounds, mixtures, quasi mixtures, and chemical reactions.	http://qsar4u.com/index.php
	SPCI	A knowledge-mining tool to retrieve SAR from chemical data sets on the basis of structural and physicochemical interpretation of QSAR models.	

OSDD	OSDD	The CRDD web portal provides computer resources related to drug discovery on a single platform.	http://crdd.osdd.net/oscadd/
SCRAMBLE'N'GAMBLE	SCRAMBLE'N'GAMBLE	To quickly and easily generate random data for statistical evaluation of QSAR models.	http://drugdesign.pl/scramble-n-gamble/
Chemistry Development Kit (CDK)	CDK	To provide functionalities like SMILES, Chemical Markup Language (CML), and MIT Design Language (MDL); structure generators; 2D diagram editing and generation; 3D geometry generation; substructure search using exact structures and Smiles ARbitrary Target Specification (SMARTS)–like queries; molecular descriptor calculation for QSAR; and fingerprint calculation.	cdk.sourceforge.net
Open Babel	Open Babel	An open-source chemical toolbox intended to be a cross platform library that is built to support interconversion between various file formats used in cheminformatics, molecular modeling, and related areas.	http://openbabel.org

(*Continued*)

Table 10.5 (*Continued*)

CheS-Mapper	CheS-Mapper	A 3D molecular viewer to visualize and explore the chemical data sets. It divides data sets into clusters of similar compounds and then arranges them in the 3D space so that their spatial proximity reflects their similarity.	http://ches-mapper.org/
DemQSAR	DemQSAR	A stand-alone Java application tool that can be used to predict the human volume of distribution (VDss) and human clearance (CL). DemQSAR integrates the open-source CDK library to compute various molecular descriptors and fingerprints, and thus the QSAR models can be built.	http://agknapp.chemie.fu-berlin.de/dempred/demqsar
VEGA Non-Interactive Client (VEGANIC)	VEGA[2]	An open-source platform that provides valid QSAR models to be used especially under the European legislation for chemical substances. VEGANIC is software under the VEGA platform.	http://amcc.it/vega_download/
OECD QSAR Toolbox	QSAR Toolbox	The key features are identification of relevant structural characteristics and potential mechanism of action of a target chemical and identification of other chemicals that have the same structural characteristics and/or mechanism of action.	http://www.oecd.org/chemicalsafety/risk-assessment/theoecdqsartoolbox.htm

OCHEM	OCHEM	A web-based platform that provides the tools for automation of typical steps necessary to create a predictive QSAR model.	https://ochem.eu/home/show.do
Chembench	Chembench	To provide a broad range of tools for data visualization and creation and validation of the predictive QSAR models and using them for virtual screening of chemical libraries.	https://chembench.mml.unc.edu/
Molecular Property Diagnostic Suite	MPDSTB 1.0.1	Conceptualized to assess and estimate the multifarious aspects of drug-likeliness of any given molecule in order to diagnose its potential application as a drug.	http://mpds.osdd.net/
Virtual Computational Chemistry Laboratory	ALOGPS 2.1	A program to predict the lipophilicity and aqueous solubility of molecules.	http://www.vcclab.org/
	ASNN	To calculate the highly predictive nonlinear neural network models.	
	E-BABEL	A molecular structure information interchange hub.	
	PNN	To produce clearly interpretable analytical nonlinear models.	
	PCLIENT	To generate more than 3000 descriptors.	
	E-DRAGON	To calculate DRAGON molecular indices.	
	PLS	To implement the original two-step descriptor selection procedure.	

are most often user-friendly tools involving the graphical user interface (GUI) [53]. Software is classified under two distinct groups: commercial and open access or open-source ones. Commercial ones are powerful and well developed under a strong industrial background but expensive, which restricts their usage to industry and research laboratories. On the other hand, open-source ones are freely available, that is, open access, which makes them easy to use, learn, and apply by beginners and connoisseurs alike. To support the scientific community in resolving the diverse problems arising in miscellaneous areas of science, it is critical that cheminformatics resources be openly accessible. A good number of resources are available in the present times from different research laboratories, universities, and industries, as well as from government funding organizations. To limit the discussion, we have reported open-source software that are directly or indirectly required for QSAR model development only in Table 10.5.

10.10 Conclusion

The success story of QSAR modeling has increased tremendously in recent years, with the growing number of chemical, pharmaceutical, and material science databases and open access software. A good number of encouraging results through applications of QSAR approaches have communicated the effectiveness of this largely implicated computational method. With the introduction of advanced chemometric tools, a hefty number of research groups have been working on the extension of the remarkable relationship between molecular structure and activity response, specifically in the fields of medicine and predictive toxicology. In this aspect, to make the QSAR models transferable and acceptable to the scientific community, especially to experimentalists, validation aspects, along with mechanistic interpretation, are of paramount importance. A wide range of validation techniques have been illustrated for detecting the capability of the QSAR models to predict the response of any new molecule entity. What is more important, a single metric may often prove to be insufficient for assessing the performance of a model and therefore the overall model quality (MQ) and predictive capacity should be considered for the selection of the ideal QSAR model for a particular endpoint. If sufficient consideration is given to the decisive issues of QSAR model validation, the models could indeed be employed successfully for virtual screening and to computationally derive reliable hits. The discussed validation metrics, the concept of the AD, randomization techniques, the prediction confidence, and the uncertainty aspect should be helpful for both computational and synthetic chemists, along with biologists, for virtual screening of chemical libraries using QSAR models. We believe that the illustrative explanation of individual metrics and validation techniques given in this chapter will be helpful for the QSAR beginners as well as for experts.

Conflict of Interest

The authors confirm that there is no conflict of interest regarding the contents of this article.

Acknowledgments

S. K. and J. L. are thankful to the National Science Foundation (NSF/CREST HRD-1547754 and NSF/RISE HRD-1547836) for financial support. K. R. wants to thank the UGC, New Delhi, for financial assistance under the UPE-II scheme.

References

1. Selassie, C. D. (2003). *Burger's Medicinal Chemistry and Drug Discovery*, Abraham, D. J., ed. (John Wiley, New York, USA), pp. 1–48.
2. Kar, S. and Roy, K. (2010). Predictive toxicology using QSAR: a perspective, *J. Indian Chem. Soc.*, **87**, 1455–1515.
3. Fujita, T. (2017). The application of classical QSAR to agrochemical research, *Int. J. Quant. Struct.-Prop. Relat.*, **2**, 1–18.
4. Dearden, J. C. (2016). The history and development of quantitative structure-activity relationships (QSARs). *Int. J. Quant. Struct.-Prop. Relat.*, **1**, 1–44.
5. Puzyn, T., Rasulev, B., Gajewicz, A., Hu, X., Dasari, T. P., Michalkova, A., Hwang, H. M., Toropov, A., Leszczynska, D. and Leszczynski, J.

(2011). Using nano-QSAR to predict the cytotoxicity of metal oxide nanoparticles, *Nat. Nanotechnol.*, **6**, 175–178.

6. Puzyn, T., Leszczynska, D. and Leszczynski, J. (2009). Toward the development of "nano-QSARs": advances and challenges, *Small*, **5**, 2494–2509.
7. Kar, S., Gajewicz, A., Puzyn, T., Roy, K. and Leszczynski, J. (2014). Periodic table-based descriptors to encode cytotoxicity profile of metal oxide nanoparticles: a mechanistic QSTR approach, *Ecotox. Environ. Saf.*, **107**, 162–169.
8. Kar, S., Gajewicz, A., Puzyn, T. and Roy, K. (2014). Nano-quantitative structure–activity relationship modeling using easily computable and interpretable descriptors for uptake of magnetofluorescent engineered nanoparticles in pancreatic cancer cells, *Toxicol. in Vitro.*, **28**, 600–606.
9. Worth, A. P., Bassan, A., De Bruijn, J., Saliner, A. G., Netzeva, T., Patlewicz, G., Pavan, M., Tsakovska, I. and Eisenreich, S. (2007). The role of the European Chemicals Bureau in promoting the regulatory use of (Q)SAR methods, *SAR QSAR Environ. Res.*, **18**, 111–125.
10. Russell, W. M. S. and Burch, R. L. (1959). *The Principles of Humane Experimental Technique* (Methuen, London).
11. Chirico, N. and Gramatica, P. (2011). Real external predictivity of QSAR models: how to evaluate it? Comparison of different validation criteria and proposal of using the concordance correlation coefficient, *J. Chem. Inf. Model.*, **51**, 2320–2335.
12. Alexander, D. L. J., Tropsha, A. and Winkler, D. A. (2015). Beware of R2: simple, unambiguous assessment of the prediction accuracy of QSAR and QSPR models, *J. Chem. Inf. Model.*, **55**, 1316–1322.
13. Cherkasov, A., Muratov, E. N., Fourches, D., Varnek, A., Baskin, I. I., Cronin, M., Dearden, J., Gramatica, P., Martin, Y. C. and Todeschini, R. (2014). QSAR modeling: where have you been? Where are you going to?, *J. Med. Chem.*, **57**, 4977–5010.
14. Consonni, V., Ballabio, D. and Todeschini, R. (2010). Evaluation of model predictive ability by external validation techniques, *J. Chemom.*, **24**, 194–201.
15. Golbraikh, A. and Tropsha, A. (2002). Beware of q^2!, *J. Mol. Graphics Modell.*, **20**, 269–276.
16. Tong, W., Xie, Q., Hong, H., Shi, L., Fang, H. and Perkins, R. (2004). Assessment of prediction confidence and domain extrapolation of two structure–activity relationship models for predicting estrogen receptor binding activity, *Environ. Health. Perspect.*, **112**, 1249–1254.

17. OECD Principles for the Validation of (Q)SARs, http://www.oecd.org/dataoecd/33/37/37849783.pdf (Accessed on: May 10, 2017).

18. OECD, Environment Directorate, joint meeting of the chemicals committee and the working party on chemicals, pesticides and biotechnology, http://www.olis.oecd.org/olis/2004doc.nsf/ (Accessed on: May 10, 2017).

19. Wold, S. and Eriksson, L. (1995). *Chemometric Methods in Molecular Design*, van de Waterbeemd, H. E., ed. (VCH, Weinheim, Germany), pp. 195–218.

20. Roy, K. (2007). On some aspects of validation of predictive QSAR models, *Expert. Opin Drug. Discov.*, **2**, 1567–1577.

21. Golbraikh, A. and Tropsha, A. (2002). Predictive QSAR modeling based on diversity sampling of experimental datasets for the training and test set selection, *J. Comput. Aided. Mol. Des.*, **16**, 357–369.

22. Roy, K. and Ambure, P. (2016). The "double cross-validation" tool for MLR QSAR model development, *Chemom. Intell. Lab. Syst.*, **159**, 108–126.

23. Mitra, I., Saha, A. and Roy, K. (2010). Exploring quantitative structure-activity relationship (QSAR) studies of antioxidant phenolic compounds obtained from traditional Chinese medicinal plants, *Mol. Simult.*, **36**, 1067–1079.

24. Roy, K., Ambure, P. and Aher, R. (2017). How important is to detect systematic error in predictions and understand statistical applicability domain of QSAR models?, *Chemom. Intell. Lab. Syst.*, **162**, 44–54.

25. Roy, K., Kar, S. and Das, R. N. (2015). *Understanding the Basics of QSAR for Applications in Pharmaceutical Sciences and Risk Assessment* (Academic Press).

26. Roy, K., Kar, S. and Das, R. N. (2015). *A Primer on QSAR/QSPR Modeling: Fundamental Concepts*, SpringerBriefs in Molecular Science (Springer).

27. Roy, K. and Kar, S. (2015). How to judge predictive quality of classification and regression based QSAR models?, in *Frontiers of Computational Chemistry*, Haq, Z. U. and Madura, J., eds. (Bentham), pp. 71–120.

28. Walkera, J. D., Carlsenb, L. and Jaworskac, J. (2003). Improving opportunities for regulatory acceptance of QSARs: the importance of model domain, uncertainty, validity and predictability, *QSAR Comb. Sci.*, **22**, 346–350.

29. Lothar, S. and Reynarowych, Z. (1982). *Applied Statistics: A Handbook of Techniques* (Springer-Verlag, New York).

30. Tropsha, A. (2010). Best practices for QSAR model development, validation, and exploitation, *Mol. Inf.*, **29**, 476–488.

31. Roy, K., Chakraborty, P., Mitra, I., Ojha, P. K., Kar, S. and Das, R. N. (2013). Some case studies on application of rm2 metrics for judging quality of quantitative structure–activity relationship predictions: emphasis on scaling of response data, *J. Comput. Chem.*, **34**, 1071–1082.

32. Roy, K., Mitra, I., Kar, S., Ojha, P. K., Das, R. N. and Kabir, H. (2012). Comparative studies on some metrics for external validation of QSPR models, *J. Chem. Inf. Model.*, **52**, 396–408.

33. Mitra, I., Roy, P. P., Kar, S., Ojha, P. K. and Roy, K. (2010). On further application of rm2 as a metric for validation of QSAR models, *J. Chemom.*, **24**, 22–33.

34. Roy, K. and Kar, S. (2014). The r_m^2 metrics and regression through origin approach: reliable and useful validation tools for predictive QSAR models (commentary on 'Is regression through origin useful in external validation of QSAR models?'), *Eur. J. Pharm. Sci.*, **62**, 111–114.

35. Consonni, V., Ballabio, D. and Todeschini, R. (2009). Comments on the definition of the Q2 parameter for QSAR validation, *J. Chem. Inf. Model.*, **49**, 1669–1678.

36. Roy, K., Das, R. N., Ambure, P. and Aher, R. B. (152). Be aware of error measures. Further studies on validation of predictive QSAR models, *Chemom. Intell. Lab. Syst.*, **152**, 18–33.

37. Tong, W., Welsh, W. J., Shi, L., Fang, H. and Perkins, R. (2003). Structure-activity relationship approaches and applications, *Environ. Toxicol. Chem.*, **22**, 1680–1695.

38. Pery, A., Henegar, A. and Mombelli, E. (2009). Maximum-likelihood estimation of predictive uncertainty in probabilistic QSAR modeling, *QSAR Comb. Sci.*, **28**, 338–344.

39. Norinder, U., Carlsson, L., Boyer, S. and Eklund, M. (2014). Introducing conformal prediction in predictive modeling. A transparent and flexible alternative to applicability domain determination, *J. Chem. Inf. Model.*, **54**, 1596–1603.

40. Clark, R. (2009). DPRESS: localizing estimates of predictive uncertainty, *J. Cheminf.*, **1**, 11.

41. Sheridan, R. P. (2013). Using random forest to model the domain applicability of another random forest model, *J. Chem. Inf. Model.*, **53**, 2837–2850.

42. Vovk, V., Gammerman, A. and Shafer, G. (2005). *Algorithmic Learning in a Random World* (Springer, New York).

43. Eklund, M., Norinder, U., Boyer, S. and Carlsson, L. (2012). Conformal prediction in QSAR, in *Artificial Intelligence Applications and Innovations: Proceedings, Part II of AIAI 2012 International Workshops: AIAB, AIeIA, CISE, COPA, IIVC, ISQL, MHDW, and WADTMB*, Iliadis, L., Maglogiannis, I., Papadopoulos, H., Karatzas, K. and Sioutas, S., eds. (Springer, Berlin, Heidelberg), pp. 166–175.

44. Papadopoulos, H. and Haralambous, H. (2011). Reliable prediction intervals with regression neural networks, *Neural Networks*, **24**, 842–851.

45. Shafer, G. and Vovk, V. (2008). A tutorial on conformal prediction. *J. Mach. Learn. Res.*, **9**, 371–421.

46. Gammerman, A., Vovk, V. and Vapnik, V. (1998). Learning by transduction, *14th Conference on Uncertainty in Artificial Intelligence*, pp. 148–155.

47. Rücker, C., Rücker, G. and Meringer, M. (2007). y-Randomization and its variants in QSPR/QSAR, *J. Chem. Inf. Model.*, **47**, 2345–2357.

48. Gramatica, P. (2007). Principles of QSAR models validation: internal and external, *QSAR Comb. Sci.*, **26**, 694–701.

49. Tetko, I. V., Bruneau, P., Mewes, H. W., Rohrer, D. C. and Poda, G. I. (2006). Can we estimate the accuracy of ADME-Tox predictions?, *Drug Discov. Today*, **11**, 700–707.

50. Roy, K. and Kar, S. (2015). Importance of applicability domain of QSAR models, in *Quantitative Structure-Activity Relationships in Drug Design, Predictive Toxicology, and Risk Assessment*, Roy, K., ed. (IGI Global), pp. 180–211

51. Hanser, T., Barber, C., Marchaland, J. F. and Werner, S. (2016). Applicability domain: towards a more formal definition, *SAR QSAR Environ. Res.*, **27**, 865–881.

52. Roy, K., Kar, S. and Ambure, P. (2015). On a simple approach for determining applicability domain of QSAR models, *Chemom. Intell. Lab. Syst.*, **145**, 22–29.

53. Ambure, P., Aher, R. B. and Roy, K. (2015). Recent advances in the open access cheminformatics toolkits, software tools, workflow environments, and databases, in *Methods in Pharmacology and Toxicology* (Springer), pp. 257–296.

Chapter 11

Green Toxicology Meets Nanotoxicology: The Process of Sustainable Nanomaterial Development and Use

Alexandra Maertens and Thomas Hartung
Center for Alternatives to Animal Testing,
Johns Hopkins Bloomberg School of Public Health,
Baltimore, Maryland, USA
amaerte1@jhu.edu

Green chemistry is a movement that seeks to design benign compounds in a manner such that waste and toxicity are reduced at every stage of production. Nanomaterials have an enormous potential to be part of the green chemistry revolution, but at the same time, there is uncertainty about how to characterize the health and safety challenges they present as well as lingering concerns about how "green" they truly are. For "green chemistry" to provide less toxic alternatives, there needs to be "green toxicology"—that is to say a toxicological paradigm that can guide the design of safer alternatives rather than our discovering hazards after a chemical has become commonplace. Early identification of hazards requires

Computational Nanotoxicology: Challenges and Perspectives
Edited by Agnieszka Gajewicz and Tomasz Puzyn

ISBN 978-981-4800-64-8 (Hardcover), 978-0-429-34137-3 (eBook)
www.jennystanford.com

a shift of the toxicological paradigm away from slow and expensive animal testing toward more rapid methods that can work in tandem with the research and development process. The need is especially acute for nanomaterials, as they present several challenges: toxicity that can be a function of both chemical and physical properties, novel toxic mechanisms, and complicated ADME properties and their sheer molecular complexity and diversity.

11.1 Green Toxicology

Green chemistry began as a movement to steer the practice of chemical research proactively toward sustainability and benign design [1]. The Twelve Principles of Green Chemistry established by Anastas and Warner can be summarized with the idea that chemistry should focus on designing less toxic products, using a production process as safe as possible, using sustainable inputs, minimizing waste, and reducing exposure.

Toxicology has traditionally been concerned with the very last stages of novel chemical or material development—or often after a product has been available for years and is in widespread use. "Green toxicology" [2] seeks to alter this paradigm and put toxicological considerations at the very early stages of chemical development—basically "front-loading" toxicology testing and making toxicology a discipline that can work side by side with chemists in the development of new toxicity.

Toxicology has faced on the one hand the Scylla of failing to foresee the toxicity of a compound before it is in widespread use and on the other hand the Charybdis of adopting the precautionary principle and assuming chemicals are hazardous until proven otherwise. A prime example of the former is the history of asbestos, which went from being a miraculous substance to being banned in the last century. Despite the fact that the warning was raised shortly after its introduction in 1898 and the first rat study that demonstrated carcinogenicity was performed in 1911, asbestos continued to be in use until finally it was banned by the EU and the US in the year 2000 [3]. On the other hand, the precautionary principle assumes hazard when in reality the great majority of chemicals in industrial use are relatively nontoxic [4]—and this presumption of hazard for new chemicals may lead to useful chemicals being unnecessarily abandoned. As an example, the case of genotoxicity—a key concern in hazard assessment due to the difficulty of identifying human carcinogens —has in some respects been well served by in vitro tests that front-load toxicity testing with excellent sensitivity [5], but the false positive rate from such tests is startlingly high. Kirkland [6] estimated that more than 90% of noncarcinogens are false positive, and although this study has one of the higher estimates of false positives, the case remains that the low specificity has the potential to consign whole classes of chemicals unnecessarily to the dustbin. As nanomaterials have been the focus of substantial popular anxiety as well as scientific uncertainty about their effects, adopting test batteries with low-specificity could lead to an overly precautious approach and therefore lost opportunities.

11.2 Green Toxicology Principles Applied to Nanomaterials

11.2.1 Principle 1: Design Benign

The ideal time to detect toxicity is before a compound has been synthesized. Yet to do this, toxicologists need to provide tools that can predict toxicity on the basis of structural features or other properties. To date, predictive toxicology has focused on two aspects— (i) absorption, distribution, metabolism, and excretion (ADME) properties that can be deduced from physiologically based pharmacokinetic (PBPK) models and (ii) qualitative or quantitative structure–activity relationship (QSARS) and read-across approaches that predict toxicity from structural features.

The many ways in which nanomaterials (including nanomedicines) interact with biological barriers represent a significant data gap for nanotoxicology. Nanomaterials have different transport properties from the equivalent bulk materials, and PBPK models need to be repurposed for nanomaterials and use nano-specific descriptors that capture traffic within tissues and cells, interaction with blood and tissue cells, tissue/blood partition coefficients, tissue concentration, and permeability through membranes [7].

While QSARS have certainly proved their merit in the pharmaceutical industry, the success has been difficult to replicate in industrial chemicals and QSARs face specific challenges when applied to nanomaterials. Pharmaceuticals are generally a single compound designed for biological activity and usually lie within a fairly narrow chemical space following the Lipinski rules. In contrast, nanomaterials are structurally diverse and can be composed of organic, metal, metal oxide, silica, or carbon nanotubes, with different core and surface coating materials. Moreover, some properties (such as agglomeration and coating material) can be modified by the biological environment of the nanomaterial. Nonetheless, QSARs have been adapted to nanomaterials (sometimes called QNARs[1]) with the addition of nano-specific descriptors that include size, shape, surface area, relaxivities, solubility, zeta potential, corona, and surface charge as well as absorption and distribution [8]. Currently, the lack of high-quality data sets [9], limited understanding of the range of possibility of mechanisms of action, and the range of nano-specific descriptors will need to be improved. All such predictive models are dependent on both a well-defined domain and adequate, high-quality data ("garbage in, garbage out" being a well-known mantra of data mining).

Read-across is currently the most common alternative to animal testing [10] and has the advantage of not requiring a large data set. As has been demonstrated with polymers, class-based approaches that take into consideration the range of molecular weights, potential impurities, and group chemicals by an understanding of mechanisms of toxicity can be successfully adapted by regulatory agencies [11]. A similar approach would likely be useful for nanomaterials, and a framework for this has been outlined by Walser [12] that groups nanomaterials into common categories on the basis of similar structural elements, size distribution, shapes, and structural element such as core and coating. Other strategies have focused on categorization according to exposure/use scenarios as well as physicochemical properties that are linked to specific biological outcomes (e.g., potential to induce acute oxidative stress and inflammation) [13].

[1]Quantitative nanostructure–activity relationships.

11.2.2 Principle 2: Test Early, Fail Safe

The pharmaceutical industry has long understood the necessity of front-loading toxicity so that new products would fail early and therefore fail cheaply [14]. This change in paradigm came about in part because of the large numbers of compounds that had been deemed safe in animal testing yet failed in human clinical trials, often because of ADME considerations, lack of efficacy, or unforeseen, off-target toxicity [15]. This required the development of rapid, screening-level tests that could provide human-relevant data.

Such in vitro tests offer substantial improvements in comparison to black box animal tests, which provide little, if any, mechanistic information and focus on high-dose exposures. In contrast, in vitro tests offer the possibility of examining effects of low-concentration exposures on cells and potentially organ systems, given the rapid growth of organ-on-a-chip systems [16, 17].

Nanomaterials specifically have several concerns that can be targeted with in vitro systems. Nanomaterial uptake is typically either from the air-blood barrier in the lungs, the intestine-blood barrier in the gastrointestinal tract, and the skin; in the instances of nanomaterials that are demonstrated to have some potential for systemic circulation, a key additional concern would be the blood-brain barrier and placental barrier [18]. All of these can potentially be modeled with in vitro systems and focusing on the realistic bioavailability would likely give a better estimate of hazard than extrapolation from toxicity studies in animals by routes such as instillation, which cause both tissue damage and stress for the animal.

At the same time, one must be mindful of the potential for such in vitro systems to introduce artifacts of their own; for example, specific nanomaterials are known to interfere in several standardized assays of oxidative stress [19]. The MTT assay and the comet assay may have substantial interference from the nanoparticles [20, 21]. Many in vitro tests have been performed with concentrations sufficient for inducing agglomeration at the surface of the cell [22], which leads to cytotoxicity due to a lack of oxygen and nutrients from the surrounding medium. Therefore, one key to the successful application of in vitro test systems that can be used with confidence as an effective screening tool is a more comprehensive, systematic attempt to understand the limits of current assay systems with respect to nanotoxicology and to adapt appropriate protocols and standard operating procedures that are specific for nanomaterials. Such standardized protocols will enable another key aspect necessary to propel in vitro predictive toxicology forward—the development and curation of extensive data sets of reliable and replicable in vitro results for both toxicokinetic features as well as the molecular mechanisms and endpoints most associated with nanomaterials (reactive oxygen species, inflammation) that can be used to develop more precise in silico tools to guide benign design.

11.2.3 Principle 3: Avoid Exposure and thus Testing Needs

Toxicologists tend to think of molecules and think less of the inputs, outputs, and processes involved in industrial chemistry. Unfortunately, this eliminates the opportunity to design a synthetic process that prevents or minimizes exposure—and in the absence of exposure, there can be no risk. A broad concept of toxicology should include an understanding of the entire industrial process, including consumer exposure and potential ecotoxicity, as well as consideration of not just the primary material but the inputs, impurities, and the likely degradation products as well. An example of the importance of these considerations is the case of fullerenes, which were initially thought to have substantial toxicity; later it was established that the toxicity was due to the solvent [23], and indeed solvents, catalysts, or impurities often contribute substantially to the toxicity of a material [24, 25].

At every step, there are likely opportunities to avoid exposure and therefore reduce testing needs—one-pot synthesis (meaning that all reactions take place in the same vessel) can limit occupational exposure to reactants and catalysts. Industrial processes can be engineered to minimize environmental dispersal, often by simple changes in workflow that simultaneously minimize waste. All materials should be designed to degrade into benign products.

11.2.4 Principle 4: Make Testing Sustainable

The sheer number of possible nanomaterials guarantees that it is not possible to conduct animal testing for every possible substance. Additionally, animal testing is both expensive and inefficient—it not only consumes a great deal of materials but also produces an enormous amount of waste. It requires amounts of material that are not practical at the research and development stage. Finally, it acts as an economic barrier to the adoption of greener technologies [26]. For testing to be sustainable, it has to substantially reduce its reliance on such an expensive and wasteful process.

Additionally, the quality of published data must be substantially improved. Recent reviews of the literature [27–29] have pointed to substantial data gaps and data reporting issues in the published literature, including failure to adequately characterize the experiments in terms of dosing regime, inappropriately high (overload) doses or insufficient attention to the dose-response curve, incomplete characterization of the physical chemical properties of the material, failure to consider potential interference in in vitro assays, and inadequate data concerning exposure scenarios. More broadly speaking, the lack of clearly delineated endpoints and agreed-upon interpretation hampers replicability and comparability of results [28]. The difficulty in publishing no-effect studies likely skews the literature [30], while on the other hand, the higher-quality, standardized, guideline-compliant in vivo studies are often performed for regulatory purposes, and the data are either not made publicly available or not available in a way that is machine readable [31]. Not only is there substantial duplication—it was recently found by the European Chemicals Agency that multiple substances had over 45 repeat Draize tests [32]—because of this, but it also means that such data are not available for computational analysis. For testing to be sustainable, it should be built on a strong foundation of quality and adopt the same approach to standardization and quality control as is already practiced in areas like analytical and clinical chemistry.

11.2.5 Principle 5: Adopt Test Strategies That Are Not Mature Enough for Regulatory Use

Regulatory agencies typically do not move quickly—a traditional validation study takes about one decade—and therefore regulatory testing regimes are not based on cutting-edge science. Given the pace at which nanomaterials science is changing, and the our steadily improving understanding of nanotoxicology, it will likely prove difficult for regulatory agencies to keep up. One example of a proactive approach is the ITS-NANO33, which provided a framework to prioritize research that was developed from the consensus of major stakeholder groups, encompassing testing needs for physicochemical characterization, exposure identification, hazard identification, and modeling, in an "intelligent testing strategy" that can assess nanomaterials as accurately and efficiently as possible. In the longer term, this opens up an opportunity for nanomaterial research and development to use a more flexible approach and begin evaluating and adopting in silico and in vitro tests that are informative yet individually too inaccurate for regulatory purposes [34, 35]. Eventually, it may prove possible to combine data from multiple tests in an "integrated testing strategy" that is flexible, maximizes predictive power, and also provides a quantitative estimate of the uncertainty in the data [36]. This capacity can then help move regulatory agencies to validate and adopt a more extensive battery of tests. If gathered systematically and shared, the data can help strengthen predictive approaches that focus on understanding the specific cellular and molecular changes that lead to adverse outcomes; such molecular-based understanding is in turn key in guiding chemists in the design of less hazardous materials.

11.2.6 Principle 6: Green Toxicology as Twenty-First-Century Toxicology

Nanotoxicology will certainly require a twenty-first-century toxicology—that is to say, toxicology that embraces systems-level thinking, computational approaches, high-throughput and high-content in vitro technologies, as well as a "big data" approach. Many of the life science disciplines have transitioned over the last two decades from a reductionist and largely qualitative science to a more quantitative approach that utilizes high-throughput, high-content data and multiscale modeling to look at biology from a more systems perspective. However, the "systems biology" perspective has proven difficult for regulatory toxicology and hazard assessment to take advantage of, and the field of toxicology has only recently begun to assemble the type of large-scale data sets that have proven crucial in many fields of medicine and molecular biology. However, in the case of nanotoxicology, such data sets will likely be essential, as the complexity of the types of materials and biological interactions would be better served by an approach that uses the insights of systems biology, computational modeling, and exploratory data mining to locate the molecular mechanisms of toxicity in perturbed biological systems rather than a reductionist approach based on "feet-up/feet-down" assays (i.e., LD_{50}) or other comparatively coarse-grained approaches. The green toxicology paradigm offers a path to twenty-first-century toxicology appropriate for the age of nanomaterials.

11.3 Going Forward

Greener toxicology for green nanomaterials will require a large-scale commitment on the part of the community. To begin with, there needs to be a coordinated effort to produce a rational grouping of nanomaterials based on physicochemical properties, hazard, and exposure properties. Secondly, there needs to be an assessment of existing tools, to understand both their limitations and their applicability domain, with an eye toward developing the most cost-effective and accurate way to characterize hazards. Thirdly, we need to better understand how to intelligently integrate our existing knowledge. Lastly, we need to develop and validate new tools that can meet the challenges posed by nanomaterials and provide high-quality data—this will likely include organ-on-a-chip systems as well as other customized solutions that can characterize different life cycle stages as well as possible biotransformations.

This is no doubt a very tall order, yet such an effort offers the opportunity for nanomaterial toxicology to be as cutting-edge in the world of hazard assessment as it is in molecular design. Conversely, if the toxicology community fails to produce a convincing body of sound data to thoroughly characterize the hazards of nanomaterials, there is both the Charybdis that a nanomaterial in widespread use will prove hazardous as well as the Scylla that nanomaterial regulations will be driven by public (mis)perceptions and regulatory agency overreaction. Good science is the only way through such a dilemma.

References

1. Anastas, P. T. and Warner, J. C. (2000). *Green Chemistry: Theory and Practice* (Oxford University Press).
2. Maertens, A., et al. (2014). Green toxicology, *ALTEX*, **31**, 243–249.
3. Gee, D. and Greenberg, M. (2001). Asbestos: from 'magic' to malevolent mineral, *Late lessons from early warnings: the precautionary principle 1896–2000*, 52–63.
4. Hartung, T. (2010). Food for thought ... on alternative methods for chemical safety testing, *ALTEX*, **27**, 3–14.
5. Walmsley, R. M. and Billinton, N. (2011). How accurate is in vitro prediction of carcinogenicity? *Br. J. Pharmacol.*, **162**, 1250–1258.
6. Kirkland, D., Aardema, M., Henderson, L. and Müller, L. (2005). Evaluation of the ability of a battery of three in vitro genotoxicity tests to discriminate rodent carcinogens and non-carcinogens I. Sensitivity, specificity and relative predictivity, *Mutat. Res.*, **584**, 1–256.
7. Li, M., Zou, P., Tyner, K. and Lee, S. (2017). Physiologically based pharmacokinetic (PBPK) modeling of pharmaceutical nanoparticles, *AAPS J.*, **19**, 26–42.
8. Raies, A. B. and Bajic, V. B. (2016). In silico toxicology: computational methods for the prediction of chemical toxicity. *Wiley Interdiscip. Rev. Comput. Mol. Sci.*, **6**, 147–172.
9. Lubinski, L., et al. (2013). Evaluation criteria for the quality of published experimental data on nanomaterials and their usefulness for QSAR modeling, *SAR QSAR Environ. Res.*, **24**, 995–1008.
10. Ball, N., et al. (2016). Toward good read-across practice (GRAP) guidance, *ALTEX*, **33**, 149–166.
11. Boethling, R. S. and Nabholz, J. V. (1996). *Environmental Assessment of Polymers under the US Toxic Substances Control Act* (United States Environmental Protection Agency).
12. Walser, T. and Studer, C. (2015). Sameness: the regulatory crux with nanomaterial identity and grouping schemes for hazard assessment, *Regul. Toxicol. Pharmacol.*, **72**, 569–571.

13. Godwin, H., et al. (2015). Nanomaterial categorization for assessing risk potential to facilitate regulatory decision-making, *ACS Nano*, **9**, 3409–3417.

14. Hartung, T. and Zurlo, J. (2012). Food for thought ... alternative approaches for medical countermeasures to biological and chemical terrorism and warfare, *ALTEX*, **29**, 251.

15. Schuster, D., Laggner, C. and Langer, T. (2005). Why drugs fail--a study on side effects in new chemical entities, *Curr. Pharm. Des.*, **11**, 3545–3559.

16. Judson, R. S., et al. (2011). Estimating toxicity-related biological pathway altering doses for high-throughput chemical risk assessment, *Chem. Res. Toxicol.*, **24**, 451–462.

17. Lee, J. B. and Sung, J. H. (2013). Organ-on-a-chip technology and microfluidic whole-body models for pharmacokinetic drug toxicity screening, *Biotechnol. J.*, **8**, 1258–1266.

18. Savolainen, K., et al. (2010). Risk assessment of engineered nanomaterials and nanotechnologies: a review, *Toxicology*, **269**, 92–104.

19. Tournebize, J., Sapin-Minet, A., Bartosz, G., Leroy, P. and Boudier, A. (2013). Pitfalls of assays devoted to evaluation of oxidative stress induced by inorganic nanoparticles, *Talanta*, **116**, 753–763.

20. Kroll, A., Pillukat, M. H., Hahn, D. and Schnekenburger, J. (2012). Interference of engineered nanoparticles with in vitro toxicity assays, *Arch. Toxicol.*, **86**, 1123–1136.

21. Magdolenova, Z., Lorenzo, Y., Collins, A. and Dusinska, M. (2012). Can standard genotoxicity tests be applied to nanoparticles? *J. Toxicol. Environ. Health A*, **75**, 800–806.

22. Hirsch, C., Roesslein, M., Krug, H. F. and Wick, P. (2011). Nanomaterial cell interactions: are current in vitro tests reliable? *Nanomedicine*, **6**, 837–847.

23. Spohn, P., et al. (2009). C60 fullerene: a powerful antioxidant or a damaging agent? The importance of an in-depth material characterization prior to toxicity assays, *Environ. Pollut.*, **157**, 1134–1139.

24. Constable, D. J. C., Jimenez-Gonzalez, C. and Henderson, R. K. (2007). Perspective on solvent use in the pharmaceutical industry, *Org. Process Res. Dev.*, **11**, 133–137.

25. Sheldon, R. A. (1992). Organic synthesis-past, present and future, *Chem. Ind.*, 903–906.

26. Bottini, A. A. and Hartung, T. (2009). Food for thought ... on the economics of animal testing, *ALTEX*, **26**, 3–16.

27. Card, J. W., Jonaitis, T. S., Tafazoli, S. and Magnuson, B. A. (2011). An appraisal of the published literature on the safety and toxicity of food-related nanomaterials, *Crit. Rev. Toxicol.*, **41**, 22–49.

28. Krug, H. F. (2014). Nanosafety research—are we on the right track? *Angew. Chem. Int. Ed.*, **53**, 12304–12319.

29. Hristozov, D. R., Gottardo, S., Critto, A. and Marcomini, A. (2012). Risk assessment of engineered nanomaterials: a review of available data and approaches from a regulatory perspective, *Nanotoxicology*, **6**, 880–898.

30. Krug, H. F. and Wick, P. (2011). Nanotoxicology: an interdisciplinary challenge, *Angew. Chem. Int. Ed. Engl.*, **50**, 1260–1278.

31. Luechtefeld, T., et al. (2016). Global analysis of publicly available safety data for 9,801 substances registered under REACH from 2008–2014, *ALTEX*, **33**, 95–109.

32. Luechtefeld, T., et al. (2016). Analysis of draize eye irritation testing and its prediction by mining publicly available 2008–2014 REACH data, *ALTEX*, **33**, 123–134.

33. Stone, V., et al. (2014). ITS-NANO--prioritising nanosafety research to develop a stakeholder driven intelligent testing strategy, *Part. Fibre Toxicol.*, **11**, 9.

34. Hartung, T. and Hoffmann, S. (2009). Food for thought ... on in silico methods in toxicology, *ALTEX*, **26**, 155–166.

35. Hartung, T., Hoffmann, S. and Stephens, M. (2013). Mechanistic validation, *ALTEX*, **30**, 119–130.

36. Hartung, T., Luechtefeld, T., Maertens, A. and Kleensang, A. (2013). Integrated testing strategies for safety assessments, *ALTEX*, **30**, 3–18.

Chapter 12

Issues for and Examples of Computational Design of "Safe-by-Design" Nanomaterials

David A. Winkler
La Trobe Institute for Molecular Science, La Trobe University, Kingsbury Drive, Bundoora, 3046, Australia
Monash Institute of Pharmaceutical Sciences, Monash University, Parkville 3052, Australia
School of Pharmacy, University of Nottingham, Nottingham NG7 2RD, UK
School of Chemical and Physical Sciences, Flinders University, Bedford Park 5042, Australia
CSIRO Manufacturing, Bayview Avenue, Clayton 3168, Australia
d.winkler@latrobe.edu.au

The interactions of nanomaterials with biological systems are exceedingly complex and multifactorial. While a detailed understanding of the myriad of processes that occur when a nanoparticle enters the body that lead to a possible adverse effect is not possible, data-driven statistical and machine learning modeling methods (nano-QSAR models) can capture enough of the complexity to help make useful predictions for new materials, thus aiding materials regulatory processes. Increased understanding of

Computational Nanotoxicology: Challenges and Perspectives
Edited by Agnieszka Gajewicz and Tomasz Puzyn

ISBN 978-981-4800-64-8 (Hardcover), 978-0-429-34137-3 (eBook)
www.jennystanford.com

the mechanisms responsible for nanotoxicity and the availability of good predictive models are now making possible a "safety-by-design" (SbD) paradigm for nanomaterial development for medical and other diverse commercial applications. Here we discuss issues limiting the application of this paradigm and provide examples of SbD.

12.1 Introduction

The dream of scientists working with nanomaterials is to have a process for reliably designing new nanomaterials with specified performances and with known, predictable, and tailorable biological properties. Such a process would allow the discovery of new materials with hitherto unknown physical or biological advantages while minimizing any adverse impacts on workers, the public, and the environment. Clearly, it is not yet possible to create such a bespoke nanomaterials design process, but experimental materials scientists and computational modelers have started to define the components for such a system. Encouragingly, significant progress has been made along the path to such a useful design paradigm, but there is still a long way to go. This chapter outlines some of the main issues that safe-by-design (SbD) processes face and how these difficulties can be overcome using a combination of experiments on the effects of nanomaterials on biological systems and advanced material modeling methods.

Bioactive molecules, such as drugs and materials more generally exist in a space of possibilities that is truly immense and essentially infinite for all practical purposes. This poses problems of how best to explore such large chemical spaces. Currently, accelerated synthesis and testing technologies and the use of structure–property models and evolutionary algorithms provide the most rational path forward. Compared to small molecules, nanomaterials have additional complexities due to the sensitivity of their properties to provenance (manufacturing processes and processing, size and shape distributions), the physicochemical transformations they may undergo (agglomeration, dissolution, erosion), and their interactions with complex biological fluids (serum, plasma, ecosystems) that modify their surfaces by binding proteins, humic substances, ions, etc. The sum total of these processes generates the so-called biologically relevant entity responsible for the positive and potentially adverse properties of nanomaterials.

Clearly, understanding the toxicity of nanomaterials is important for human and environmental health and safety as well as public acceptance, given the rapidly increasing number of products containing them [1]. Although there were more than 6000 publications on nanotoxicology in 2014, some key questions are still unanswered and some established paradigms need to be challenged or even replaced [2].

Otrowski et al. recently reviewed the state of knowledge about nanotoxicology, an important step in understanding the health and environmental implications of widespread use of nanomaterials. They used bibliometric techniques to analyze the current nanotoxicological literature and found a wide range of disciplines and subfields were involved in nanomaterial research. In particular, they found that in vitro testing in the absence of an exposure pathway emphasizes acute toxicity and mortality rather than chronic exposure and morbidity. Relatively little research has been reported on nanomaterials in real consumer products, particularly regarding their environmental fate.

Fadeel and Garcia-Bennett strongly advocated a more careful and broad study of the potential toxicities of nanoparticles used in biomedical applications in particular [3]. They argued that there needs to be a much clearer understanding of mechanisms of nanomaterial-induced toxicities and a detailed assessment of the factors influencing the biocompatibility and toxicity of nanoparticles. They reviewed recent developments related to inorganic nanoparticles used in nanomedicine, focusing on synthesis, physicochemical properties, and cytotoxicity of these nanoparticles. They identified interactions between nanomaterials and the immune system as being particularly important. They extended their discussions to the importance of size and surface area of nanoparticles relative to that of cells and essential cellular machinery, such as the ribosome, and described in detail the importance of high-throughput synthesis and assessment methods to screen a large range of nanoparticle types [4]. These data are essential for the development of artificial intelligence–based computational models that can leverage scarce data to make biological activity predictions for new materials not yet synthesized or tested. Such models allow new materials to be triaged for possible adverse biological effects, assisting regulatory agencies to assess nanomaterials for safety more efficiently.

This theme was also taken up by Warheit [5]. He was also concerned about the dearth of nanotoxicity data and additionally provided some useful guidance on common misconceptions in nanoparticle structure–property relationships. Some of these misconceptions are (i) nanoparticles are always more toxic than bulk particles of a similar or identical composition, (ii) particle size and surface area are the key factors influencing nanotoxicity, and (iii) all forms of a given nanomaterials particles have similar toxicity profiles. If insufficient data are available and computational models are not robust or widely applicable enough, then regulatory agencies and the public will remain concerned about the impact of nanomaterials on health and the environment. The precautionary principle may lead to the establishment of special, restrictive regulations for commercialization of products containing nanomaterials. He did not support contemporary views that in silico or in vitro cell culture assays are sufficient to identify health hazards in vivo.

The specific promise and challenge of developing robust in silico models for regulatory uses was reviewed by Clark et al. [6]. They acknowledged the great complexity of quantifying the risks posed by nanomaterials because of the wide range of factors that influence the toxicity of a given nanomaterial. Shape, size, and their distributions; core chemical composition and surface chemistry; and interaction with complex biological fluids are only a few of the factors affecting toxicity. They also advocated rapid and efficient screening methods for evaluating the potential hazards of nanomaterials. Given sufficient experimental data to train them, predictive toxicity models can provide a valuable element to the armory of regulators. Similar themes were developed by Culha and Altunbek [7] and Autrup [8].

12.2 Biocorona: Biologically Relevant Entity

It was recognized early in the development of nanoparticle research that the materials do not interact with cells and other biological entities in a pristine state. Most biological fluids are very complex mixtures of small molecules, proteins, peptides, and ions, for example.

The "corona" of molecules attached to the surface of nanoparticles is dynamic, dependent on the particular biological compartment the materials transit, and dependent on the competition between thermodynamic and kinetic processes [9]. More abundant proteins with lower affinity for the nanoparticle tend to bind initially, but these are slowly replaced by less abundant, higher affinity proteins. Understanding these processes and predicting them is an extremely important part of understanding the biological interactions and potential toxicities of the biologically relevant entity, the combination of the particle and its corona (Fig. 12.1). The mechanisms and complexity of corona formation, resulting in the biologically relevant entity (what the biosystems "see" and respond to), have been reviewed by Yang et al. [10]. Clearly, understanding how nanoparticles interact with proteins is crucial for both applications and safety.

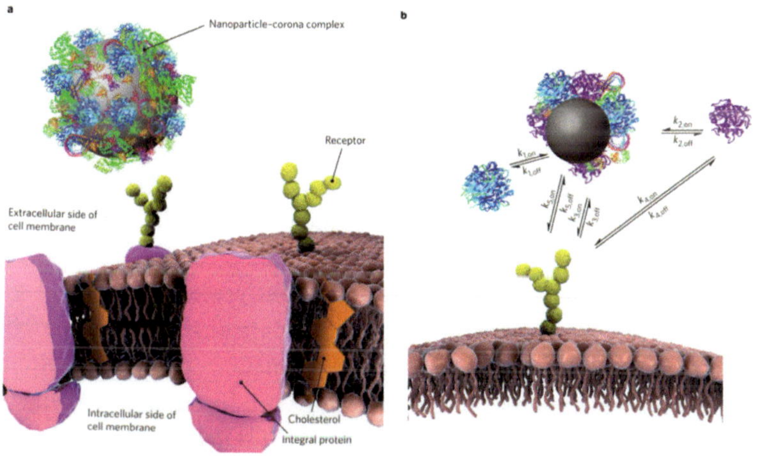

Figure 12.1 Nanoparticle corona showing dynamic interaction between proteins and nanoparticles driven by thermodynamic and kinetic processes and the interaction of the biologically relevant entity with a cell. Reprinted by permission from Springer Customer Service Centre GmbH: Springer Nature, *Nature Nanotechnology*, Ref. [11], Copyright (2012).

Experimental and computational methods have been developed to study this important process. Cedervall et al. used isothermal titration calorimetry and surface plasmon resonance (SPR) to study the interactions of proteins with nanoparticles [9]. Tassa and colleagues used high-throughput SPR to study the interaction between a single protein target and a structurally related series of targeting ligands whose affinity varied over a 4500-fold range, at protein densities that reflect endogenous receptor densities [12]. Similar approaches can be used to assess the affinity of a single type of nanoparticle for a large number of proteins, potentially generating large data sets that can be used to generate in silico corona models.

The explicit computational modeling of corona is very complex, but initial promising results have been reported by Li et al. They used coarse-grained molecular dynamics (MD) methods to study the interactions between spherical, negatively charged citrate-covered silver nanoparticles and 15 common apolipoproteins [13]. They also characterized biocorona formation using electron microscopy imaging and circular dichroism spectroscopy and found decreased alpha-helix and increased beta-sheet content in proteins when bound to the nanoparticle. Chen et al. used high-throughput differential scanning fluorometry and circular dichroism to study the thermal stability and denaturation of proteins in the corona surrounding nanoparticles. They showed that nanoparticles display a particle/chemical duality and do not play a conventional ligand role in protein stabilization. This behavior affected nanoparticle solubility and size [14]. This research team also used an automated, high-throughput, and temperature-controlled dynamic light scattering method and MD to study thermostability and reversibility of silver nanoparticle-protein binding [15]. They used branched polyethyleneimine coated–silver nanoparticles interacting with like-charged lysozyme and oppositely charged alpha lactalbumin in their experiments.

With sufficient experimental information on the affinities and abundances of serum and plasma proteins it should be possible in principle to develop a mathematical model that predicts the corona composition through time for a given biological compartment. However, such a useful outcome is not yet in the immediate horizon due to a lack of comprehensive data on protein-nanoparticle affinities.

12.3 System Complexity, Reproducibility, Data Generation, and Curation

12.3.1 Complexity of Nanoparticle–Biology Interactions

As the previous section shows, the interaction of diverse types of nanoparticles with the biological milieu is very complex. There are a wide variety of compositions and forms of nanomaterials now being studied or even incorporated into products, with a myriad of physicochemical properties, such as particle size, shape, agglomeration state, crystal structure, chemical composition, surface area, and surface properties [16]. Recent studies of the effects of physiochemical properties on physiological interactions between nanomaterial toxicology and target biological areas have been reviewed [17].

To add to this complexity, there are also several hundred small and macromolecular entities in biological or environment fluids that interact to a greater or lesser extent with nanoparticles. To this complexity can be then added the interactions of the biologically relevant entity—nanoparticle plus corona—with cells, biological membranes, lipids, organs, and organisms. The situation is much more complex than with pharmaceutical agents, where the molecule is usually delivered to a specific biological target, such as a receptor or an enzyme, essentially unchanged and easily characterized. Issues involved with nanomaterial characterization have been highlighted recently by Nel and coworkers [18].

The ability of nanoparticles to elicit immune or inflammatory processes in vivo is also under intense scrutiny [19]. For example, cobalt and nickel nanoparticles cause inflammation, while silver nanoparticles reduce allergic inflammation. Some fibrous nanomaterials (e.g., high-aspect-ratio carbon nanotubes) can cause lung and other cellular damage that leads to the formation of precancerous lesions. David et al. called for additional research into the relationship between nanoparticle properties and immunotoxicity and described several mechanism by which nanoparticles can suppress or otherwise modulate the immune system [20]. Complement activation, platelet activation, hemolysis, and immune stimulation processes affected by circulating nanomaterials were also reviewed recently. David et al. advocated materials and assay standardization, and harmonization of research methods and data between international regulatory agencies, to provide better-informed regulatory and preclinical assessment of nanomaterials.

12.3.2 Experimental Issues

Apart from the complexity of the interaction of nanoparticles with the biological environment, an important and complex issue is identifying the type of rapid in vitro assays most suitable to describe the ultimate effect of nanomaterials on biological systems. Zebrafish and their embryos have been invaluable for this purpose., allowing the development of high-throughput screening and in silico data transformation tools that accelerate hazard ranking [21]. They allow rapid evaluation of cellular injury by reactive oxygen species production, intracellular calcium flux, mitochondrial depolarization, and plasma membrane permeability using automated epifluorescence microscopy. Observed patterns of nanoparticle biological responses have been clustered using self-organizing maps. George et al. concluded that high-throughput screening, in silico data handling, and zebrafish assays are useful for rapid toxicological screening [21].

An important part of the process of safe regulation of nanomaterials is an accurate assessment of risk. A recent study by Warheit and Donner queried the significance of risk determinations in the absence of complete data and stressed the importance of establishing standardized test guidelines [22]. They critically assessed the ability of nanotoxicology research articles to address health and safety issues for workers and consumers. They concluded that the potential risk from exposures to nanomaterials is highly exaggerated and that no nano-specific adverse effects had been convincingly demonstrated at that time. Furthermore, results from such standardized studies, and even "no effect" study findings, provide critically useful input for assessing safe levels of exposure. Low and no-effect results are, of course, essential for the development of robust computational models linking nanoparticle properties to adverse biological effects. Unfortunately, much of these valuable data are not published or are omitted in published work in favor of those results that show higher levels of adverse effects.

Misra et al. have tackled the complexity of nanoparticle dissolution and its importance in adverse biological effects [23]. They showed that two nominally identical nanoparticles could have totally different dissolution behaviors when surface-modified. The fraught issue of nanoparticle assays has been discussed by Ong and colleagues [24]. The unique properties of nanomaterials increase the risk of interference in analytical methods, but most published papers do not take interference into account. Ong et al. found that the nature and degree of interference cannot be predicted from current knowledge of nanomaterials.

Aggregation and agglomeration is also a very common and important problem with nanoparticles [25]. This phenomenon has generated toxicological effects in biological and ecological systems as it changes the nature of the biologically relevant entity and the solubility properties of nanoparticles. It may also affect the important property of nanoparticle biopersistence [16].

The use of biomarkers for toxicity and high-throughput screening using traditional and next-generation sequencing approaches has been described by Klaper et al. [26]. They caution that the most common endpoint investigated in nanotoxicity experiments is oxidative stress, sometimes a temporary response with no negative outcome. Alternative relevant pathways can be identified by biomarkers or global gene expression studies.

12.3.3 Data Availability, Reliability, and Processing Issues

Hackenberg recognized that the knowledge base of potential human health hazards of nanomaterials is very incomplete [27]. There is a dearth of knowledge on the mechanisms by which nanoparticles are taken up by the body. Some enter via the lungs or the gastrointestinal tract, but transcutaneous penetration is unlikely in healthy skin. However, isotopic labeling studies have shown that metal ions, if not nanoparticles, from sunscreens do penetrate the skin [28, 29]. Interactions of nanoparticles with cells involve primarily cellular uptake, intracellular fate, degradation, and cytotoxicity. Key factors influencing these processes were reviewed recently by Deng and Gao [30].

One of the main limitations to the derivation of widely applicable, useful models of the adverse effects of nanoparticles is the dearth of relevant biological data. Such data are expensive and time-consuming to collect but are essential for the development of data-driven models based on machine learning methods. Oksel et al. have reviewed the current availability of nanostructure-biological activity data [31]. They concluded that the available nanotoxicity data are insufficient for generating robust nano-QSAR models in the majority of cases. Oksel et al. also expressed concern over the paucity of nanostructure-biological activity data [31] as the quality of nano-QSAR models is clearly directly related to the quality of data used to train them. Krug and Nau [32] also noted that many published studies do not provide usable toxicological data, particularly for computational modeling, although they are often highly cited or appear in the popular press. Gajewicz also argued for widely available, systematically varied libraries of nanomaterials assessed in multiple toxicity assays to provide reliable data for nano-QSAR models with significant domains of applicability (spaces over which models can predict properties of new materials) and experimental data sufficient for the application and validation of such models [33].

The exponential increase in research into the potential toxicity of nanoparticles has seen a concomitant increase in the literature, and automated ways of extracting information from papers are beginning to emerge. Natural language processing can partially automate the informatics tasks associated with engineered nanomaterials (ENMs) and their physicochemical and biological properties, provenance, and exposure effects. A paper by Lewinski and McInnes identified nine useful natural language processing packages: NanoPort, NanoMapper, TechPerceptor, a text mining framework, a nanodevice analyzer, a clinical trial document classifier, nanotoxicity searcher, NanoSifter, and NEIMiner [34]. The literature on nanomaterials risk analysis can also be mined for the keyword "co-occurrence" to identify trends in analysis research [35].

The need for rigor in the way data on nanoparticles and their biological effects are collected and stored is discussed by Marchese Robinson and coworkers [36]. They advocated the use of an ISA-TAB-Nano-based data collection framework to support the generation of data-driven models of potential adverse effects of nanoparticles. The issues still to be addressed when applying this approach include identifying which (meta)data should be reported and how these should be recorded. They defined data collection templates, business rules that extend the generic ISA-TAB-Nano specification and code to parsing and integration of data sets. An effective understanding of the toxicological impact of nanoparticles requires the consideration of relevant factors, including the temperature of the target tissue, the plasma gradient, cell shape, interfacial effects, and personalized protein corona [37]. Given the extremely large size of experimental and materials composition and property spaces, design of experiments and combinatorial experiments, coupled with automated methods, is required to address a significant portion of these spaces and to gain a better perspective on interactions between these parameters. A systematic approach to studying the factors influencing nanoparticle toxicity has been advocated recently by Baskin et al. [38].

The mechanisms by which nanomaterials affect cells, signaling pathways, and, ultimately, in vivo responses are clearly extremely complex. Unravelling the mechanistic steps in these processes will take a very long time and extensive resources but is ultimately very important in the rational design of nanomaterials for real applications that incorporate SbD principles. Sizochenko et al. used causation inference methods within structure–activity relationship modeling to understand the underlying structure in nanotoxicity data sets and suggest mechanisms of nanotoxicity for the two types of cells [39].

12.3.4 Social Issues and Medical Applications

The very important social aspects of technology are sometimes overlooked. Consumers have problems locating reliable and balanced information on the benefits and risks of nanotechnology [40]. Communicating scientific facts to the public is complex, and transmission of hazard and risk data related to safety of nanomaterials is particularly challenging. As products containing nanomaterials become more common, there will be increasing public interest in reliable and easy-to-understand sources of information, such as in the DaNa (Data and Knowledge on Nanomaterials) project. Websites such as www.nanoobjects.info also provide accessible information on properties and applications of materials to the public and scientists.

The literature is now growing on the deliberate manipulation of the surfaces of nanoparticles to change their biological and physicochemical properties to improve nanosafety or target specific cell types for therapeutic or diagnostic purposes. This topic will be expanded in a later section on in silico approaches to SbD and nanomedicine.

12.4 QSAR Modeling for Nanomaterials

The use of statistical or machine learning methods to generate computational models linking nanomaterials' physicochemical properties to their biological effects is a relatively recent but very active area of research. These studies aim to leverage the limited amount of experimental data to allow predictions to be made of biological properties of new nanomaterials, particularly for triaging materials for regulatory purposes. They can also elucidate important properties of nanomaterials that correlate most strongly with biological effects, especially those that are adverse for human or environmental health. Published research in the field has been comprehensively reviewed in other chapters of this monograph, so only a few key reviews are listed here.

Magnetic nanoparticles used in medicine, largely cancer diagnostics and therapy, are one type of nanomaterial that has received relatively little attention from computational modelers. Research in this field has been reviewed very recently by Winkler [41]. The outcomes of a COST-sponsored workshop on the application of QSAR methods to nanomaterials for regulatory purposes has also been reviewed [42]. Wolfram et al. have reviewed the use of models to understand the safety of nanoparticles used in medicine [43].

A QSAR study that is specifically targeting SbD was reported by Le et al. [44]. They probed the influence of particle size, aspect ratio, doping type, doping concentration, and surface coating on biological effects of nanoparticles. Oksel et al. also demonstrated how decision tree methods can aid in the interpretability of nano-QSAR models [45]. Recent advances and unresolved issues in the application of computational modeling to the prediction of the biological effects of nanomaterials were also discussed in reviews in the last two years [46]. The lack of experiments that alter the properties of nanomaterials in a systematic way, and the need for experimental design, has prompted recent calls for this to be addressed in order to significantly advance the nano-QSAR field [47]. Wang et al. reported a new statistical method, stochastic kriging with qualitative factors, to model exposure-response data for nanoparticles [48]. This allows integration of data from multiple sources to generate accurate information from limited data sets.

12.5 Read-Across

One of the most important issues hindering the development and adoption of nano-QSAR models is the lack of widely available, systematically varied experimental data from libraries of nanomaterials that are required to train and validate such models [38, 49]. This also importantly impacts the utility of models developed with the available data sets, as the domains of applicability (and confidence with which predictions can be made of properties of new nanomaterials that differ from the training data) are relatively small. One way of leveraging the sparse data available and partially filling data gaps is to use the concept of read-across. Read-across relies on a key assumption that chemicals or materials that are structurally similar (or physicochemically similar in the case of materials) have similar biological properties. Grouping similar materials in this way means that the biological impact (e.g., toxicity) of one or more source materials can be used to make predictions of the same endpoint for new or experimentally untested target materials [50]. Gajewicz proposed a simple and transparent read-across algorithm for a prescreening hazard assessment of nanomaterials, making it possible to leverage existing limited data more effectively [33]. She provided limited validation of the value of this approach using three real-world, sparse nanotoxicity data sets. She proposed a new quantitative read-across approach based on linear algebra and the widely used unsupervised pattern recognition method, principal component analysis [33, 51]. As Fig. 12.2 shows, simple linear interpolation or extrapolation of neighboring data points allows an estimate of the likely properties of missing data. Her studies confirmed the applicability and usefulness of this read-across algorithm for preliminary hazard assessment of nanomaterials. She recommended the method for triaging nanomaterials for further testing and risk assessment. However, the basic similarity assumptions may not be true in all cases and difficult to quantify for some complex materials.

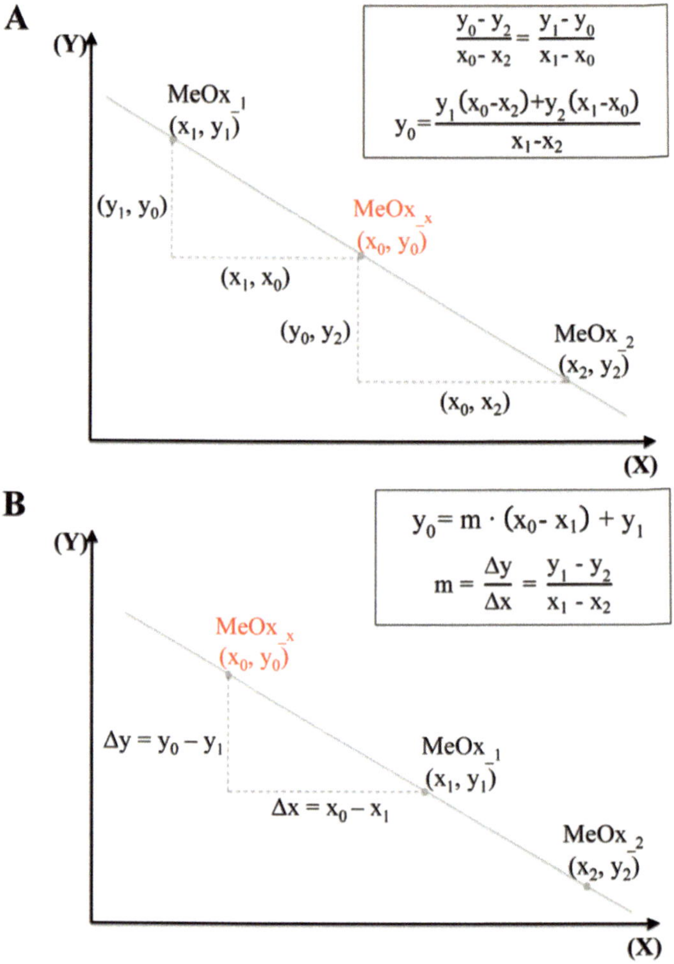

Figure 12.2 Using one-point-slope interpolation (A) and two-point extrapolation (B) to predict properties of related materials. Reproduced from Ref. [51] by permission of The Royal Society of Chemistry. https://creativecommons.org/licenses/by-nc/3.0/.

The data sets considered in two recent studies consisted of cytotoxicity to *Escherichia coli* of a set of 17 metal oxide nanoparticles characterized by a single descriptor ΔH_{Me+} and the toxicity of metal oxide nanoparticles to a human keratinocyte cell line characterized by two descriptors ΔH^c and χ^c. The success in using read-across for the single nanoparticle type and biological endpoint is encouraging, and the extent to which it is useful across a wider range of nanomaterials and biological endpoints is yet to be confirmed. However, it is a very useful step along the road to eliminate the subjectivity of current expert assessment of read-across properties.

12.6 Toward Safety-by-Design

There are an increasing number of published research studies into modeling of the beneficial and adverse properties of nanomaterials, largely using machine learning and other statistical methods, some using physics-based approaches like molecular modeling and dynamics. This topic has been comprehensively covered by previous chapters, so only a few key examples will be given in this section. The main focus is cases where computational approaches have been used to try to achieve safety by design (SbD).

Given the immensity of the materials space that is theoretically encompassed by nanomaterials, it is clearly impossible to explore this entire space to find new nanomaterials with useful properties and lack of significant adverse effects on biological systems. While the types of accelerated experimental testing and materials synthesis described in a previous section will generate very useful data for building models and regulating the use of nanomaterials, the ultimate aim of nanosafety research is the development of methods for engineering SbD materials. This implies the use of robust and predictive models and nanotoxicity mechanistic information to discover materials that have optimum useful properties while remaining relatively benign to biological systems. Estimating potential risks of nanotechnology as markets develop can ensure success of the technology by allowing responsible use without exposing the consumers to unacceptable risk [52]. Several national nanotechnology initiatives have been developed to support safe, responsible application of nanotechnology. For example, the

National Institutes of Health developed the Nano Health and Safety Enterprise Initiative speed development of ENMs that are safe by design. The initiative aims to discover key interactions of ENMs with biological systems and the resulting physiological responses relevant to human exposure. The project requires tight integration between materials science, basic biology, pathobiology, and informatics.

The number of case studies of SbD reported in the literature is still relatively small but increasing. For example, Donaldson and coworkers studied the pulmonary hazards of high-aspect-ratio nanoparticles (HARNs) to enable future SbD [53]. The most common HARNs, carbon nanotubes, are being used on an increasingly large scale, with higher human exposure. The review described the significant hazard associated with other fibers, such as asbestos, and disclosed a robust structure–toxicity model that identified thin, long, and biologically persistent fibers as the largest contributor to pathogenic effects. Most HARNs conform to the model, facilitating the intentional design of safer HARNs. A recent paper described the types of information that are needed to be generated to realize an SbD framework [54]. Such a process requires well-characterized materials, nanoinformatics to store and model data, a good understanding of nano-bio interaction mechanisms, and relevant biological assays that are indicative of likely in vivo responses, and studies have increasingly identified the importance of genomics and systems biology. Early success in improving nanoparticle SbD has included the use of iron doping to reduce the solubility of labile nanomaterials like ZnO and to alter their redox properties [44, 55].

A recent transatlantic think tank on toxicology concluded there is no well-accepted path to effective hazard identification for ENMs. They identified the need for improved nanomaterial characterization; more realistic toxicity testing protocols using whole animal models, such as zebrafish or *Caenorhabditis elegans*; and methods to probe functional pathways and toxicity modes of action. The group also identified a need for reliable, open, curated databases that allow simple integration of data with existing databases, simplifying sharing of information. However, this requires the establishment of an agreed nanomaterials markup language [56] and ontology (e.g., http://www.enanomapper.net).

To this end, the first substance database to support SbD, eNanoMapper, was reported in 2014. This defines a computational infrastructure for toxicological data management of ENMs based on open standards and ontologies [57]. eNanoMapper provides nanosafety collaborators with transparent data sharing, data analysis, and useful computational toxicology models for ENMs. The database uses flexible data storage, semantic web technologies, and open-source components and web services. The authors favored an ontology-supported data model, describing the materials and measurements in a consistent manner.

Lynch and coworkers developed a paradigm for clustering nanomaterials into classes on the basis of physicochemical properties that provides insight into mechanisms of nanomaterials toxicity [58]. A fundamental understanding of the mechanisms of ENM interaction with living systems and the environment is still lacking, hampering efforts to regulate these materials. Lynch et al. advocated identifying critical properties correlating with adverse effects in biological systems, allowing the development of computational models predicting toxicities of related ENMs (nano-QSAR and read-across approaches) that allow adverse physicochemical features to be avoided or minimized in ENM production. To achieve this, more systematically varied libraries of ENMs are required, a point also stressed very recently by Bai et al. [47]. Clearly, physicochemical properties of pristine ENMs are usually correlated and cannot be varied independently, and nanomaterials age, transform, and dissolve in the environmental or under biological conditions, further complicating the establishment of robust predictive models for SbD [59].

Chen et al. [60] reiterated that the categorization of ENM hazards is severely hindered by the lack of accessible toxicity data. They strongly supported the role of nano-QSAR modeling as a tool to leverage scarce and expensive experimental toxicity data and minimize the use of experimental animals. They developed nano-QSAR models to assist in the hazard categorization of ENMs for regulatory purposes, which could also be used for SbD. Their decision tree models could correctly classify the global LC_{50} values of > 70% of the samples in the training (320 ENMs) and test sets (80 ENMs), potentially expediting the identification of ENM hazards and aiding regulatory decisions. A French project, SERENADE [61], also applied the SbD process to the synthesis of safer nanoproducts. It combined knowledge and methods from a range of disciplines to achieve a useful SbD. Another data-driven strategy for SbD nanoparticles was reported by Brehm and colleagues working on a European FP7-funded project [62]. They recognized that safety testing of a large number of synthetically accessible nanomaterials is unfeasible and computational modeling approaches are needed to identify problematic nanomaterials in a time- and cost-effective way. The project developed a comprehensive computational framework to rank the environmental and health impacts of ENMs. They discussed issues related to data generation, nano-specific descriptors, nano-QSAR modeling, hazard ranking, and risk assessment.

A very recent paper describes how SbD can draw on and learn from long-established principles within drug discovery and development industry [63]. The paper stressed that SbD is not precisely defined for ENMs. The authors reviewed how SbD in the European nanomaterials research community compared to safety assessment in the pharmaceutical industry. They concluded that SbD was only one of a series of complex issues, so products containing ENMs still require safety testing and regulation. Although risk reduction is needed, current SbD approaches treat safety as an inherent material property that reduces societal issues to technical problems.

An example of a potential SbD nanomedicine approach was reported recently by Le et al. [44]. They showed how chemical modification of the surface of nanomaterials could dramatically alter the cell targeting profile, making the delivery of nanomaterials and their therapeutic or diagnostic cargo to specific cell types intentional, while avoiding bystander cells. They could selectively deliver nanoparticles into four different kinds of cancer cell. Furthermore, the relationships between the surface chemistry and the delivery profile could be described by a machine learning model that could be used to improve the targeting profile and design new materials with higher safety.

12.7 Nanomedicine Implications

The application of nanotechnology to medicine has been accelerating faster than research into identifying adverse biological effects of nanomaterials. The pharmaceutical and biotechnology industries have a sophisticated regulatory and safety enterprise in operation that has evolved and refined over a very long time. As most drugs are either ingested or injected, the safety and efficacy of the materials is of paramount importance, unlike manufactured nanomaterials, where entry into the body is adventitious, accidental, or occupational.

Nanotechnology promises important opportunities for medicine: targeted delivery of drugs to specific organs or cell types, more efficient delivery of drugs, reduced side effects, and improved imaging and diagnostics. These opportunities have stimulated the dynamic growth of medical applications of nanotechnology over the past decade. The growth in nanomedicine has generated advances in bioengineering, regenerative medicine, and the development of medical devices [64].

However, new drugs, materials, diagnostics, and devices also provide new challenges due to their interactions with the body on the cellular, subcellular, and molecular levels. The opportunities and challenges associated with the development of multifunctional nanomaterials and nanotheranostics (therapeutic diagnostics) have been reviewed by Radomska et al. [64]. Applications of nanotechnology in medicine are hampered by suboptimal in vitro models, lack of in vitro–in vivo correlations, variability within in vitro protocols, and issues with material purity and characterization [65]. Most of the in vitro toxicological data are not sufficiently accurate to allow a decision to be made on whether nanomaterials for drug delivery and medical device implants are safe. New techniques, such as microbioreactors, tissue engineering, advanced imaging methods, and high-content analysis, should help resolve these issues by enabling the procurement of larger amounts of more accurate and reproducible data, thus improving the translation of nanotechnology to the market.

Finding the optimum balance between safety and efficacy, cost and efficiency, and benefit and adverse effects is very complex. Recent reviews have summarized these opportunities and obstacles to the development of nanopharmaceuticals for human use [66]. Although, nanotechnology promises many benefits, its potential dangers require thorough investigation by regulatory and scientific bodies. Regulatory bodies such as the US Food and Drug Administration, Therapeutic Good Administration, European Medicine Agency, Environmental Protection Agency, National Institute for Occupational

Safety and Health, and World Health Organization are developing regulatory frameworks to ensure public health.

Examples of a bespoke, targeted nanomedicine approach was reported recently by Le et al. [44], discussed above, and Liu et al. [67]. The study by Liu et al. used a combinatorial library of surface-modified gold nanoparticles and a machine learning approach to understand and control the specific and nonspecific interactions between the nanoparticles and a common human enzyme, acetylcholinesterase. This work illustrated the insight and benefits that a synergistic blend of experimental combinatorial synthesis, biological testing, and computational modeling can achieve.

12.8 Conclusions and Perspective

It is clear that the field of modeling and predicting the biological properties of nanomaterials has advanced rapidly over the last decade or so by leveraging the considerable body of knowledge on small-molecule and materials modeling and design. Although physics-based models of interactions between nanomaterials and biological entities are still very challenging and largely limited to idealized model systems, improvements in coarse-grained MD calculations and the inevitable increase in computing power will gradually relax the limitations on these types of calculations and simulations. One of the primary benefits of this will be a better molecular-level understanding of how biology interacts with materials.

Data-driven machine learning/artificial intelligence (AI) models will still be the mainstay for predicting the biological effects of nanomaterials, especially for regulatory purposes where these QSAR methods are well understood and have been used for decades to regulate industrial chemicals. Several paradigm-shifting technologies will see these methods grow in power and popularity. Firstly, the rapid increase in automation in laboratories will allow the generation of much larger and more chemically diverse libraries of materials. At the same time, improved bioassays, again increasingly roboticized, will generate a broader range of biological responses, increasingly exploiting new "omics" technologies. These much larger data sets will significantly improve the performance of data-driven nano-QSAR models because the domain of applicability of the models will be larger and the number of relevant biological endpoints will be expanded. AI methods will also make the task of inferring useful mechanistic information from large and diverse data sets easier. In particular, some of the new deep learning methods may provide a substantial improvement in quantitative modeling of biological effects of nanomaterials and also in inferring mechanisms of toxicity from the data.

Improved predictive models and mechanistic knowledge will make the task of designing nanomaterials that are SbD easier and more efficient. It will also become clearer which types nanomaterials pose the greatest risk and which physicochemical properties are the largest drivers of adverse biological effects. This will allow nanomaterials to be regulated efficiently so the dual aims of ensuring the safety of workers, the public, and the environment and effective commercial exploitation without undue barriers can be achieved.

Acknowledgments

The author acknowledges the financial support of CSIRO Manufacturing and the Australian Government National Enabling Technologies Scheme (NETS). The contributions of my research team, Frank Burden, Tu Le, Vidana Epa, are also gratefully acknowledged.

References

1. Ostrowski, A. D., Martin, T., Conti, J., Hurt, I. and Harthorn, B. H. (2009). Nanotoxicology: characterizing the scientific literature, 2000–2007, *J. Nanopart. Res.*, **11**(2), 251–257.
2. Gebel, T., Foth, H., Damm, G., Freyberger, A., Kramer, P.-J., Lilienblum, W., Roehl, C., Schupp, T., Weiss, C., Wollin, K.-M. and Hengstler, J. G. (2014). Manufactured nanomaterials: categorization and approaches to hazard assessment, *Arch. Toxicol.*, **88**(12), 2191–2211.
3. Fadeel, B. and Garcia-Bennett, A. E. (2010). Better safe than sorry: understanding the toxicological properties of inorganic nanoparticles manufactured for biomedical applications, *Adv. Drug Delivery Rev.*, **62**(3), 362–374.
4. Feliu, N. and Fadeel, B. (2010). Nanotoxicology: no small matter, *Nanoscale*, **2**(12), 2514–2520.

5. Warheit, D. B. (2010). Debunking Some misconceptions about nanotoxicology, *Nano Lett.*, **10**(12), 4777–4782.
6. Clark, K. A., White, R. H. and Silbergeld, E. K. (2011). Predictive models for nanotoxicology: current challenges and future opportunities, *Reg. Toxicol. Pharmacol.*, **59**(3), 361–363.
7. Culha, M. and Altunbek, M. (2011). Nanotoxicology: how to test the safety of engineered nanomaterials? *Curr. Opin. Biotechnol.*, **22**, S28–S29.
8. Autrup, H. (2013). Safety of engineered nanomaterials, *Clin. Toxicol.*, **51**(4), 254–255.
9. Cedervall, T., Lynch, I., Lindman, S., Berggard, T., Thulin, E., Nilsson, H., Dawson, K. A. and Linse, S. (2007). Understanding the nanoparticle-protein corona using methods to quantify exchange rates and affinities of proteins for nanoparticles, *Proc. Natl. Acad. Sci. U. S. A.*, **104**(7), 2050–2055.
10. Yang, S.-T., Liu, Y., Wang, Y.-W. and Cao, A. (2013). Biosafety and bioapplication of nanomaterials by designing protein-nanoparticle interactions, *Small*, **9**(9–10), 1635–1653.
11. Monopoli, M. P., Aberg, C., Salvati, A. and Dawson, K. A. (2012). Biomolecular coronas provide the biological identity of nanosized materials, *Nat. Nanotechnol.*, **7**(12), 779–786.
12. Tassa, C., Duffner, J. L., Lewis, T. A., Weissleder, R., Schreiber, S.,L., Koehler, A. N. and Shaw, S. Y. (2010). Binding affinity and kinetic analysis of targeted small molecule-modified nanoparticles, *Bioconjugate Chem.*, **21**(1), 14–19.
13. Li, R., Chen, R., Chen, P., Wen, Y., Ke, P. C. and Cho, S. S. (2013). Computational and experimental characterizations of silver nanoparticle-apolipoprotein biocorona, *J. Phys. Chem. B*, **117**(43), 13451–13456.
14. Chen, P., Seabrook, S. A., Epa, V.,C., Kurabayashi, K., Barnard, A. S., Winkler, D. A., Kirby, J. K. and Ke, P. C. (2014). Contrasting effects of nanoparticle binding on protein denaturation, *J. Phys. Chem. C*, **118**(38), 22069–22078.
15. Wang, B., Seabrook, S. A., Nedumpully-Govindan, P., Chen, P., Yin, H., Waddington, L., Epa, V. C., Winkler, D. A., Kirby, J. K., Ding, F. and Ke, P. C. (2015). Thermostability and reversibility of silver nanoparticle-protein binding, *Phys. Chem. Chem. Phys.*, **17**(3), 1728–1739.
16. Lai, D. Y. (2015). Approach to using mechanism-based structure activity relationship (SAR) analysis to assess human health hazard potential of nanomaterials, *Food Chem. Toxicol.*, **85**, 120–126.

17. Shin, S. W., Song, I. H. and Um, S. H. (2015). Role of physicochemical properties in nanoparticle toxicity, *Nanomater*, **5**(3), 1351–1365.

18. Nel, A. E., Brinker, C. J., Parak, W. J., Zink, J. I., Chan, W. C. W., Pinkerton, K. E., Xia, T., Baer, D. R., Hersam, M. C. and Weiss, P. S. (2015). Where are we heading in nanotechnology environmental health and safety and materials characterization? *ACS Nano*, **9**(6), 5627–5630.

19. Ali, A., Suhail, M., Mathew, S., Shah, M. A., Harakeh, S. M., Ahmad, S., Kazmi, Z., Alhamdan, M. A. R., Chaudhary, A., Damanhouri, G. A. and Qadri, I. (2016). Nanomaterial induced immune responses and cytotoxicity, *J. Nanosci. Nanotechnol.*, **16**(1), 40–57.

20. David, C. A. W., Owen, A. and Liptrott, N. J. (2016). Determining the relationship between nanoparticle characteristics and immunotoxicity: key challenges and approaches, *Nanomedicine*, **11**(11), 1447–1464.

21. George, S., Xia, T., Rallo, R., Zhao, Y., Ji, Z., Lin, S., Wang, X., Zhang, H., France, B., Schoenfeld, D., Damoiseaux, R., Liu, R., Lin, S., Bradley, K. A., Cohen, Y. and Nal, A. E. (2011). Use of a high-throughput screening approach coupled with in vivo zebrafish embryo screening to develop hazard ranking for engineered nanomaterials, *ACS Nano*, **5**(3), 1805–1817.

22. Warheit, D. B. and Donner, E. M. (2015). How meaningful are risk determinations in the absence of a complete dataset? Making the case for publishing standardized test guideline and 'no effect' studies for evaluating the safety of nanoparticulates versus spurious 'high effect' results from single investigative studies, *Sci. Tech. Adv. Mater.*, **16**(3).

23. Misra, S. K., Dybowska, A., Berhanu, D., Luoma, S. N. and Valsami-Jones, E. (2012). The complexity of nanoparticle dissolution and its importance in nanotoxicological studies, *Sci. Total Environ.*, **438**, 225–232.

24. Ong, K. J., MacCormack, T. J., Clark, R. J., Ede, J. D., Ortega, V. A., Felix, L. C., Dang, M. K. M., Ma, G., Fenniri, H., Veinot, J. G. C. and Goss, G. G. (2014). Widespread nanoparticle-assay interference: implications for nanotoxicity testing, *PLoS One*, **9**(3), e90650.

25. Sayes, C. M. (2014). The relationships among structure, activity, and toxicity of engineered nanoparticles, *Kona Powder Part. J.*, **31**, 10–21.

26. Klaper, R., Arndt, D., Bozich, J. and Dominguez, G. (2014). Molecular interactions of nanomaterials and organisms: defining biomarkers for toxicity and high-throughput screening using traditional and next-generation sequencing approaches, *Analyst*, **139**(5), 882–895.

27. Hackenberg, S. (2014). Risk assessment of nanoparticles in consumer products, *HNO*, **62**(6), 432–438.

28. Gulson, B., McCall, M., Korsch, M., Gomez, L., Casey, P., Oytam, Y., Taylor, A., McCulloch, M., Trotter, J., Kinsley, L. and Greenoak, G. (2010). Small amounts of zinc from zinc oxide particles in sunscreens applied outdoors are absorbed through human skin, *Toxicol. Sci.*, **118**(1), 140–149.

29. Gulson, B., McCall, M., Korsch, M., Gomez, L., Casey, P., Oytam, Y., Taylor, A., McCulloch, M., Trotter, J., Kinsley, L. and Greenoak, G. (2011). Small amounts of zinc from zinc oxide particles in sunscreens applied outdoors are absorbed through human skin (vol 118, pg 140, 2010), *Toxicol. Sci.*, **120**(2), 530–530.

30. Deng, J. and Gao, C. (2016). Recent advances in interactions of designed nanoparticles and cells with respect to cellular uptake, intracellular fate, degradation and cytotoxicity, *Nanotechnology*, **27**(41).

31. Oksel, C., Ma, C. Y. and Wang, X. Z. (2015). Current situation on the availability of nanostructure-biological activity data, *SAR QSAR Environ. Res.*, **26**(2), 79–94.

32. Krug, H. F. and Nau, K. (2017). Reliability for nanosafety research, *Chem. Ingen. Tech.*, **89**(3), 215–223.

33. Gajewicz, A. (2017). Development of valuable predictive read-across models based on "real-life" (sparse) nanotoxicity data, *Environ. Sci. Nano*, **4**(6), 1389–1403.

34. Lewinski, N. A. and McInnes, B. T. (2015). Using natural language processing techniques to inform research on nanotechnology, *Beilstein J. Org. Chem.*, **6**, 1439–1449.

35. Erbis, S., Ok, Z., Isaacs, J. A., Benneyan, J. C. and Kamarthi, S. (2016). Review of research trends and methods in nano environmental, health, and safety risk analysis, *Risk Anal.*, **36**(8), 1644–1665.

36. Marchese Robinson, R. L., Cronin, M. T. D., Richarz, A.-N. and Rallo, R. (2015). An ISA-TAB-Nano based data collection framework to support data-driven modelling of nanotoxicology, *Beilstein J. Org. Chem.*, **6**, 1978–1999.

37. Azhdarzadeh, M., Saei, A. A., Sharifi, S., Hajipour, M. J., Alkilany, A. M., Sharifzadeh, M., Ramazani, F., Laurent, S., Mashaghi, A. and Mahmoudi, M. (2015). Nanotoxicology: advances and pitfalls in research methodology, *Nanomedicine*, **10**(18), 2931–2952.

38. Baskin, I. I., Winkler, D. and Tetko, I. V. (2016). A renaissance of neural networks in drug discovery, *Expert Opin. Drug Discovery*, **11**(8), 785–795.

39. Sizochenko, N., Rasulev, B., Gajewicz, A., Mokshyna, E., Kuz'min, V. E., Leszczynski, J. and Puzyn, T. (2015). Causal inference methods to assist in mechanistic interpretation of classification nano-SAR models, *RSC Adv.*, **5**(95), 77739–77745.

40. Nau, K., Bohmer, N., Kuehnel, D., Marquardt, C., Paul, F., Steinbach, C. and Krug, H. F. (2016). The DANA(2.0) knowledge base on nanomaterials - communicating current nanosafety research based on evaluated literature data, *J. Mater. Ed.*, **38**(3–4), 93–108.

41. Winkler, D. A. (2017). Computational modelling of magnetic nanoparticle properties and in vivo responses, *Curr. Med. Chem.*, **24**(5), 483–496.

42. Winkler, D. A., Mombelli, E., Pietroiusti, A., Tran, L., Worth, A., Fadeel, B. and McCall, M. J. (2013). Applying quantitative structure-activity relationship approaches to nanotoxicology: current status and future potential, *Toxicology*, **313**(1), 15–23.

43. Wolfram, J., Zhu, M., Yang, Y., Shen, J., Gentile, E., Paolino, D., Fresta, M., Nie, G., Chen, C., Shen, H., Ferrari, M. and Zhao, Y. (2015). Safety of nanoparticles in medicine, *Curr. Drug Targets*, **16**(14), 1671–1681.

44. Le, T. C., Yin, H., Chen, R., Chen, Y., Zhao, L., Casey, P. S., Chen, C. and Winkler, D. A. (2016). An experimental and computational approach to the development of ZnO nanoparticles that are safe by design, *Small*, **12**(26), 3568–3577.

45. Oksel, C., Winkler, D. A., Ma, C. Y., Wilkins, T. and Wang, X. Z. (2016). Accurate and interpretable nanoSAR models from genetic programming-based decision tree construction approaches, *Nanotoxicology*, **10**(7), 1001–1012.

46. Winkler, D. A. (2016). Recent advances, and unresolved issues, in the application of computational modelling to the prediction of the biological effects of nanomaterials, *Toxicol. Appl. Pharmacol.*, **299**, 96–100.

47. Bai, X., Liu, F., Liu, Y., Li, C., Wang, S., Zhou, H., Wang, W., Zhu, H., Winkler, D. A. and Yan, B. (2017). Toward a systematic exploration of nano-bio interactions, *Toxicol. Appl. Pharmacol.*, **323**, 66–73.

48. Wang, K., Chen, X., Yang, F., Porter, D. W. and Wu, N. (2014). A new stochastic kriging method for modeling multi-source exposure-response data in toxicology studies, *ACS Sustain. Chem. Eng.*, **2**(7), 1581–1591.

49. Gajewicz, A. (2017). What if the number of nanotoxicity data is too small for developing predictive Nano-QSAR models? An alternative

read-across based approach for filling data gaps, *Nanoscale*, **9**(24), 8435–8448.

50. OECD (2014). Guidance on grouping of chemicals. Series on testing and assessment No. 194 Organisation of Economic Cooperation and Development, Paris, France, 2014. Organisation of Economic Cooperation and Development, Paris.
51. Gajewicz, A., Jagiello, K., Cronin, M. T. D., Leszczynski, J. and Puzyn, T. (2017). Addressing a bottle neck for regulation of nanomaterials: quantitative read-across (nano-QRA) algorithm for cases when only limited data is available, *Environ. Sci. Nano*, **4**(2), 346–358.
52. Tinkle, S. S. (2010). Maximizing safe design of engineered nanomaterials: the NIH and NIEHS research perspective, *Wiley Interdiscip. Rev. Nanomed. Nanobiotechnol.*, **2**(1), 88–98.
53. Donaldson, K., Murphy, F., Schinwald, A., Duffin, R. and Poland, C. A. (2011). Identifying the pulmonary hazard of high aspect ratio nanoparticles to enable their safety-by-design, *Nanomedicine*, **6**(1), 143–156.
54. Roca, C. P., Rallo, R., Fernandez, A. and Giralt, F. (2013). Nanoinformatics for safe-by-design engineered nanomaterials, in *Towards Efficient Designing of Safe Nanomaterials: Innovative Merge of Computational Approaches and Experimental Techniques*, Leszczynski, J. and Puzyn, T., eds. (RSC), pp. 89–107.
55. Xia, T. A., Zhao, Y., Sager, T., George, S., Pokhrel, S., Li, N., Schoenfeld, D., Meng, H. A., Lin, S. J., Wang, X., Wang, M. Y., Ji, Z. X., Zink, J. I., Madler, L., Castranova, V., Lin, S. and Nel, A. E. (2011). Decreased dissolution of ZnO by iron doping yields nanoparticles with reduced toxicity in the rodent lung and zebrafish embryos, *ACS Nano*, **5**(2), 1223–1235.
56. Kaufman, J. G. and Begley, E. F. (2003). MatML: a data interchange markup language, *Adv. Mater. Process.*, **161**(11), 35–36.
57. Jeliazkova, N., Doganis, P., Fadeel, B., Grafstrom, R., Hastings, J., Jeliazkov, V., Kohonen, P., Munteanu, C. R., Sarimveis, H., Smeets, B., Tsiliki, G., Vorgrimmler, D. and Willighagen, E. (2014). The first eNanoMapper prototype: a substance database to support safe-by-design, in *2014 IEEE International Conference on Bioinformatics and Biomedicine*, Zheng, H., Hu, X., Berrar, D., Wang, Y., Dubitzky, W., Hao, J. K., Cho, K. H. and Gilbert, D., eds.
58. Lynch, I., Weiss, C. and Valsami-Jones, E. (2014). A strategy for grouping of nanomaterials based on key physico-chemical descriptors as a basis for safer-by-design NMs, *Nano Today*, **9**(3), 266–270.

59. Borges, T., Silva, M. J. and Louro, H. (2015). Moving towards a safe by design approach for ENM: linking ENM relevant properties to toxicological concerns, *Toxicol. Lett.*, **238**(2), S35–S35.

60. Chen, G., Peijnenburg, W. J. G. M., Kovalishyn, V. and Vijver, M. G. (2016). Development of nanostructure-activity relationships assisting the nanomaterial hazard categorization for risk assessment and regulatory decision-making, *RSC Adv.*, **6**(57), 52227–52235.

61. Bottero, J. Y., Rose, J., de Garidel, C., Masion, A., Deutsch, T., Brochard, G., Carriere, M., Gontard, N., Wortham, H., Rabilloud, T., Salles, B., Dubosson, M., Cathala, B., Boutry, D., Ereskovsky, A., Auplat, C., Charlet, L., Heulin, T., Frejafon, E. and Lanone, S. (2017). SERENADE: safer and ecodesign research and education applied to nanomaterial development, the new generation of materials safer by design, *Environ. Sci. Nano*, **4**(3), 526–538.

62. Brehm, M., Kafka, A., Bamler, M., Kuhne, R., Schuurmann, G., Sikk, L., Burk, J., Burk, P., Tamm, T., Tamm, K., Pokhrel, S., Madler, L., Kahru, A., Aruoja, V., Sihtmae, M., Scott-Fordsmand, J., Sorensen, P. B., Escorihuela, L., Roca, C. P., Fernandez, A., Giralt, F. and Rallo, R. (2017). An integrated data-driven strategy for safe-by-design nanoparticles: the FP7 MODERN project, *Adv. Exp. Med. Biol.*, **947**, 257–301.

63. Hjorth, R., van Hove, L. and Wickson, F. (2017). What can nanosafety learn from drug development? The feasibility of "safety by design", *Nanotoxicology*, **11**(3), 305–312.

64. Radomska, A., Leszczyszyn, J. and Radomski, M. W. (2016). The nanopharmacology and nanotoxicology of nanomaterials: New opportunities and challenges, *Adv. Clin. Exp. Med.*, **25**(1), 151–162.

65. Bhattacharjee, S. and Brayden, D. J. (2015). Development of nanotoxicology: implications for drug delivery and medical devices, *Nanomedicine*, **10**(14), 2289–2305.

66. Koopaei, N. N. and Abdollahi, M. (2016). Opportunities and obstacles to the development of nanopharmaceuticals for human use, *Daru J. Pharm. Sci.*, **24**, 23.

67. Liu, Y., Winkler, D. A., Epa, V. C., Zhang, B. and Yan, B. (2015). Probing enzyme-nanoparticle interactions using combinatorial gold nanoparticle libraries, *Nano Res.*, **8**(4), 1293–1308.

Index